Neuromethods

Series Editor
Wolfgang Walz
University of Saskatchewan
Saskatoon, SK, Canada

For further volumes:
http://www.springer.com/series/7657

Biomarkers for Preclinical Alzheimer's Disease

Edited by

Robert Perneczky

Department of Psychiatry and Psychotherapy,
Ludwig-Maximilians-Universität München,
Munich, Germany

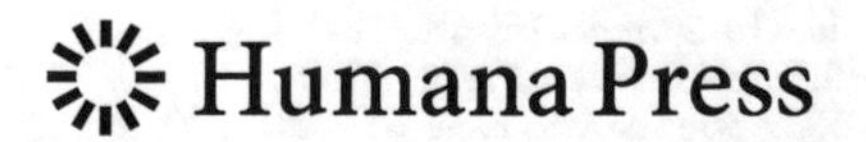

Editor
Robert Perneczky
Department of Psychiatry and Psychotherapy
Ludwig-Maximilians-Universität München
Munich, Germany

ISSN 0893-2336 ISSN 1940-6045 (electronic)
Neuromethods
ISBN 978-1-4939-9255-3 ISBN 978-1-4939-7674-4 (eBook)
https://doi.org/10.1007/978-1-4939-7674-4

Printed on acid-free paper

This Humana Press imprint is published by Springer Nature
The registered company is Springer Science+Business Media, LLC
The registered company address is: 233 Spring Street, New York, NY 10013, U.S.A.

Series Preface

Experimental life sciences have two basic foundations: concepts and tools. The *Neuromethods* series focuses on the tools and techniques unique to the investigation of the nervous system and excitable cells. It will not, however, shortchange the concept side of things as care has been taken to integrate these tools within the context of the concepts and questions under investigation. In this way, the series is unique in that it not only collects protocols but also includes theoretical background information and critiques which led to the methods and their development. Thus it gives the reader a better understanding of the origin of the techniques and their potential future development. The *Neuromethods* publishing program strikes a balance between recent and exciting developments like those concerning new animal models of disease, imaging, in vivo methods, and more established techniques, including, for example, immunocytochemistry and electrophysiological technologies. New trainees in neurosciences still need a sound footing in these older methods in order to apply a critical approach to their results.

Under the guidance of its founders, Alan Boulton and Glen Baker, the *Neuromethods* series has been a success since its first volume published through Humana Press in 1985. The series continues to flourish through many changes over the years. It is now published under the umbrella of Springer Protocols. While methods involving brain research have changed a lot since the series started, the publishing environment and technology have changed even more radically. Neuromethods has the distinct layout and style of the Springer Protocols program, designed specifically for readability and ease of reference in a laboratory setting.

The careful application of methods is potentially the most important step in the process of scientific inquiry. In the past, new methodologies led the way in developing new disciplines in the biological and medical sciences. For example, Physiology emerged out of Anatomy in the nineteenth century by harnessing new methods based on the newly discovered phenomenon of electricity. Nowadays, the relationships between disciplines and methods are more complex. Methods are now widely shared between disciplines and research areas. New developments in electronic publishing make it possible for scientists that encounter new methods to quickly find sources of information electronically. The design of individual volumes and chapters in this series takes this new access technology into account. Springer Protocols makes it possible to download single protocols separately. In addition, Springer makes its print-on-demand technology available globally. A print copy can therefore be acquired quickly and for a competitive price anywhere in the world.

Saskatoon, Canada *Wolfgang Walz*

Preface

The constantly increasing number of individuals with dementia due to Alzheimer's disease (AD) poses a significant financial and emotional burden on the affected families and the global society. Therapeutic strategies designed to treat symptoms or alter the disease course have so far failed to make a positive impact, despite billions of US dollars of R&D investments and massive efforts of the industry and academic communities. These failures in treatment have led many to believe that symptomatic AD, including its earliest clinical stages, is resistant to drug interventions.

The recently increased focus on biomarkers to diagnose early disease is fuelled by the hope of identifying a therapeutic window, in which the brain is still largely intact and therefore amenable to treatment effects. The ability of biomarkers to adequately define the preclinical (i.e., at risk of AD dementia) disease stage may ultimately allow novel or repurposed drugs to finally achieve clinically meaningful results for the affected individuals and to help to prevent dementia and associated disability.

This book discusses the usefulness of established biomarkers (imaging, fluid, and genetic) to detect preclinical AD, providing detailed protocols for state-of-the-art biomarker measurement and analysis. It is also explained how cutting-edge technology is used to develop novel improved biomarkers for the earliest, presymptomatic stages of AD. Our book aims to provide a comprehensive overview specifically of biomarkers for the earliest detectable disease stages. All previous publications cover biomarkers for clinical, symptomatic AD.

This publication comprises five parts: In *Part I* (Chaps. 1 and 2) we explain why AD is one of the major challenges for the global societies and healthcare systems, which highlights the urgent need for improved approaches for early diagnosis enabling disease prevention. In *Part II* (Chaps. 3–5) clinical and research concepts are presented, which are important to improve early recognition of AD. *Part III* (Chaps. 6–10) provides a comprehensive overview of methods currently used in the AD diagnostic work-up, and it is explained how these methods can be applied to preclinical disease. In *Part IV* (Chaps. 11–15) cutting-edge technology innovations and their value for early AD diagnosis are discussed. Finally, in *Part V* (Chaps. 16–18), we consider important ethical considerations in relation to biomarker-based early diagnosis, and we also discuss the meaningfulness of biomarker endpoints in clinical AD research.

The book is targeted at individuals with an interest in the use of advanced biomarker strategies to significantly improve the early diagnosis of AD, and thereby to accelerate the development of effective, disease-modifying drugs. This includes researchers, clinicians, and those interested in regulatory and medical affairs, both from academia and industry. We wish to present biomarker development approaches as a strategy for the study of AD with the hope and expectation that the results will translate into more effective diagnosis and treatment and improved public health policies. We expect this book to complement other excellent volumes and monographs on AD that cover basic science or clinical aspects of the disease.

London, UK *Robert Perneczky*

Contents

Contributors

MARINA ÁVILA-VILLANUEVA • *Alzheimer Disease Research Unit, CIEN Foundation, Carlos III Institute of Health, Queen Sofía Foundation Alzheimer Center, Madrid, Spain*

PANAGIOTIS ALEXOPOULOS • *Department of Psychiatry, University Hospital of Patras, University of Patras, Patras, Greece; Department of Psychiatry and Psychotherapy, Klinikum rechts der Isar, Technical University of Munich, Munich, Germany*

JOHANNES ATTEMS • *Institute for Neuroscience, Newcastle University, Newcastle upon Tyne, UK*

THORSTEN BARTSCH • *Department of Neurology, Memory Disorders and Plasticity Group, Dementia and Alzheimer's Clinic, University Hospital Schleswig-Holstein, Kiel, Germany*

KATHARINA BRONNER • *Department of Psychiatry and Psychotherapy, Technische Universität München, Munich, Germany*

JAMES CAMERON • *Centre for Clinical Brain Sciences, University of Edinburgh, Edinburgh, UK*

PAUL EDISON • *Division of Brain Sciences, Department of Medicine, Neurology Imaging Unit, Imperial College London, Hammersmith Hospital, London, UK; Department of Psychological Medicine, Cardiff University School of Medicine, Cardiff, UK*

MIGUEL A. FERNÁNDEZ-BLÁZQUEZ • *Alzheimer Disease Research Unit, CIEN Foundation, Carlos III Institute of Health, Queen Sofía Foundation Alzheimer Center, Madrid, Spain*

PETER FUHR • *Section of Clinical Neurophysiology, Department of Neurology, Hospital of the University of Basel, Basel, Switzerland*

SERGE GAUTHIER • *The McGill University Research Centre for Studies in Aging, McGill University, Montreal, QC, Canada*

PANTELEIMON GIANNAKOPOULOS • *Department of Psychiatry, Faculty of Medicine, University of Geneva, Geneva, Switzerland; Division of Institutional Measures, Medical Direction, University Hospitals of Geneva, Belle-Idée, Chêne-Bourg, Switzerland*

OLIVER GRANERT • *Department of Neurology, Memory Disorders and Plasticity Group, Dementia and Alzheimer's Clinic, University Hospital Schleswig-Holstein, Kiel, Germany*

TIMO GRIMMER • *Centre for Cognitive Disorders, Department of Psychiatry and Psychotherapy, Klinikum rechts der Isar, Technical University of Munich, Munich, Germany*

LEA GRINBERG • *Department of Neurology, University of California, San Francisco, CA, USA*

MICHEL J. GROTHE • *German Center for Neurodegenerative Diseases (DZNE), Rostock, Germany*

CHRISTIAN HABECK • *Cognitive Neuroscience Division, Department of Neurology, College of Physicians and Surgeons, Columbia University, New York, NY, USA*

SVEN HALLER • *Affidea Carouge Radiologic Diagnostic Center, Geneva, Switzerland; Department of Surgical Sciences, Radiology, Uppsala University, Uppsala, Sweden; Department of Neuroradiology, University Hospital Freiburg, Freiburg, Germany; Department of Neuroradiology, Faculty of Medicine, University of Geneva, Geneva, Switzerland*

JOHANNES HAMANN • *Department of Psychiatry and Psychotherapy, Technische Universität München, Munich, Germany*
FLORIAN HATZ • *Section of Clinical Neurophysiology, Department of Neurology, Hospital of the University of Basel, Basel, Switzerland*
PETER HÄUSSERMANN • *Department of Geriatric Psychiatry, LVR Klinik Köln, Academic Teaching Hospital, University of Cologne, Köln, Germany*
HELMUT HEINSEN • *Universidade de Sao Paulo Faculdade de Medicina, Sao Paulo, Brazil; Department of Psychiatry, Psychosomatics and Psychotherapy, Julius-Maximilians-University, Würzburg, Germany*
AMANDA J. HESLEGRAVE • *Department of Molecular Neuroscience, UCL Institute of Neurology, London, UK*
WENDY E. HEYWOOD • *Centre for Translational Omics, University College London Institute of ChildHealth, London, UK*
INGO KILIMANN • *German Center for Neurodegenerative Diseases (DZNE), Rostock, Germany; Department of Psychosomatic Medicine, University of Rostock, Rostock, Germany*
ALEXANDER F. KURZ • *Department of Psychiatry and Psychotherapy, Klinikum rechts der Isar, Technische Universität München, Munich, Germany*
NICOLA T. LAUTENSCHLAGER • *Academic Unit for Psychiatry of Old Age, Department of Psychiatry, The University of Melbourne and NorthWestern Mental Health, Mental Health, Melbourne, Australia*
XIAOFENG LI • *The McGill University Research Centre for Studies in Aging, McGill University, Montreal, QC, Canada; Department of Neurology, The Second Affiliated Hospital of Chongqing Medical University, Chongqing, People's Republic of China*
TOM MACGILLIVRAY • *Centre for Clinical Brain Sciences, University of Edinburgh, Edinburgh, UK*
KIRSTY E. MCALEESE • *Institute for Neuroscience, Newcastle University, Newcastle upon Tyne, UK*
SARAH MCGRORY • *Centre for Cognitive Ageing and Cognitive Epidemiology, University of Edinburgh, Edinburgh, UK*
MIGUEL MEDINA • *Alzheimer Disease Research Unit, CIEN Foundation, Carlos III Institute of Health, Queen Sofía Foundation Alzheimer Center, Madrid, Spain; CIBERNED (Network Center for Biomedical Research in Neurodegenerative Diseases), Universidad Autónoma de Madrid, Madrid, Spain*
KEVIN M. MILLS • *Centre for Translational Omics, University College London Institute of ChildHealth, London, UK*
KOK PIN NG • *The McGill University Research Centre for Studies in Aging, McGill University, Montreal, QC, Canada; Department of Neurology, National Neuroscience Institute, Singapore, Singapore*
THARICK A. PASCOAL • *The McGill University Research Centre for Studies in Aging, McGill University, Montreal, QC, Canada*
TOM PEARSON • *Centre for Clinical Brain Sciences, University of Edinburgh, Edinburgh, UK*
ROBERT PERNECZKY • *Department of Psychiatry and Psychotherapy, Ludwig-Maximilians-Universität München, Munich, Germany*
JENNIFER GRACE PERRYMAN • *Ludwig-Maximilians-University Munich, Munich, Germany*

CHENGXUAN QIU • *Aging Research Center, Department of Neurobiology, Care Sciences and Society, Karolinska Institutet-Stockholm University, Stockholm, Sweden*
CRISTELLE RODRIGUEZ • *Division of Institutional Measures, Medical Direction, University Hospitals of Geneva, Belle-Idée, Chêne-Bourg, Switzerland*
PEDRO ROSA-NETO • *The McGill University Research Centre for Studies in Aging, McGill University, Montreal, QC, Canada*
CÉCILIA SAMIERI • *INSERM, Bordeaux Population Health Research Center, UMR 1219, University of Bordeaux, Bordeaux, France*
CHAIDO SIRINIAN • *Clinical and Molecular Oncology Laboratory, Division of Oncology, Medical Faculty, School of Health Sciences, University of Patras, Patras, Greece*
YAAKOV STERN • *Cognitive Neuroscience Division, Department of Neurology, College of Physicians and Surgeons, Columbia University, New York, NY, USA*
STEFAN TEIPEL • *German Center for Neurodegenerative Diseases (DZNE), Rostock, Germany; Department of Psychosomatic Medicine, University of Rostock, Rostock, Germany*
LAUREN WALKER • *Institute for Neuroscience, Newcastle University, Newcastle upon Tyne, UK*
JUNFANG XU • *Research Center for Public Health, School of Medicine, Tsinghua University, Beijing, People's Republic of China*
DAVIDE ZANCHI • *Department of Psychiatry (UPK), University of Basel, Basel, Switzerland*
HENRIK ZETTERBERG • *Department of Molecular Neuroscience, UCL Institute of Neurology, London, UK; Clinical Neurochemistry Laboratory, Department of Psychiatry and Neurochemistry, Institute of Neuroscience and Physiology, The Sahlgrenska Academy, University of Gothenburg, Mölndal, Sweden*

Part I

Alzheimer's Disease as a Societal Challenge

Chapter 1

Worldwide Economic Costs and Societal Burden of Dementia

Junfang Xu and Chengxuan Qiu

Abstract

Dementia is the principal cause of functional dependence and institutionalization in the elderly and has already posed tremendous economic burden on the aging society. Thus, dementia has been recognized by the World Health Organization as a global public health priority. Here, we briefly introduce the conceptual and methodological issues of the cost-of-illness studies (COI) of dementia, summarize the major literature regarding the economic costs of dementia, and identify some key issues that need be addressed in future COI studies of dementia. We identified 17 COI studies that estimated the costs of dementia in different countries, in which the annual total costs per patient with dementia varied from US$2935 to US$64168. Differences in methodology, data sources, cost categories assessed, and severity of patients employed in the COI studies are likely to contribute to the substantial variations in cost estimates of dementia. Thus, to increase the comparability of findings from COI studies, the methodological issues need to be harmonized and standardized, and consensus on the methodological principles in the COI studies would be essential. Further well-designed COI studies of dementia could provide critical evidence to help develop health policy and improve allocation of healthcare resources.

Key words Dementia, Aging, Care Costs, Economic Burden, Societal Burden, Cost-of-Illness Studies

1 Introduction

Dementia is defined as a clinical syndrome characterized by progressive deteriorations in multiple cognitive domains (e.g., memory and executive function) that are severe enough to interfere with daily social and professional functioning [1]. Clinically, Alzheimer's disease is the most common form of dementia, followed by vascular dementia, accounting for 50–70% and 15–20% of all dementia cases, respectively. However, population-based neuroimaging and autopsy-verified studies have revealed that pure Alzheimer's dementia or pure vascular dementia cases are virtually rare, whereas a large majority of dementia cases are attributable to mixed vascular and neurodegenerative pathology in the brain. Dementia, as a clinical syndrome in general, is a major cause of functional dependence and institutionalization in older people, which has enormous economic and societal impact on the aging society.

Robert Perneczky (ed.), *Biomarkers for Preclinical Alzheimer's Disease*, Neuromethods, vol. 137, https://doi.org/10.1007/978-1-4939-7674-4_1,

2 Dementia: A Priority of Global Public Health

Population aging is a worldwide universal phenomenon. Thus, dementia or Alzheimer's disease as a strongly age-dependent disorder has posed huge impact on public health, healthcare, and social service systems in all countries across the world. The World Alzheimer Report 2015 estimated that worldwide approximately 47.5 million people were living with dementia, and the number is projected to reach 75 million in 2030 and 132 million in 2050 [2]. The global costs of dementia care were estimated to increase from US$604 billion in 2010 (~1.0% of global GDP) to US$818 billion in 2015 (~1.1% of global GDP). The growth of economic costs is faster than that of prevalence of the disease. Thus, dementia has been recognized by WHO and Alzheimer's Disease International (ADI) as a global public health priority [3]. The 2013 G8 Dementia Summit in London brought together world leaders from major industrial countries, research community, industry, and charities to discuss what can be done to stimulate greater investment in innovative dementia research, to accelerate the development of preventive and treatment approaches for dementia, and to improve quality of life for people with dementia. In March 2015, WHO hosted the first ministerial conference on global action against dementia aiming to raise awareness of the social and economic burden caused by dementia and to highlight the potential that this huge burden can be reduced only if the world collectively commits to placing dementia high on the global public health agenda.

3 Estimation of Dementia Costs: Conceptual and Methodological Issues

The economic burden of a disease is often defined by cost-of-illness (COI) studies, in which the overall economic impact of the disease on individuals, caregivers, and the society can be estimated [4]. Quantifying dementia costs is highly relevant for policy making and resource allocation as well as for determining the cost-effectiveness of early diagnosis, treatment, and intervention approaches for dementia [5]. Furthermore, it can motivate policy makers to develop strategic action plan and to prioritize policy in social care and proper management of dementia. The economic burden of an illness largely depends on how the illness is defined [5]. In most of the COI studies of dementia, dementia has been clinically diagnosed according to the *Diagnostic and Statistical Manual of Mental Disorders* criteria, fourth edition (DSM-IV) [1], and the National Institute of Neurological and Communicative Disorders and Stroke and the Alzheimer's Disease and Related Disorders Association criteria (NINCDS-ADRDA) are used for defining Alzheimer's disease [6]. The severity of dementia and the living conditions should also be specified as economic costs of dementia are varied considerably with these factors [5, 7–10]. Moreover, to estimate the economic costs of a disease, it is necessary to specify the study approaches, perspectives, and cost components.

3.1 The Cost-of-Illness Approaches

The COI approaches are usually used to estimate dementia costs where epidemiological (e.g., number of patients with dementia) and economic or cost data are integrated. The prevalence-based and incidence-based approaches are the two main types of COI studies [11]. The prevalence-based approach is based on the prevalence of a disease at a given year to estimate economic burden of the disease. The total costs are estimated by multiplying the total number of patients with dementia in a given year with the mean cost per patient in the same year [11]. In the incidence-based approach, the total costs are calculated by multiplying the total number of patients with newly diagnosed (incident) dementia in a given year with the lifetime costs of these patients [12]. The prevalence-based approach has been frequently used to estimate dementia costs for two reasons. First, most cost data of dementia available in the real world cover costs of a short period of time rather than the lifetime. Second, dementia, and Alzheimer's disease in particular, is an irreversible chronic health condition with an insidious onset, and prevalence could measure burden of the disease in the society [13].

3.2 Estimation of Dementia Costs from Different Perspectives

The economic burden of dementia can be analyzed from three perspectives: family, healthcare provider, and the society [5]. The economic burden from a societal perspective refers to the costs of all resources consumed or lost due to dementia, irrespective of who the payer is [14]. In this regard, costs from the societal perspective are most relevant for decision makers whose main interest is the welfare of the society as a whole. A healthcare provider may be concerned with only the costs of a hospital or nursing home in order to optimize healthcare delivery of the hospital or nursing home within a given budget [14]. The costs from a family perspective include those paid by the family due to the treatment or care for dementia [14]. The cost components of different perspectives can be varied, but it is important to include each cost item only once regardless of the perspectives to avoid double counting.

3.3 Cost Components in Cost-of-Illness Studies of Dementia

There are four cost components in COI studies of dementia [14]: (a) direct medical costs, (b) direct nonmedical costs, (c) indirect costs, and (d) intangible costs (Fig. 1). Direct medical costs include all healthcare expenditures used for dementia, including goods and service costs that are related to diagnosis and treatments of dementia such as hospitalization and outpatient services. Direct nonmedical

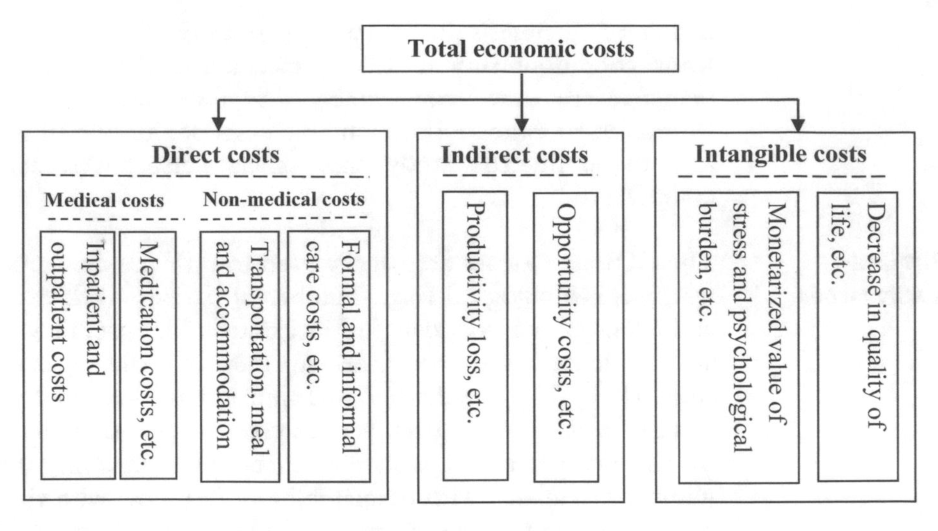

Fig. 1 Main cost components in the cost-of-illness studies of dementia

costs are incurred by outpatient visits or inpatient hospitalizations, including costs of transportation and accommodation for patient and family members during these visits, as well as costs of formal care at nursing home and informal care. Informal care cost often refers to the opportunity costs (i.e., working time or leisure time loss) of unpaid care provided by family members, relatives, or friends [15]. Although it is sometimes considered to be part of indirect costs, informal care cost is usually included in the direct nonmedical costs because it replaces formal care services that have fallen into this category [16]. Indirect costs included the costs of productivity loss due to premature death and disability. Intangible costs mean the monetary value of suffering from depression or upset caused by dementia.

In most of the COI studies of dementia, costs of certain items or components were not included in the estimates of total costs due to lack of consensus on how to value these costs. For example, direct nonmedical costs for home safety modifications, indirect costs due to premature mortality, intangible costs, and costs of crime committed by patients with dementia are usually not included in cost estimates [17].

3.4 Estimation of Costs of the Main Cost Components

3.4.1 Direct Costs

The bottom-up approach and top-down approach are the two methods that are commonly used to estimate the direct costs of dementia [11]. In the bottom-up approach, cost data by individual cost items are collected through questionnaire survey or review of medical records. Then, estimates of the total costs can be obtained by multiplying the mean cost per patient with the number of patients derived from prevalence studies. The top-down method uses national or regional statistics on the total costs of diseases to separate out the costs of a certain disease (e.g., dementia) [11].

However, it is often difficult to tease out the cost of a certain illness from the national or regional databases because information on different diseases is usually sparse in the statistics. Thus, the bottom-up approach is more frequently used in the COI studies of dementia. However, a potential disadvantage is that costs for dementia may be overestimated due to double-counting costs because patients with dementia are likely to have multiple coexisting chronic health conditions (e.g., heart disease, stroke, and depression). Thus, adjusting for the presence of these coexisting conditions is important in estimating the costs due to dementia alone.

3.4.2 Informal Care Costs

Informal care provided by family members, relatives, or friends has a significant impact on the societal costs of dementia [18]. Informal caregiving, in the form of assistance with basic activities of daily living (ADLs), is an important component of the support required by patients with dementia, yet it is unclear how to attribute the monetary cost to informal caregiver's time. Indeed, it is complicated to quantify the informal care costs owing to lack of consensus in methodological issues. First, quantifying caregiver's time can be problematic. The time spending on informal care often depends on the needs of care in self-care ADLs and instrumental ADLs. The cost is much lower when only considering the cares in basic ADLs compared with cares in both basic and instrumental ADLs, as well as time spending in supervision [19]. Moreover, dementia patient and the caregiver may both involve the same activities such as shopping together [20]. Second, estimating informal care cost is also complicated. Most of the previous studies have applied two different methods to assessing informal care costs [20–22]: the opportunity cost approach and the replacement cost approach. To estimate the opportunity cost, the best alternative use of the caregiver's time is taken to indicate the value of that time spent in care. If a caregiver gives up paid work to care, the lost earnings are estimated. Quantifying the costs of caregiver's time for retired people can be challenging because there are no obvious market prices [16]. The replacement cost approach assumes that the informal caregiver's inputs should be calculated according to the cost of replacing their care inputs with those of a professional caregiver [16].

3.4.3 Indirect Costs

The appropriate method for estimating indirect costs remains debatable [5]. The common approaches are the human capital, the willingness-to-pay, and the friction cost methods [23]. In the human capital approach, the average earnings by age and gender are combined with time lost from work or years of working life lost due to premature death to estimate unrealized earnings [24]. However, human capital approach with average wage may be subject to capital market imperfections. The willingness-to-pay approach, in contrast, incorporates both lost productivity and the intrinsic value of life to estimate the average costs that an individual would be willing to pay for an additional year of life. The friction cost method assumes that production losses are confined to the period needed to replace the "sick" worker [25].

Of note, the national and international guidelines for COI studies advocate that the past or future costs of a disease should be discounted to the present costs, usually by applying a discount rate of 3–5%. Most COI studies in the USA and Europe employ 3% or 3.5% as the discount rate [26, 27]. The following equation is used to estimate costs:

$$C_a = C_t \sum_{t}^{n=1} (1+r)^{-n},$$

where C_a denotes the cost in present year a, C_t represents the cost in year t, r refers to the discount rate, and n implies the time interval between year a and year t [5].

In addition, there are always uncertainties when estimating costs of a disease because COI studies are often based on a set of assumptions and data sources [28]. Sensitivity analysis can be carried out to assess the impact of variations of certain factors on the cost estimates. In the sensitivity analysis of dementia costs, we could assess the impact on cost estimates of variations in factors such as prevalence, medical costs, average wages, productivity weight, time spending on informal care, and the proportion of patients with dementia who receive informal and formal care [24, 29–31].

3.5 Sources of Cost Data

The sources of cost data in COI studies include the following. First, we can collect cost data through questionnaire survey that includes costs by individual items, where direct and indirect costs can be estimated. When designing the questionnaire, cost indicators can be relatively comprehensive. However, data collected through the survey may be subject to recall bias. Accurate and precise data may be obtained through follow-up survey, but it can be expensive and time consuming [23]. Moreover, the questions and guides in the questionnaire may vary across studies, thus resulting in inaccuracies of the data [23]. Second, cost data can be derived from electronic medical records, which usually include accurate and reliable information on medical expenses. However, medical records often include data of direct medical costs, and the costs may vary depending on available facilities of hospitals or medical institutes. Thus, cost data from certain medical institutes may not be representative of the target population [23]. Finally, relevant data needed for cost estimates can also be obtained from literature or government statistics. This is usually convenient and timesaving, but it may not capture all the utilizations, and the representativeness should also be kept in mind [32].

4 Economic and Societal Burden of Dementia: A Brief Summary of Literature

We searched PubMed, Google Scholar, Web of Science, and China National Knowledge Infrastructure databases for studies estimating the economic costs of Alzheimer's disease and other forms of dementia that were published since the 2000s. Table 1 provides a summary of major literature regarding economic burden of dementia in 2000–2017. A total of 17 studies were identified. There are substantial variations in the estimates of total costs per dementia patient, ranging from US$2935 (China) to US$64168 (USA) [24, 33–48], largely depending on methodologies, perspectives of cost estimation, data sources, cost categories, and setting and severity of patients with dementia. Moreover, there are regional differences in economic costs of dementia, such that dementia causes higher economic costs in high-income countries (HICs) compared with low- and middle-income countries (LMICs). To reduce future burden of dementia, national and global strategic action plan to promote primary and secondary prevention of dementia as well as to improve social care of dementia is critical [24].

5 Future Perspectives

First, methodological issues need to be reported in detail in future COI studies of dementia, because an insufficient description of the methodology may lead to misunderstandings or misinterpretations of the results. The methodological limitations should also be fully discussed. These issues are particularly relevant given that consensus on the methodology of the COI studies forms a basis for estimating and comparing the economic burden of a disease across studies from different regions. Moreover, the indicators selected in the study should be combined with local conditions. For example, the disability-adjusted life years (DALYs) index needs to be adjusted according to the local economic index, age weight, disability weight, and life expectancy. In addition, the clinical stages and severity of dementia should be taken into consideration when estimating the costs of dementia.

Second, cost data from large samples with long-term follow-ups are preferred for estimating dementia costs. Current COI studies have been limited by small sample size [15] and a relatively short period of follow-ups [49]. Moreover, survival benefits from treatment, dynamic changes in prevalence of risk factors for dementia (e.g., hypertension and diabetes), increases in care costs (e.g., use of medical devices), and hiring professional caregivers are likely to affect the economic burden of dementia. However, these data are generally missing, which limits the use of advanced approaches (e.g., microsimulation approach).

Table 1
A brief summary of literature regarding economic costs of dementia, 2000–2017

Study, country	Perspectives of cost estimation	Costs by categories (US$)			
		Direct medical care	Direct nonmedical care	Indirect cost informal care	Total costs
Rigaud et al. [33], France	Societal	6663	5632	18,858	31,153
Scuvee-Moreau et al. [34], Belgium	NA	8218	5222	1233	14,673
Beeri et al. [35], Israel	Societal	3974	9326	6593	19,893
Jönsson et al. [36], Nordic region	Societal	3113	7167	3757	14,038
Francois et al. [37], Finland	Societal	NA	NA	NA	33,333
Wolstenholme et al. [38], UK	NA	1818	30,650	NA	32,468
Livingston et al. [39], UK	Societal	NA	NA	NA	35,287
Lopez-Bastida et al. [40], Spain	Societal	4848	2306	29,884	37,881
Coduras et al. [41], Spain	Societal	4744	5798	12,016	22,558
Mesterton et al. [42], Sweden	Societal	3155	39,373	4428	46,956
Kiencke et al. [43], Germany	Healthcare payer	11,786	NA	NA	11,786
Zencir et al. [44], Turkey	NA	2128	NA	1364	3492
Wang et al. [45], China	NA	863	431	163	2935[a]
Xu et al. [24], China	Societal	1152	4779[b] 4612[c]	88	6020[d] 5852[e]
Hurd et al. [46], USA	Societal	33,329	NA	30,839[f] 14,591[g]	64,168[f] 47,920[g]
Suh et al. [47], Korea	Societal	4394	445	655	11,389
Allegri et al. [48], Argentina	Societal	3389	2488	1832	7709

[a]Net costs. NA: not available
[b]Informal care
[c]Nursing home care
[d]Cost for patients living at home
[e]Cost for patients living at nursing home
[f]Valued according to replacement cost
[g]Valued according to cost of forgone wages

Finally, the potential impact of secular trends in dementia occurrence on future societal and economic burden of the disease needs to be clarified [50]. Currently, all estimates of dementia costs have been based on the assumption of a stable or increased prevalence over time. However, recent studies have suggested a declining trend in occurrence of dementia in some HICs (e.g., UK, Sweden, and USA) [51–54]. In contrast, an increase in the age-specific prevalence of dementia has been reported in Asia (e.g., Japan and China) or LMICs. As prevalence of dementia is the second most important driver of societal costs of dementia after population aging [2], changes in dementia occurrence are likely to have significant impact on the total number of people with dementia in the community and thus have a huge impact on the societal burden. The impacts of time trends in dementia occurrence on global societal costs are also complicated by the fact that the number of dementia cases would increase to a greater extent in LMICs (~250%) than in HICs (~116%) [29]. Therefore, further research to clarify the impact of time trends in dementia occurrence on societal costs is urgently needed, in which up-to-date epidemiological data and cost data should be systematically reviewed and analyzed using the standardized COI approaches. This will help develop health policy and allocate resources.

Acknowledgment

Chengxuan Qiu received research grants from the Swedish Research Council (VR), the Swedish Research Council for Health, Working Life and Welfare (FORTE), and Karolinska Institutet, Stockholm, Sweden.

References

1. American Psychiatric Association (2000) Diagnostic and statistical manual of mental disorders, 4th edn. American Psychiatric Association, Washington DC
2. Alzheimer's Disease International (ADI) (2015) World Alzheimer report 2015, the global impact of dementia: an analysis of prevalence, incidence, cost and trends. Alzheimer's Disease International (ADI), London. http://www.worldalzreport2015.org/downloads/world-alzheimer-report-2015.pdf. Accessed 10 Jun 2017
3. World Health Organization and Alzheimer's Disease International (2012) Dementia: a public health priority. World Health Organization, Geneva. http://apps.who.int/iris/bitstream/10665/75263/1/9789241564458_eng.pdf?ua=1. Accessed 10 June 2017
4. Quentin W, Riedel-Heller S, Luppa M, Rudolph A, König HH (2010) Cost of illness studies of dementia: a systematic review focusing on stage dependency of costs. Acta Psyciatr Scand 121:243–259
5. Costa N, Derumeaux H, Rapp T et al (2012) Methodological considerations in cost of illness studies on Alzheimer disease. Health Econ Rev 15:18
6. McKhann G, Drachman D, Folstein M, Katzman R, Price D, Stadlan EM (1984) Clinical diagnosis of Alzheimer's disease: report of the NINCDS-ADRDA work group under the auspices of Department of Health and Human Services Task Force on Alzheimer's disease. Neurology 34:939–944
7. Hope T, Keene J, Gedling K, Fairburn CG, Jacoby R (1998) Predictors of institutionaliza-

tion for people with dementia living at home with a carer. Int J Geriatr Psychiatry 13: 682–690

8. Small GW, McDonnell DD, Brooks RL, Papadopoulos G (2002) The impact of symptom severity on the cost of Alzheimer disease. J Am Geriatric Soc 50:321–327
9. Mauskopf J, Racketa J, Sherrill E (2010) Alzheimer's disease: the strength of association of costs with different measures of disease severity. J Nutr Health Aging 14:655–663
10. Herman N, Tam DY, Balshaw R, Sambrook R, Lesnikova N, Lanctôt KL (2010) The relation between disease severity and cost of caring for patients with Alzheimer disease in Canada. Can J Psychiatr 55:768–775
11. Wimo A, Jönsson L, Fratiglioni L et al (2016) The societal costs of dementia in Sweden 2012 – relevance and methodological challenges in valuing informal care. Alzheimers Res Ther 8:59
12. Girmens JF, Koerber C, Miadi-Fargier H, Wittrup-Jensen KU (2013) Costs of incidence-based central retinal vein occlusion in France. Value Health 16:A504–A504
13. Andlin-Sobocki P, Jönsson B, Wittchen HU, Olesen J (2005) Cost of disorders of the brain in Europe. Eur J Neurol 12(Suppl 1):1–27
14. Leung GM, Yeung RY, Chi I, Chu LW (2003) The economics of Alzheimer disease. Dement Geriatr Cogn Disord 15:34–43
15. Olesen J, Gustavsson A, Svensson M, Wittchen HU, Jönsson B (2012) The economic cost of brain disorders in Europe. Eur J Neurol 19:155–162
16. Wimo A, Jönsson L, Bond J, Prince M, Winblad B, Alzheimer Disease International (2013) The worldwide economic impact of dementia 2010. Alzheimers Dement 9:1–11.e3
17. Deb A, Thornton JD, Sambamoorthi U, Innes K (2017) Direct and indirect cost of managing Alzheimer's disease and related dementias in the United States. Expert Rev Pharmacoecon Outcomes Res 17:189–202
18. Schneider J, Hallam A, Islam MK et al (2003) Formal and informal care for people with dementia: variations in costs over time. Ageing Soc 23:303–326
19. Prince M, Bryce R, Albanese E et al (2013) The global prevalence of dementia: a systematic review and metaanalysis. Alzheimers Dement 9:63–75.e2
20. Jönsson L, Wimo A (2009) The cost of dementia in Europe: a review of the evidence, and methodological considerations. PharmacoEconomics 27:391–403
21. McDaid D (2001) Estimating the costs of informal care for people with Alzheimer's disease: methodological and practical challenges. Int J Geriatr Psychiatry 16:400–405
22. Nordberg G, Wimo A, Jönsson L et al (2007) Time use and costs of institutionalised elderly persons with or without dementia: results from the Nordanstig cohort in the Kungsholmen project: a population based study in Sweden. Int J Geriatr Psychiatry 22:639–648
23. Xu JF (2016) Study on the economic burden of mental disorders and the influence of health insurance-based on patients in Shandong province. Shandong University, Jinan
24. Xu J, Wang J, Wimo A, Fratiglioni L, Qiu C (2017) The economic burden of dementia in China, 1990–2030: implications for health policy. Bull World Health Organ 95:18–26
25. Oostdam N, Bosmans J, Wouters MG, Eekhoff EM, Mechelen WV, Poppel MNV (2012) Cost-effectiveness of an exercise program during pregnancy to prevent gestational diabetes: results of an economic evaluation alongside a randomised controlled trial. BMC Pregnancy Childbirth 12:64
26. Gold MR, Russel LB, Siegel JE, Daniels N, Weinstein MC (1996) Cost effectiveness in health and medicine. JAMA 276:1172–1177
27. Luengo-Fernandez R, Leal J, Gray A, Sullivan R (2013) Economic burden of cancer across the European Union: a population-based cost analysis. Lancet Oncol 14:1165–1174
28. Xu J, Wang J, Wimo A, Qiu C (2016) The economic burden of mental disorders in China, 2005–2013: implications for health policy. BMC Psychiatry 16:137
29. Wimo A, Guerchet M, Ali GC et al (2017) The worldwide costs of dementia 2015 and comparisons with 2010. Alzheimers Dement 13:1–7
30. Wimo A, Winblad B, Jönsson L (2010) The worldwide societal costs of dementia: estimates for 2009. Alzheimers Dement 6:98–103
31. Wang H, Gao T, Wimo A, Yu X (2010) Caregiver time and cost of home care for Alzheimer's disease: a clinic-based observational study in Beijing, China. Ageing Int 35:153–165
32. Zhang YY (2012) Study on the economic burden of inpatients with hepatitis B owning basic medical insurance for urban workers in Tianjin. Tianjin University, Tianjin
33. Rigaud AS, Fagnani F, Bayle C, Latour F, Traykov L, Forette F (2003) Patients with Alzheimer's disease living at home in France: costs and consequences of the disease. J Geriatr Psychiatry Neurol 16:140–145

34. Scuvee-Moreau J, Kurz X, Dresse A (2002) The economic impact of dementia in Belgium: results of the National Dementia Economic Study (NADES). Acta Neurol Belg 102:104–113
35. Beeri MS, Werner P, Adar Z, Davidson M, Noy S (2002) Economic cost of Alzheimer disease in Israel. Alzheimers Dis Assoc Disord 16:73–80
36. Jönsson L, Eriksdotter Jö nhagen M, Kilander L et al (2006) Determinants of costs of care for patients with Alzheimer's disease. Int J Geriatr Psychistry 21:449–459
37. Francois C, Sintonen H, Sulkava R, Rive DB (2004) Cost effectiveness of memantine in moderately severe to severe Alzheimer's disease: a Markov model in Finland. Clin Drug Invest 24:373–384
38. Wolstenholme J, Fenn P, Gray A, Keene J, Jacoby R, Hope T (2002) Estimating the relationship between disease progression and cost of care in dementia. Br J Psychiatry 181:36–42
39. Livingston G, Katona C, Roch B, Guilhaume C, Rive B (2004) A dependency model for patients with Alzheimer's disease: its validation and relationship to the costs of care. The LASER-AD study. Curr Med Res Opin 20:1007–1016
40. Lopez-Bastida J, Serrano-Aguilar P, Perestelo-Perez L, Olivamoreno J (2006) Social-economic costs and quality of life of Alzheimer disease in the Canary Islands, Spain. Neurology 67:2186–2191
41. Coduras A, Rabasa I, Frank A et al (2010) Prospective one-year cost-of-illness study in a cohort of patients with dementia of Alzheimer's disease type in Spain: the ECO study. J Alzheimers Dis 19:601–615
42. Mesterton J, Wimo A, By A, Langworth S, Winblad B, Jönsson L (2010) Cross sectional observational study on the societal costs of Alzheimer's disease. Curr Alzheimer Res 7:358–367
43. Kiencke P, Dietmar D, Grimm C, Rychlik R (2011) Direct costs of Alzheimer disease in Germany. Eur J Health Econ 12:533–539
44. Zencir M, Kuzu N, Beşer NG, Ergin A, Catak B, Sahiner T (2005) Cost of Alzheimer's disease in a developing country setting. Int J Geriatr Psychiatry 20:616–622
45. Wang G, Cheng Q, Zhang S et al (2008) Economic impact of dementia in developing countries: an evaluation of Alzheimer-type dementia in shanghai, China. J Alzheimers Dis 15:109–115
46. Hurd MD, Martorell P, Langa KM (2013) Monetary costs of dementia in the United States. N Engl J Med 369:489–490
47. Suh GH, Knapp M, Kang CJ (2006) The economic costs of dementia in Korea, 2002. Int J Geriatr Psychiatry 21:722–728
48. Allegri RF, Butman J, Arizaga RL et al (2007) Economic impact of dementia in developing countries: an evaluation of costs of Alzheimer-type dementia in Argentina. Int Psychogeriatr 19:705–718
49. Cavallo MC, Fattore G (1997) The economic and social burden of Alzheimer disease on families in the Lombardy region of Italy. Alzheimer Dis Assoc Disord 11:184–190
50. Hurd MD, Martorell P, Langa K (2015) Future monetary costs of dementia in the United States under alternative dementia prevalence scenarios. J Popul Ageing 8:101–112
51. Matthews FE, Arthur A, Barnes LE et al (2013) A two-decade comparison of prevalence of dementia in individuals aged 65 years and older from three geographical areas of England: results of the cognitive function and ageing study I and II. Lancet 382:1405–1412
52. Matthews FE, Stephan BC, Robinson L et al (2016) A two decade dementia incidence comparison from the cognitive function and ageing studies I and II. Nat Commun 7:11398
53. Satizabal CL, Beiser AS, Chouraki V, Chêne G, Dufouil C, Seshadri S (2016) Incidence of dementia over three decades in the Framingham heart study. N Engl J Med 374:523–532
54. Langa KM, Larson EB, Crimmins EM et al (2017) A comparison of the prevalence of dementia in the United States in 2000 and 2012. JAMA Intern Med 177:51–58

Chapter 2

Epidemiology and Risk Factors of Alzheimer's Disease: A Focus on Diet

Cécilia Samieri

Abstract

Alzheimer's disease (AD) appears to result from a highly dynamical, time-dependent cascade of neuropathological processes, which involve the balance between two antagonist pathways: an accumulation of lesions in the brain (evolving progressively over decades), balanced by a variety of cellular and tissular compensatory mechanisms. Preclinical research has indicated that several nutritional factors have the potential to modulate many components of this complex system; dietary interventions may thus constitute promising strategies for the prevention of AD. Extensive epidemiological research has been conducted on nutrition, diet, and AD dementia over the past two decades. This chapter reviews existing epidemiological literature on nutrition and cognitive aging and AD dementia, with a specific focus on observational, prospective studies, including both clinical outcomes and brain imaging biomarkers. Most studies investigated candidate foods and nutrients, including "good" fats (e.g., marine long-chain omega-3 polyunsaturated fatty acids provided by fish intake), antioxidant nutrients (i.e., vitamins C and E, carotenoids, and polyphenols, found in plant products), and B vitamins (e.g., folate provided by green leafy vegetables). Overall, associations between individual nutrients and both clinical outcomes and brain imaging biomarkers have been relatively inconsistent. Healthy dietary patterns such as the Mediterranean diet, which combine several healthy foods and nutrients, have been more consistently related to AD outcomes. Still, most randomized controlled trials on nutritional supplements for the prevention of AD have remained inconclusive so far. Future research should leverage innovative approaches such as "omics" methods to explore dietary exposures, novel imaging markers to capture prodromal AD, and longitudinal statistical approaches to model long-term trajectories of risk factors and AD markers, to better understand how and when diet may shape the brain and prevent AD.

Key words Alzheimer's disease, Dementia, Epidemiology, Population-based, Diet, Nutrition, Vitamins, Fatty acids, Clinical trial

1 Introduction

Dementia is the most frequent disease associated with brain aging. Approximately 36 million of persons were affected by dementia in 2010 worldwide, and with the gradual aging of populations, this prevalence is expected to dramatically increase in the next decades, to reach possibly around 131 million persons by 2050 [1]. Alzheimer's disease (AD) is the most frequent form of dementia,

Robert Perneczky (ed.), *Biomarkers for Preclinical Alzheimer's Disease*, Neuromethods, vol. 137,
https://doi.org/10.1007/978-1-4939-7674-4_2,

accounting for approximately two thirds of cases. No curative treatment for AD has been identified yet, and therapeutic research has been extremely disappointing so far. Sporadic AD (the form which occurs in older age and represents 99% of all cases) appears to result from a complex interplay between genetic and environmental risk factors. The strongest genetic risk factor for sporadic AD is the ε4 allele of the apolipoprotein E gene (APOEε4). Furthermore, about 20 single nucleotide polymorphisms have been moderately associated with disease risk in genome-wide association studies [2]. Environmental risk factors for AD and dementia include cardio-metabolic risk factors, psychosocial factors, and lifestyle [3]. Recent epidemiological findings from several large prospective studies across Europe and the USA have established a decrease in the incidence of dementia in the last two decades [4, 5]. Improvements of educational level and in the management of cardiovascular risk factors have been proposed as a plausible explanation for such a decrease of incidence, suggesting that lifestyle-based preventive interventions aimed at improving cardiovascular health may be a promising strategy to lower the burden of dementia.

However, there is now strong evidence that pathophysiological mechanisms evolve silently over years, if not decades, before clinical diagnosis of dementia in the late life; it is thus likely that prevention may be effective only through a long-course perspective, targeting individuals in the preclinical phase of the disease [3]. The mechanisms underlying AD dementia appear complex and are still not fully understood. They involve a progressive accumulation in the brain of several pathophysiological alterations, including (1) Aβ proteins which aggregate extracellularly in the form of plaques, (2) intracellular production of hyper-phosphorylated Tau proteins in the form of tangles, and (3) neuro-inflammation which occurs through activation of microglia. The interaction of these pathological events will eventually lead to neurodegeneration and related cognitive decline; however, the nature of such interactions and their chronology are still under debate [3]. The primary theory of AD focuses on the accumulation of amyloid plaques as the main neuropathological culprit of the disease. According to the "amyloid cascade hypothesis," brain Aβ accumulation, which results from increased production of Aβ, decreased clearance, or both, may further precipitate the production of Tau (also found in normal aging) and may increase neuro-inflammation caused by microglial cells, which leads to neurodegeneration. An alternative theory views accumulation of Aβ and Tau as two parallel processes, causing AD and enhancing each other's toxic effects [6]. Overall, Aβ appears necessary but not sufficient to the development of clinical dementia. Still, the reasons for brain Aβ accumulation remain partly unknown, and the factors which dictate how brain health will respond to amyloid/Tau pathology have still to be elucidated. A key additional player may be vascular brain health. Indeed, age-related alterations of the neurovasculature which constitutes the blood-brain barrier (also known as the "blood-brain barrier breakdown") may both increase the permeability of the brain to various neurotoxic substances, cells, and pathogens (bacteria and viruses, e.g., herpes virus, which can trigger amyloid deposition [7]) and strongly hinder the clearance of Aβ from the brain [8]. Early dysfunction of the vascular system may thus be the first step that predispose to the accumulation of amyloid, along with reducing the oxygen and metabolite supply (sugar, nutrients, etc.) to the brain. Another important player, which adds to the complexity of the disease, relies on compensatory mechanisms, generally described as cognitive and brain reserve. Cognitive and brain reserve enable the brain, thanks to a more developed tissue (i.e., more numerous neurons and/or higher synaptic density) and alternative neuronal circuits, to cope with the presence of lesions for a long time before the expression of clinical symptoms. Hence, higher educational level (a proxy of cognitive reserve) has been associated with delayed cognitive decline and AD dementia in epidemiological studies [9].

In summary, AD dementia appears to result from a highly dynamical neuropathological system based on several processes (amyloid production and clearance, Tau production, neuro-inflammatory processes, neurovascular health, and cognitive/brain reserve), of which interactions over 10–20 years will determine cognitive trajectories and the risk to develop dementia in the late life. Furthermore, the neurovascular unit (made of the microvascular endothelium defining the blood-brain barrier and astrocytes, pericytes, neurons, and microglial cells) is a key player of this dynamic network, and each of the elements forming the neurovascular unit has a strong plasticity (e.g., neurovascular plasticity, neuronal/synaptic plasticity, microglia plasticity)—thus allowing the system to adapt for a long time before falling down. Interestingly, lifestyle factors, in particular nutrition, have the potential to modulate the plasticity of each of these elements over the life course. Furthermore, selected nutrients may have a specific role against the accumulation of amyloid, Tau, and vascular lesions in AD.

In the context of a disease evolving over years with no curative treatment, prevention is considered one of the most promising strategies to face dementia and AD [10]. Understanding the early influence and interactions of non-modifiable risk factors (e.g., gender, genetics, age) and modifiable risk factors (e.g., physical activity, diet, and cognitive stimulation) for dementia in population-based samples is considered as a research priority to reduce the global burden of dementia by 2025 [10]. In this chapter, we will provide an overview of current knowledge on nutrition and diet for the prevention of dementia and AD through an epidemiological perspective. We will focus on prospective observational studies, in which exposure precedes outcome, and we will put less emphasis on cross-sectional studies which assess exposure and outcome concomitantly, because prospective studies address temporality of associations (a pillar of causality). We will start with studies on clinical outcomes (i.e., dementia/AD diagnosis and cognitive decline) and continue with most recent studies on AD biomarkers (i.e., assessment of brain atrophy, vascular health, and amyloid load using brain imaging) which give access to subtle, preclinical impairments in early phases of the neuropathological continuum underlying AD. Nutritional exposures will be examined through a bottom-up approach, addressing first the most important individual candidates for brain health and second a global approach of nutrition through dietary patterns, which take into account potential additive and/or synergistic effects of the food matrix on health. Furthermore, we will examine both food/nutrient intake data gained from dietary surveys and nutrient biomarkers, which are less prone to measurement error and incorporate interindividual variability in biodisponibility. The chapter will end with an overview on main findings from randomized controlled trials and brain aging outcomes, discussing potential reasons for inconsistencies with observational studies and providing a few insights on potential future innovations in the field.

2 Nutrition, Diet, and Alzheimer's Disease Clinical Outcome

2.1 Nutrient and Food Candidates for Brain Health

Several nutrients and foods have neuroprotective effects which have been largely established in preclinical research. Overall, these nutrient/food candidates fall into three main categories: (1) "good" fats, i.e., unsaturated fats, in particular polyunsaturated fats from the omega-3 family (n-3 PUFAs), mainly provided by vegetable oils (for precursors, e.g., alpha linolenic acid [ALA]) and by fish (for the long-chain derivatives, e.g., eicosapentaenoic acid [EPA] and docosahexaenoic acid [DHA] which are directly useful for brain structure and function); (2) antioxidant nutrients, including vitamins E and C, carotenoids, and polyphenols, provided by fruits, vegetables, vegetable oils (e.g., olive oil), and plant-derived beverages (coffee, tea, wine); (3) other vitamins involved in various neuroprotective mechanisms, including B vitamins (folate or vitamin B9, particularly abundant in green leafy vegetables, and vitamin B12 provided by animal products) and vitamin D, found in fatty fish. Each of these nutrients has been individually related to the risk of AD clinical outcomes in prospective observational studies.

2.1.1 "Good" Fats: Omega-3 Polyunsaturated Fats and Fish

Lipids represent 50–60% of total brain weight, mostly under the form of PUFAs in phospholipids of neuronal membranes or in the gain of myelin, and the long-chain (LC) n-3 PUFA DHA represents up to 60% of esterified fatty acids in neuronal membranes. DHA is essential to the development of the mammalian brain—in particular for the hippocampus, a key structure for memory which is altered in both aging and AD neuropathology. Rodent studies have demonstrated that LC n-3 PUFA preferentially accumulate in the hippocampus and are involved in neuronal transmission, neurogenesis, synaptic plasticity, learning, and memory [11]. Furthermore, hippocampal DHA levels decrease with age and in AD, and this decrease is associated with reduced hippocampal-dependent spatial learning memory ability [12]. EPA and DHA have anti-inflammatory properties which may help lower neuroinflammation associated with normal brain aging and AD. DHA may also exert anti-amyloid effects [13]. Hence, an adequate supply of essential LC n-3 PUFA in the diet (mainly from fish) may be critical for the maintenance of cognition and memory with aging.

Some observational studies have found that intakes of fish, or higher concentrations of LC n-3 PUFA in blood, may be related to a lower risk of dementia and to slower cognitive decline [14–18] in older subjects. However, inconsistent findings have been reported, especially with cognitive decline [19]. In 2016, a meta-analysis of 21 prospective cohort studies including 181,580 participants and 4438 dementia cases followed for 2–21 years concluded that each 1-serving/week increment of dietary fish was associated with a 5% lower risk of dementia (95% confidence interval [CI], 1–10%) and a 7% lower risk of AD (95% CI, 5–10%). Dietary DHA was also significantly associated with lower risks of dementia and AD, while there was no significant association with dietary EPA or blood EPA and DHA biomarkers [20]. Likewise, blood ALA, the precursor of the LC n-3 PUFA found in vegetable oils (colza, soybean, walnut), was not associated with dementia risk—which may not be surprising given that the conversion of ALA to derivatives EPA and DHA is considered very limited in humans [21].

Recently, the Rush Memory and Aging cohort reported, among 286 older participants (aged 90 years old on average), a cross-sectional association between higher intake of fish and lower AD neuropathology after brain autopsy at death (an average 8 years later); yet the association was limited to carriers of the APOEε4 allele [22, 23]. There is a biological rationale for a vulnerability of APOEε4 carriers to lower DHA status, especially in early AD stages, before marked neurodegeneration [24]. First, APOEε4 may decrease circulating DHA by increasing DHA hepatic catabolism (through increased affinity to LDL and VLDL receptors). Second, APOEε4 is associated with a breakdown of the blood-brain barrier and may decrease the transfer of DHA to the brain (by altering the efficacy of the Mfsd2a receptor). Third, APOEε4 is associated with hypolipidated lipoproteins and may thus compromise the transport of DHA in the brain. However, interactions between fish and APOEε4 carrier status in relation to AD clinical outcomes have been inconsistent in epidemiological studies so far, with studies reporting association limited to APOEε4 carriers

[22, 23, 25, 26], others reporting association in APOEε4 noncarriers [27–30], and most studies reporting no interaction [20]. Moreover, an inverse association between low serum DHA and cerebral amyloidosis was found independent of the APOEε4 genotype in a small study of older non-demented persons [31].

2.1.2 Antioxidant Nutrients and Fruits, Vegetables, and Specific Antioxidant-Rich Compounds (Olive Oil, Red Wine, Coffee, Tea, Cocoa)

As n-3 PUFA, antioxidant nutrients (vitamins E and C, carotenoids, and polyphenols) and their main food sources (fruits, vegetables, and specific polyphenol-rich foods and beverages, e.g., olive oil, wine, coffee, tea, cocoa) have been investigated for more than 15 years in relation to brain aging outcomes in epidemiological research. Antioxidant nutrients may help lower oxidative stress associated with AD neuropathology, thereby counteracting oxidation of easily peroxidable brain LC PUFA (e.g., DHA). In particular, vitamin E, provided by seed oils and nuts, represents the major lipid-soluble antioxidant found in cells; the main function of vitamin E is to prevent the peroxidation of membrane phospholipids [32]. Carotenoids, found in colored fruits and vegetables, are precursors of retinoids, key signaling molecules for synaptic plasticity [33]. As all liposoluble vitamins (i.e., vitamins A, D, E, and K), vitamins A and E may affect the generation and clearance of Aβ both by direct effects and indirectly by altering the cellular lipid homeostasis [32]. Aside from the "classical" antioxidant vitamins, polyphenols have been increasingly investigated over the last 5 years in relation to brain health, since their contribution to the total antioxidant capacity of the human diet appears much larger than that of vitamins [34] and some polyphenols have powerful biological effects for brain function and brain health. For example, selected flavonoids have strong vascular beneficial properties and improve vasoactivity both in the periphery and brain (e.g., flavan-3-ols [35], flavanols [36]). A human clinical study demonstrated that 3-month supplementation with flavanols from cocoa (catechins and epicatechins) in healthy adults improved cognition and enhanced hippocampal vascular plasticity at magnetic resonance imaging (MRI) [36]. Beyond vascular beneficial effects, certain polyphenols are found in brain areas involved in learning and memory and improve neuronal signaling and cognitive behavior (i.e., anthocyanins [37]), and some species appear to exert specific anti-amyloid effects. For example, resveratrol, found in grape and red wine, increases Aβ clearance in vitro [38]; however, the very high doses necessary to reach such biological effects may be impossible to reach from diet. Non-resveratrol polyphenols such as anthocyanin and quercetin (a flavonol) metabolites [e.g., quercetin-3-O-glucuronide] appear to cross the blood-brain barrier and may decrease Aβ aggregation and improve learning memory in mice [39, 40].

Epidemiological studies have primarily investigated the relation of dietary intakes of antioxidant foods to the risk of dementia or cognitive decline. Most studies on the overall consumption of fruit and vegetables found an association between higher intakes of vegetables and lower risk of AD or less cognitive decline, while association with fruit intake was less consistent [41]. Furthermore, a low-to-moderate consumption of alcoholic beverages was associated with a lower risk of dementia in several large prospective studies [42]; although all alcoholic beverages may be protective (consistently with a neuroprotective role of ethanol at very moderate doses), a stronger association was found with red wine, which provides, in addition to ethanol, a large variety of polyphenols. As red wine, olive oil (a specific feature of the Mediterranean diet) contains many polyphenol species (e.g., tyrosol) and may contribute to protect the brain from age-related diseases. In the Three-City (3C) study, intensive use of olive oil for cooking and seasoning was related to lower risk of stroke and less decline in visual memory [43, 44]. However, epidemiological research on olive oil and brain health has been very limited so far and deserves further investigation. Likewise, coffee and tea intakes were related to lower cognitive decline or dementia risk in a few studies; yet inconsistent findings were also reported, especially from prospective studies [45], and overall, these associations need to be further explored in large-scale studies.

Intakes of individual antioxidant nutrients were also examined in relation to dementia outcomes. However, studies on carotenoids and vitamins were hampered by the inclusion of supplement users, who have considerably higher intakes of antioxidants, and high-dose supplements may be more harmful than beneficial [46]. A meta-analysis of seven prospective studies on dietary intakes (excluding supplements) of vitamins E and C and β-carotene and AD risk found that moderate to high intakes (defined as being in the second to last quantile of exposure distribution) of vitamins E and C were related to a 24% (95% CI 16–33%) and a 17% (95% CI 6–28%) lower risk of AD, respectively, whereas the risk reduction associated with β-carotene (−12%) was only borderline significant (95% IC −27%, +3%) [47]. With antioxidant nutrients examined through blood biomarkers, findings have been relatively inconsistent so far. Lower plasma vitamin E was associated with higher cognitive impairment/dementia/AD risk in several prospective cohorts [48–52], but not all [53], and there is evidence that the association between vitamin E and cognitive decline/dementia is influenced by concurrent cholesterol levels [49, 51, 52]. Among carotenoids, plasma lutein was the single specie modestly inversely associated with dementia risk over 10 years in the 3C study [54], whereas in the MacArthur Studies of Successful Aging, serum β-carotene was associated with lower cognitive decline but only among APOEε4 carriers [55]. In contrast, in the Nurses' Health Study, none of the individual carotenoid or vitamin E species measured in plasma was associated with cognitive decline over 4 years [56].

Thus overall, epidemiological evidence for an association between lower intakes of carotenoids and antioxidant vitamins in dementia and AD has been mixed in existing literature.

Furthermore, promising, yet limited epidemiological research has been conducted on polyphenols and brain aging outcomes. Inverse associations were found between higher intakes of total flavonoids [57–60], flavonols [61], and anthocyanidins found in berries (blueberry, strawberry, raspberry) [60] and lower risk of dementia or lower cognitive decline. An Italian study found an association between total urinary polyphenols and lower cognitive decline over 3 years [62]. However, more research is needed both at the preclinical stage to better understand the complex bioavailability and metabolism of polyphenols and at the clinical stage to identify most promising targets for prevention.

2.1.3 Vitamins B and D

B vitamins, including folate (vitamin B9), vitamin B6, and vitamin B12, have a fundamental role in brain development and functioning and have been extensively examined in relation to age-related brain diseases [63]. B vitamins regulate homocysteine (Hcy) levels, and hyperhomocysteinemia is an established risk factor for cardiovascular diseases and dementia [64]. Hyperhomocysteinemia may affect brain function through both vascular mediation and neurotoxicity [65]. B vitamins may also have a direct effect on the brain, independently of Hcy: a lower status in B vitamins decreases S-adenosylmethionine, a major intermediate of methylation reactions which are involved in synaptic transmission and epigenetic regulation [66]. Moreover, tetrahydrofolate, the active form of folate, is involved in DNA replication which is critical for adult hippocampal neurogenesis [67]; direct mechanistic effects of B vitamins on the brain independent of Hcy may therefore be particularly relevant to folate. Prospective epidemiological studies have found associations between higher intakes of B vitamins and a lower risk of dementia and AD; four studies have reported associations with dietary folate (but not with vitamins B6 and B12) [68–71], although null [72–74] and even harmful associations (when folate was combined with low vitamin B12 status) [74] were also reported. Findings were also inconsistent in studies using blood status in B vitamins (see for review [75]). Importantly, most studies were conducted in the USA, where fortification of all flour and uncooked cereal grain product in folic acid became mandatory in 1998, leading to a subsequent dramatic increase in folate status which may have biased observational findings. Moreover, it is possible that folate is protective for the brain in lower intake ranges and become inefficient—and even detrimental for those with low B12 status—at higher ranges (as those found in North American population under folic acid fortification policy). In the French 3C study in France (a country where folic acid fortification is not mandatory), an inverse association was found between higher folate intake and lower dementia risk over 10 years [71].

As B vitamins, vitamin D has a major role in brain development and functioning; yet, beyond its well-documented role in calcium homeostasis and bone health, vitamin D has gained interest only quite recently for the prevention of age-related brain diseases [76]. Vitamin D results from endogenous synthesis after sun exposure and to a lower extent (about 20%) from intake of animal products including fatty fish, eggs, and dairies. The prevalence of vitamin D deficiency is high worldwide; in a systematic review of studies including 168,000 individuals from 44 countries, 37% of the participants had 25(OH)D (the active form of vitamin D) seric concentrations below 20 ng/mL (50 nmol/L; the threshold for hypovitaminosis D [i.e., insufficiency or deficiency] according to the definition of the World Health Organization and the US Institute of Medicine) [77], with older persons particularly at risk. Vitamin D is critical for brain function as it participates in neurotransmission (vitamin D receptors which are present in several brain areas, including regions involved in learning and memory). Furthermore, various neuroprotective properties have been attributed to vitamin D in preclinical studies, including modulation of immune response and of inflammatory and oxidative pathways; promotion of neurogenesis, synaptogenesis, and Aβ clearance; and prevention of neuronal death [76]. Several prospective observational studies have reported an association between higher vitamin D and lower cognitive decline, specifically of executive functions, whereas associations with episodic memory have been less consistent [78]. Prospective research on the risk of dementia has been limited to date. A meta-analysis of five observational studies including a total of 18,639 participants that compared to individuals with deficient vitamin D status (25(OH)D $<$ 20 nmol/L) to older persons with sufficient vitamin D (25(OH)D $>$ 50 nmol/L) found a 54% higher risk of dementia associated with vitamin D deficiency [79]. However, this meta-analysis did not include the most recent strong association of low vitamin D to cognitive decline and the risk of dementia and AD in the 3C study [80] and the null findings from two large cohorts, the Framingham Heart Study [81] and the Uppsala Longitudinal Study of Adult Men [82].

In summary, 20 years of preclinical and epidemiological research have established a strong scientific rationale for a role of a least seven families of nutrients (n-3 PUFA, vitamins E and C, carotenoids, polyphenols, vitamins B and D) in normal brain function and for the prevention of age-related brain diseases. Still, all these nutrients are contained in foods and consumed within a diet; moreover, they may have synergistic—or at least additive—effects on metabolic pathways underlying normal brain function and pathological brain aging. Thus, a higher level of data integration must consider dietary patterns in relation to health and disease.

2.2 Nutrient and Dietary Patterns

Dietary patterns incorporate all dietary habits in a global appraisal of diet and may be more relevant to brain health and diseases than nutrients taken in isolation. Indeed, many nutrients exert biological effects on the brain (as developed above); yet the combination of these nutrients in the diet may have a stronger effect than any nutrient considered individually. Alternatively, individual nutrients may be not sufficiently powerful to protect the brain from age-related diseases, and global approaches of diet may be more promising. Moreover, nutrients may act in synergy (although the concept of synergy of the food matrix has remained largely hypothetical to date). For example, antioxidant nutrients (e.g., carotenoids, vitamin D) may contribute to decrease the peroxidation of PUFAs in neuronal membranes of the aging brain [83, 84]. Accordingly, regular intake of fruits and vegetables (the main sources of antioxidant nutrients) appeared necessary to observe a cognitive benefit of fish consumption (a source of n-3 PUFAs and vitamin D) in the 3C study [28]. As done earlier with cardiovascular diseases, many cohort studies have thus examined the associations between dietary patterns and brain aging outcomes.

2.2.1 The Mediterranean Diet: A Landmark Model of Healthy Diet and Brain Health

Most convincing research on overall diet quality and brain health has certainly been obtained with the Mediterranean diet (MeDi). The MeDi, now considered as a landmark model for a healthy diet, has been defined as the traditional diet from Mediterranean countries; it is characterized by high intakes of plant foods (fruits, vegetables, cereals, and legumes), low intakes of meats and dairies, moderate intakes of fish and wine during meals, and olive oil as a main source of added fats [85]. Primary findings for a protective role of the MeDi on AD [86] and cognitive aging [87] were provided by the US cohort WHICAP and the French 3C study. These initial findings were replicated in several populations (see for review [85]); a meta-analysis combining five cohort studies found that subjects with higher MeDi adherence (i.e., in the last tertile of distribution of a score of adherence to the MeDi) had a 33% lower risk (95% CI 19–45%) of mild cognitive impairment or AD [88]. However, inconsistent results were also reported, especially with cognitive decline in the non-Mediterranean North American population (e.g., the MeDi was associated with higher average cognitive status, but not cognitive decline, in two large US cohorts [89, 90]). A Mediterranean diet score adapted to the Australian diet was associated with less decline in executive function over 3 years, but only in carriers of the APOEε4 allele [91].

2.2.2 Other Approaches of Food Quality and Brain Health

Other indices of diet quality have been investigated in relation to dementia outcomes. For example, among 27,860 subjects from 40 countries, a modified Alternative Healthy Eating Index, comprising seven components (including vegetables, fruits, nuts and soybean proteins, whole grain, deep-fried foods, ratio of fish to meat and egg, and alcohol), was associated with risk of a ≥3-point decline in

Mini-Mental State Examination score over 5 years [92]. The Dietary Approaches to Stop Hypertension (DASH) diet, which differs from the MeDi in that it specifies low consumption of saturated fat and commercial pastries and sweets and higher consumption of low-fat dairies, was associated with lower cognitive decline in the Memory and Aging Project [93], but not in the Nurses' Health Study [94]. With the objective to conceptualize a diet pattern score more specific to brain health, Morris et al. recently proposed the MIND (Mediterranean-DASH Intervention for Neurodegenerative Delay) diet, a hybrid of the Mediterranean-DASH diets. Similar to the Mediterranean and DASH diets, the MIND diet score emphasizes plant-based foods and limited intakes of animal and high saturated fat foods but uniquely specifies the consumption of berries and green leafy vegetables and does not specify high fruit, high dairy, or high potato consumption. The MIND diet score has 15 dietary components including ten brain healthy food groups (green leafy vegetables, other vegetables, nuts, berries, beans, whole grains, fish, poultry, olive oil, and wine) and five unhealthy food groups (red meats, butter and stick margarine, cheese, pastries and sweets, and fried/fast food). Higher adherence to the MIND diet was associated with lower cognitive decline and a lower risk of AD in the Memory and Aging Project [95, 96]; yet replications in other cohorts are warranted to assess the external validity of this diet for the protection of the brain in non-US populations.

2.2.3 Exploratory Nutrient/Dietary Patterns and Brain Health

Exploratory approaches have also been utilized to identify nutrient/dietary patterns related to brain health without formulating any a priori hypothesis on diet-to-brain relationships. Cluster analysis, principal component analysis, and partial least squares analyses (and a variant, the reduced rank regression) have been employed to characterize dietary patterns in various populations (see for review [97, 98]). Overall, findings from these studies are difficult to compare because the patterns obtained with exploratory approaches are constitutively specific to the studied population, and limited research has explored longitudinal associations with dementia outcomes so far. Studies have often interpreted their patterns as "healthy" or "prudent" versus "unhealthy" or "western" diets, yet definitions have varied largely across populations and associations with cognitive decline/dementia were relatively inconsistent. For example, in the Whitehall II study, the "prudent" diet, characterized by higher intakes of vegetables, fruits, dried legume, and fish, was related to lower odds of cognitive impairment, while the "western" pattern (sweetened desserts, chocolates, fried food, processed meat, pies, refined grains, high-fat dairy products, margarine) was related to higher odds of cognitive deficit; however, adjustment for education attenuated these cross-sectional associations [99]. In the Australian Imaging, Biomarkers and Lifestyle (AIBL) study, the "prudent" pattern obtained by factor analysis (characterized by high intakes of vegetables, fruits, nuts, whole grains) was not associated with cognitive decline over 3 years, while the "western" pattern (red and processed meats, refined grains, poultry, potatoes, sweets, margarine, high-fat dairies) was associated with increased decline in visuospatial performances, but only among APOEε4 allele noncarriers [91]. In the Newcastle 85+ Study, three patterns were identified by cluster analysis; two of them, i.e., pattern 1 associated with high intakes of red meat, potatoes, and gravy and pattern 2 associated with high intake of butter, were associated with lower cognitive performances at baseline. However, no association was found with cognitive decline over 5 years [100].

More convincing findings were obtained in the Swedish National study on Aging and Care-Kungsholmen (SNAC-K) [101] and in the US cohort Washington Heights-Inwood Columbia Aging Project (WHICAP) [102]. For example, in WHICAP, a "healthy" pattern, identified with reduced rank regression to reflect a diet rich in n-3 PUFA, n-6 PUFA, vitamin E, and folate, but poor in SFA and vitamin B12, was characterized by higher intakes of nuts, fish, tomatoes, poultry, cruciferous vegetables, fruits, and dark and green leafy vegetables and lower intakes of high-fat dairy, red meat, organ meat, and butter; in multivariate models, a higher score to this dietary pattern was strongly associated with a decreased risk of AD over 4 years [102]. Finally, we recently reported in a subsample from the French 3C study ($N = 666$) a strong association between a nutrient biomarker pattern combining several nutrient deficiencies, including low blood status vitamin D, low carotenoids, and low PUFAs (versus high SFAs) and the risk of dementia over 12 years. Compared with individuals in the first quintile of the pattern score, participants in the highest quintile of score had an approximately fourfold increased risk of dementia in multivariate models—an association that was stronger in magnitude than any of the associations between the individual nutrient biomarkers and dementia [103].

Overall, although the exploratory approaches developed above are uniquely adapted to identify novel food and nutrient combinations predictive of dementia, their lack of reproducibility across populations has limited their utility for etiological research so far.

3 Nutrition, Diet, and Alzheimer's Disease Brain Imaging Biomarkers

With a disease evolving over decades before clinical symptoms become apparent, biomarkers have been a core feature of research on dementia and AD [104, 105]. Biomarkers are not only useful in clinical research to capture early signs of disease in preclinical stages, but they are also highly relevant to etiological research, as they may give important insights on potential mechanisms underlying associations between risk/prevention factors and cognitive aging and dementia. For example, when applied to research on diet and dementia, the use of biomarkers may inform on whether associations of certain dietary factors to dementia risk may be mediated by pathways acting on lowering brain amyloid specifically, on preserving brain structure by avoiding neurodegeneration more generally, and/or on protecting the brain vasculature from age-related damage.

AD biomarkers have been divided into (1) pathophysiological markers, reflecting AD pathology at any point on the disease continuum, and (2) topographical and/or prognostic markers, reflecting downstream damage (e.g., neuronal loss associated with neurodegeneration, white matter microstructural alterations associated with both neurodegeneration and brain vascular injury) [106]. Pathophysiological biomarkers assess the presence of amyloid and Tau pathology and include both cerebrospinal fluid biomarkers and imaging biomarkers obtained with PET (positron emission tomography) scans (i.e., PET amyloid and PET Tau). Although not specific to AD pathology, biomarkers of cerebral small vessel disease may also be considered as part of pathophysiological biomarkers of AD [106]. Moreover, topographical biomarkers are exclusively assessed through brain imaging and include markers of macro- and microstructural changes in the brain at MRI (whole brain and hippocampal atrophy, cortical thickness obtained with volumetric MRI, alterations of white matter and gray matter microstructure obtained with diffusion tensor imaging [DTI]) and markers of hypometabolism of neocortical regions using fluorodeoxyglucose (FDG)-PET [106]. We will focus here on brain imaging biomarkers (including markers of cerebral small vessel disease), which have been studied in relation to nutrition and diet only recently (very limited research has been conducted on diet and cerebrospinal fluid biomarkers, and there has been no valid plasma biomarker to date). We will address studies by order of importance in the literature in relation to nutrition and diet, by describing, first, studies on MRI topographical biomarkers reflecting neuronal loss; second, studies on pathophysiological markers of brain vascular health; and, last, the very few studies on nutrition and brain amyloid. Because most of imaging-based research has been cross-sectional, cross-sectional studies will be included in this paragraph. Relevant studies have been reviewed and summarized in Table 1.

3.1 Nutrition, Diet, and Topographical Imaging Biomarkers (Reflecting Macro-/Microstructural and/or Metabolic Changes Associated with Neuronal Loss and/or Brain Vascular Injury)

3.1.1 Brain Atrophy

Among individual dietary candidates for brain health, fish intake/LC n-3 PUFA exposures have certainly been among the most studied in relation to brain volumes in previous research (Table 1). We have reviewed six large epidemiological studies, mostly cross-sectional, on the relationships between fish intake or blood status in EPA and/or DHA and volumes of the whole brain, the hippocampus, and/or the medial temporal lobe (which includes hippocampus and other structures of the limbic system, in particular the amygdala). Overall, studies have been inconsistent though. Two cross-sectional analyses from the Cardiovascular Health Study reported null associations [107, 108] (Table 1 *lines 1–2*); a longitudinal retrospective analysis from the ADNI found relationships limited to APOE4 carriers (Table 1 *line 3*); whereas three other cohorts [109, 110] (*lines 5–6*) (including a single longitudinal

Table 1
Epidemiological studies on the relationship between nutrition, diet, and brain imaging biomarkers of AD dementia (longitudinal studies are underlined in gray)

Line	Type of imaging biomarker Study	Study (1st and last author, year, country, cohort name)	Design (cross-sectional [CS] versus longitudinal [L]; N; age at baseline [minimum, mean or range, whatever available])	Nutritional Exposure	Main findings (based on multivariate models)
	Topographical imaging biomarkers (reflecting macro/microstructural and/or metabolic changes associated with neuronal loss and/or brain vascular injury)				
	Brain atrophy				
1	***Atrophy of the whole brain/hippocampus/ enthorinal cortex/medial temporal lobe***	Virtanen and Mozaffarian; 2008; USA; Cardiovascular Health Study [107]	CS; N=4,128 participants without known cerebrovascular disease; ≥65 years	Fish intake	Fish intake **not** associated with sulcal or ventricular grades (visual grading), markers of brain atrophy
2		Virtanen and Mozaffarian; 2013; USA; Cardiovascular Health Study [107,108]	CS; N=2,293 participants without known cerebrovascular disease; ≥65 years	Plasma Phospholipid n-3 PUFA	Higher plasma phospholipid long-chain n-3 PUFA (EPA+DPA+DHA) **not** associated with sulcal or ventricular grades (visual grading), markers of brain atrophy
3		Daiello and Ott for the Alzheimer's Disease Neuroimaging Initative (ADNI); 2015; USA; ADNI [30]	**L** (retrospective cohort; exposure an outcomes modeled concomitantly over 4 years); N=819; 55-90 years	Fish oil use (mostly of moderate to long duration, ie, 70% had used fish oil during > 1 year before baseline)	Fish oil use associated with lower of both the whole brain (for normal controls and MCI participants) and the hippocampus (for normal controls and AD patients), and less cognitive decline but only among APOE4 non-carriers
4		Samieri and Barberger-Gateau; 2012; France; Three-City Study [111]	**L** (volume change over 4 years); N=281; 72.3 years	Plasma long-chain n-3 PUFA at baseline	Plasma EPA associated with lower atrophy in part of the right medial-temporal lobe; anatomical changes linked with slower cognitive decline
5		Tan and Seshadri; 2012; USA; the Framingham Offspring Study [109]	CS; N=1,575 without dementia; 67 years	EPA and DHA in red blood cells	Lower DHA level (Q1 versus Q2-Q4) associated with lower total brain volume
6		Pottala and Harris; 2014; USA; Women's Health Initiative Memory Study (WHIMS) [110]	CS (with brain imaging 8 years after blood draw); N=1,111 women; 65 to 80 years	EPA and DHA in red blood cells	Higher EPA+DHA in red blood cells (ie, omega-3 index) associated with larger brain volume and larger hippocampal volume
7		Bowman and Quinn; 2012; USA; the Oregon Brain Aging Study cohort [125]	CS; N=42 (with brain imaging) with CDR<0.5; 87 years	Nutrient biomarker patterns (based on a set of 30 biomarkers) at baseline	A nutrient biomarker pattern "BCDE" characterized by higher plasma vitamins B (B1, B2, B6, folate, B12), C, D and E associated with higher total brain volume
8		Titova and Benedict; 2013; Sweden; Prospective Investigation of the Vasculature in Uppsala Seniors [114]	CS (with brain imaging 5 years after diet assessment); N=194; 70.1 years	Mediterranean diet score at baseline	Higher Mediterranean diet score **not** associated total brain volume 5 years later (higher meat intake significantly associated with worse volume and worse cognition).
9		Gu and Scarmeas; 2015; USA; Washington Heights/Hamilton Heights Inwood Columbia Aging Project (WHICAP) [113]	CS; N=674; 80.1 years	Mediterranean diet score	Higher Mediterranean diet score associated with larger total brain volumes (especially high fish and low meat); associations were equivalent to 5 years of aging.
10		Pelletier and Samieri, 2015, Three-City Study [115]	CS (with brain imaging 9 years after diet assessment); N=146; 73 years	Mediterranean diet score at baseline	Higher Mediterranean diet score **not** associated with grey matter or white matter brain volumes in any region
11		Luciano and Deary; 2017; Scotland; Lothian Birth Cohort of 1936 [116]	**L** (volume and corticial thickness change over 3 years, with baseline brain imaging 3 years after diet assessment); N=401 (346 for cortical thickness)	Mediterranean diet score	Higher Mediterranean diet score **not** associated with total brain volumes 3 years and 6 years after diet assessment, but significantly associated with lower atrophy over 3 years (no specific association with fish or meat found)

(continued)

12	***Cortical thickness***	Walhovd and Fjell; 2014; Norway; Cognition and Plasticity through the Lifespan project [112]	L (change over 3.6 years, with blood biomarkers assessed at the second MRI visit); N=92; 53.8 years (range 23.5 to 87.8 years)	Several exposures including plasma long-chain n-3 PUFA and vitamin D	Higher DHA associated with less cortical thinning (preceding biomarker assessment) in the left middle and superior temporal cortex; higher vitamin D associated with less cortical thinning in right lateral prefrontal cortex
13		Mosconi and de Leon; 2014; USA; study at New York University school of Medicine [117]	CS; N=52; 54 years	Mediterranean diet score	Higher Mediterranean diet score associated with larger cortical thickness in the entorhinal, orbitofrontal and posterior cingulate regions
14		Gu and Scarmeas; 2015; USA; Washington Heights/Hamilton Heights Inwood Columbia Aging Project (WHICAP) [113]	CS; N=674; 80.1 years	Mediterranean diet score	Higher Mediterranean diet score **not** associated with cortical thickness (association of fish to greater cortical thickness only).
15		Staubo and Roberts, 2017, USA, Mayo Clinic Study of Aging [118]	CS; N=672; 79.8 years	Mediterranean diet score	Higher Mediterranean diet score associated with larger cortical thickness in many regions (especially high legume and fish and low carbohydrate and sugar), but not significantly with an AD signature cortical thickness
16		Luciano and Deary; 2017; Scotland; Lothian Birth Cohort of 1936 [116]	L (change over 3 years, with baseline brain imaging 3 years after diet assessment); N=401 (346 for cortical thickness)	Mediterranean diet score	Higher Mediterranean diet score **neither** associated with cortical thickness 3 years and 6 years after diet assessment **nor** with cortical thickness change over 3 years
Brain microstructure					
17	***White matter microstructure (DTI)***	Gu and Brickman, 2016, USA, WHICAP [119]	CS (with brain imaging 5.1 years after diet assessment); N=239; 84.1 years	Nutrient intake patterns (based on a set of 24 nutrients)	A pattern characterized by higher intakes of PUFA (n-3 and n-6) and vitamin E associated with higher average fractional anisotropy; this association with white matter microstructure mediated the relations of the pattern to memory and cognition
18		Pelletier and Samieri, 2015, Three-City Study [115]	CS (with brain imaging 9 years after diet assessment); N=146 not demented at baseline; 73 years	Mediterranean diet score at baseline	Higher Mediterranean diet score associated with higher fractional anisotropy and lower mean diffusivity in many white matter regions; no significant association with brain volumes
Brain hypometabolism					
19		Mosconi and de Leon; 2014; USA; study at New York University school of Medicine [121]	CS; N=49 cognitively normal individuals (70% with family history of late-onset AD and 39% carriers of the APOE4 allele); 54 years (range 25 to 72 years)	Intakes of 10 candidate nutrients	Higher intake of β carotene and folate from food sources associated with higher brain metabolism in AD regions of interest at ^{18}F-Fluorodeoxyglucose (FDG)-PET
20		Berti and Mosconi; 2015; USA; study at New York University school of Medicine [122]	CS; N=52 cognitively normal individuals (69% with family history of late-onset AD and 39% carriers of the APOE4 allele); 54 years (range 25 to 72 years)	Nutrient intake patterns (based on 35 nutrients)	Several nutrient intake patterns associated with higher brain metabolism in AD regions: - A pattern with higher intakes of vitamin B12, vitamin D and Zinc (also associated with lower amyloid load and higher grey matter volume) - A pattern with lower intakes of saturated and trans fats, cholesterol and sodium (also

(continued)

					associated with higher grey matter volume) - A pattern with higher MUFA and PUFA intakes - A pattern with higher carotenoids, vitamins A and C, fibers
	Pathophysiological imaging biomarkers				
	Markers of cerebral small vessel disease				
21	***Brain infarcts***	Scarmeas and Brickman; 2011; WHICAP [126]	CS (with brain imaging 5.8 years after diet assessment); N=707; 80.3 years	Mediterranean diet score	Higher Mediterranean diet score associated with reduced burden of brain infarcts
22	***Subclinical infarcts***	Virtanen and Mozaffarian; 2008; USA; Cardiovascular Health Study [107]	CS and L (2 MRI 5 years apart); N=4,128 for CS and N=1,124 for L analyses, without known cerebrovascular disease; ≥65 years	Fish intake	Consuming tuna/other fish ≥3 times/week associated with lower prevalence of subclinical infarcts (ischemic lesions ≥3mm) compared to <once/month; **non-significant tend** for association with incident subclinical infarcts
23		Virtanen and Mozaffarian; 2013; USA; Cardiovascular Health Study [107,108]	CS and L (2 MRI 5 years apart); N=2,293 for CS and N=1,056 for L analyses, without known cerebrovascular disease; ≥65 years	Plasma Phospholipid n-3 PUFA	Higher plasma phospholipid long-chain n-3 PUFA (EPA+DPA+DHA) associated with lower prevalence of subclinical infarcts; **no** association with incident subclinical infarcts
24	***White matter hyperintensities***	Tan and Seshadri; 2012; USA; the Framingham Offspring Study [109]	CS; N=1,575 without dementia; 67 years	EPA and DHA in red blood cells	Higher DHA level associated with lower white matter hyper-intensities volume but **not** after adjustment for the full set of potential confounders; no association with silent cerebral brain infarcts
25		Pottala and Harris; 2014; USA; Women's Health Initiative Memory Study (WHIMS) [110]	CS (with brain imaging 8 years after blood draw); N=1,111 women; 65 to 80 years	EPA and DHA in red blood cells	EPA+DHA in red blood cells (ie, omega-3 index) **not** associated with ischemic lesion volume (represented by both diffuse small-vessel disease and white matter hyperintensities)
26		Bowman and Quinn; 2012; USA; the Oregon Brain Aging Study cohort [125]	CS; N=42 (with brain imaging) with CDR<0.5; 87 years	Nutrient biomarker patterns (based on a set of 30 biomarkers) at baseline	A nutrient biomarker pattern characterized by higher plasma long-chain n-3 PUFA associated with lower white matter hyper-intensities volume, **but only** among participants without depression
	Biomarkers of brain amyloid				
27		Mosconi and de Leon; 2014; USA; study at New York University school of Medicine [121]	CS; N=49 cognitively normal individuals (70% with family history of late-onset AD and 39% carriers of the APOE4 allele); 54 years (range 25 to 72 years)	Intakes of 10 candidate nutrients	Higher intake of vitamin B12, vitamin D, n-3 PUFA from food sources associated with lower amyloid load in AD regions at ^{11}C-Pittsburgh Compound B (PIB)-PET
28		Berti and Mosconi; 2015; USA; study at New York University school of Medicine [122]	CS; N=52 cognitively normal individuals (69% with family history of late-onset AD and 39% carriers of the APOE4 allele); 54 years (range 25 to 72 years)	Nutrient intake patterns (based on 35 nutrients)	A nutrient intake pattern characterized by higher intakes of vitamin B12, vitamin D and Zinc associated with lower amyloid load
29		Yassine and Chui; USA; Aging Brain Study [31]	CS; N=61 cognitively normal individuals; 77 years (range 67 to 88 years)	Blood DHA	Serum DHA inversely related to amyloid load using ^{11}C PIB-PET

report from the Three-City (3C) study [111] (*line 4*)) found significant associations of higher blood LC n-3 PUFA status to higher brain volumes (/lower brain atrophy in 3C). In complement to these studies on brain volumes, two studies reported interesting associations between fish/DHA and cortical thickness. The Cognition and Plasticity through the Lifespan project found a longitudinal association between higher blood DHA and less cortical thinning in part of the left temporal cortex [112] (*line 12*). However, the study was strongly limited by a biomarker assessment performed after ascertainment of cortical thinning. Furthermore, in a cross-sectional analysis from the large MRI sample of the WHICAP, higher fish intake was the single component of the MeDi score associated with greater cortical thickness [113] (*line 14*). Thus, the relation of fish/LC n-3 PUFAs to cortical thickness (a subtle marker of neuronal loss) certainly deserves further research.

Aside from studies on fish/n-3 PUFAs, a number of cohorts investigated overall diet through adherence to the MeDi; however, as with fish, inconsistent results were found in relation to brain atrophy. For example, both the Swedish Prospective Investigation of the Vasculature in Uppsala Seniors and the 3C study found no association of the MeDi to brain volumes [114, 115] (*lines 8 and 10*). Likewise, in the Lothian Birth Cohort, there was no association between the MeDi and brain volumes assessed 3 years apart; yet adherence to the MeDi was related to lower brain atrophy between the two time points [116] (*line 10*). Only the WHICAP reported a significant association between the MeDi and total brain volumes [113] (*line 9*). More recent studies on cortical thickness have not been more consistent than those which targeted volumes. Two US cohorts, including the large sample ($N > 500$) from the Mayo Clinic Study of Aging, reported a cross-sectional relation between the MeDi and larger cortical thickness [117, 118] (*lines 13 and 15*). In contrast, none of the two large WHICAP and Lothian Birth Cohort studies found any significant association of the MeDi to cortical thickness [113, 116] (*lines 14 and 16*).

3.1.2 Alterations of Brain Microstructure

Studies on diet and brain microstructure, which have been extremely limited to date, yielded promising findings. Both the WHICAP and the 3C study found associations of (1) a nutrient pattern with higher PUFAs and vitamin E and (2) the MeDi, respectively, to preserved white matter structure [115, 119] (Table 1 *lines 17 and 18*). Interestingly, in both studies, the reported associations were accompanied by strong relations with cognitive function, suggesting that alterations in brain connectivity may serve as a mediator in the relationship between low diet quality and cognitive aging. Moreover, in a small clinical trial ($N = 65$), 26 weeks of supplementation with EPA + DHA (2.2 g/day) in older persons led to an improvement of white matter microstructure compared to placebo [120] (intervention study not shown in Table 1). Together, these findings suggest that brain connectivity may be a key neuroanatomical substrate for the effect of foods and nutrients on brain health. Larger epidemiological studies are warranted to better understand how diet may influence brain connectomics.

3.1.3 Brain Hypometabolism

As with cerebral microstructure, studies on diet and brain hypometabolism are at their infancy. Two small studies from the NY University School of Medicine reported associations between nutrient intake patterns and higher brain metabolism in AD regions of interest [121, 122] (Table 1, *lines 19 and 20*). However, in both studies the number of nutrients investigated was almost superior to the number of subjects included, and study samples were highly selected. These pilot studies certainly deserve validation in larger samples.

3.2 Nutrition, Diet, and Pathophysiological Imaging Biomarkers

3.2.1 Markers of Cerebral Small Vessel Disease

The term cerebral small vessel disease refers to a group of pathological processes with various etiologies that affect the small arteries, arterioles, venules, and capillaries of the brain. The causes are multiple and include cerebral amyloid angiopathy, hypertension, and vascular risk factors in general. As a consequence of small vessel disease, various subcortical lesions develop in brain parenchyma, including lacunar infarcts, white matter lesions, large hemorrhages, and microbleeds [123]. These lesions reflecting small vessel disease can be detected at MRI; as their prevalence dramatically increases with age, they are commonly considered part of the spectrum of vascular-related brain injury associated with brain aging. For example, white matter hyperintensities (WMH), defined as white matter lesions appearing as hyperintensities on T2-weighted images at MRI, range from 11–21% around 64 years old to 94% at age 82 [124].

Assuming that diet, which has a primary influence on vascular risk factors, may exert a beneficial role on the brain vasculature, several of the largest cohorts on diet and dementia have examined the relations between nutritional factors and MRI markers of cerebral small vessel disease. Primary studies were published a decade ago; still, existing literature has remained scarce (mostly limited to fish/n-3 PUFA) and, as with brain atrophy, generally inconsistent (Table 1). The Cardiovascular Health Study was among the first to examine the relation of fish/LC n-3 PUFA status to subclinical infarcts; despite significant cross-sectional associations and a very large sample size ($N > 1000$), findings were null for the prospective relationships with incident lesions [107, 108] (Table 1 *lines 22 and 23*). In addition, two large US cohorts ($N = > 1000$) failed to evidence any relation of blood LC n-3 PUFA to WMH volume [109, 110] (*lines 24 and 25*). A relation between a nutrient biomarker pattern reflecting higher plasma LC n-3 PUFA and lower WMH volume was reported in a small sample from the Oregon Brain

Aging Study Cohort [125]; however, the association was limited to participants without depression, with a high probability of chance finding in such a small sample (*line 26*). One of the rare positive findings pertains to the WHICAP, in which a higher MeDi score was associated with reduced burden of brain infarcts in a cross-sectional analysis [126] (*line 21*). Despite this positive finding, overall, one may conclude that no consistent relationship between diet and conventional MRI markers of cerebral small vessel disease has been found to date. This is not to say that nutrition and diet do not influence brain health through preservation of brain vasculature; more convincing findings may be obtained with more subtle markers of neurodegeneration and brain vascular injury, such as markers of white matter microstructure alterations, as developed above.

3.2.2 Markers of Brain Amyloid

PET amyloid reflects the neuropathological hallmark of AD and is thus considered the "king" biomarker in AD research. However, given its elevated cost, virtually no research has been made on diet and PET amyloid so far. Recently, the Atherosclerosis Risk in Communities (ARIC) study found a relationship between midlife obesity and brain amyloid deposition 20 years later, suggesting a strong, lifelong influence of diet-related risk factors on AD neuropathology [127]. However, the dietary factors likely to influence amyloid load have remained understudied to date. One of the single studies has been conducted by the NY University School of Medicine, which recently published two pilot studies on nutrients and brain amyloid [121, 122] (Table 1, *lines 27 and 28*). In one of these studies, which included 49 cognitively normal individuals with high familial and genetic risk for AD, higher intakes of vitamin B12, vitamin D, and n-3 PUFA were related to lower amyloid load in regions with typical amyloid accumulation in AD. Moreover, a small cross-sectional study from the Aging Brain Study reported a correlation between higher blood DHA and lower amyloid load among cognitively normal older adults (*line 29*). Added to the important findings on midlife obesity and brain amyloid load, these promising preliminary findings with nutrients call for additional large-scale studies on diet and brain amyloid in future research.

4 Conclusion: Toward Innovative Approaches for Future Epidemiological Research on Nutrition, Diet, and Alzheimer's Disease

Epidemiological research on diet and dementia conducted over the past 15–20 years has provided a considerable insight into the impact of lifestyle on brain health. Many large-scale observational studies have confirmed the importance of several nutrients and foods for maintaining optimal brain structure and function with aging, as suggested by preclinical studies. However, despite the amount of positive findings gained from preclinical and observational epidemiological research, surprisingly, randomized controlled trials which have tested the efficacy of supplementation in individual nutrients (mostly n-3 PUFAs, vitamin E, and B vitamins) generally failed to demonstrate any clinical efficacy on cognition in humans (see a number of meta-analyses published recently [128–132]). Several reasons may explain the discordance between observational and interventional studies in diet and brain aging. Observational studies have well-recognized limitations (e.g., measurement error and residual confounding); however, randomized trials in AD dementia have also a number of limitations, mainly due to the complexity of a disease which evolves silently over decades, and there may be several methodological explanations for the failure of most trials to demonstrate the clinical efficacy of nutrients in brain aging. Methodological limitations of existing trials may include (1) a too late time window for intervention (i.e., most trials have included mild-to-moderate AD patients, while those most likely to benefit from a nutritional intervention may need to be relatively preserved from extended and irreversible neurodegeneration), (2) a too short duration of intervention (i.e., many trials tested less than 1 year of supplementation in AD dementia, while long-term preventive interventions are certainly more relevant for a disease which potentially evolves over decades), and (3) failure to take into account baseline nutritional exposures (i.e., those most likely to benefit from supplementation of a given nutrient may have low intakes of that nutrient in their habitual diet).

It is also possible that single nutrients, which have been generally modestly associated with cognitive decline or the risk of dementia in observational studies, are not sufficiently active to modulate brain aging and that a combination of nutrients with complementary biological properties (e.g., LC n-3 PUFA and antioxidant nutrients) may be more effective for the prevention of AD dementia. Accordingly, promising observational findings were obtained with overall healthy diets (e.g., the MeDi and the MIND diets [96, 133]) or with nutrient patterns [103]. Recently, a very large US trial ($N > 5000$) failed to evidence the benefit of long-term multivitamin supplementation (vitamins E, C, and β-carotene) on cognitive aging [134]; however, the authors acknowledged that the population of male health professional included in the trial may be too well-nourished to benefit from a multivitamin. It is also possible that the combination of antioxidant vitamins tested in this large trial may not be sufficient and that a broader set of nutrients may be more relevant. For example, Nutricia developed the Souvenaid (Fortasyn Connect) supplement, including 11 nutrients involved in synaptic synthesis (uridine monophosphate; choline; phospholipids; EPA; DHA; vitamins E, C, B12, and B6; folic acid; and selenium), for the prevention of AD dementia. However, after promising results of a 2-year phase II trial in mild-to-moderate AD

[135], Souvenaid failed to lower cognitive decline as an add-on of usual AD medication in a phase III study (N = 527) [136] and did not improve decline in global cognition in prodromal dementia after 24- month supplementation (N=382); still, benefit was observed in important secondary outcomes including hippocampal atrophy [137]. In contrast with the mitigated findings obtained so far with the two nutrient combinations cited above, the Spanish PREDIMED (Prevención con Dieta Mediterránea) study evidenced the efficacy, for individuals with elevated cardiovascular risk, of adopting a Mediterranean diet supplemented with virgin olive oil or nuts for 5 years for age-related cognitive decline [138]. These significant interventional findings may certainly constitute a strong positive signal for the relevance of healthy diet/nutrient combinations in the prevention of AD dementia.

Thus, there is a large room for nutritional epidemiology in future research on AD prevention. The challenges posed by research on behavioral and biological risk factors of AD are multiple and may be successfully addressed using innovative approaches such as "omics" methods, cutting-edge imaging biomarkers, and advanced statistical methods for the modeling of big data and biomarker trajectories. The first challenge to face with will be to determine the optimal combination of foods/nutrients for prevention. It may include some of the candidate nutrients with established associations with brain aging (e.g., vitamin D, a very promising candidate yet understudied in combination with other nutrients). The optimal food/nutrient combination may also incorporate some novel dietary bioactives. For example, interesting molecules for prevention could be identified in the food metabolome—defined as the thousands of nutrients/non-nutrient metabolites present in biofluids and tissues that directly derive from food digestion. The food metabolome is a novel research area which promises to give important insights on the role of diet in brain health.

Beyond the composition of the active compound(s), another important challenge for future research will be to fully elucidate the optimal target population for a preventive intervention. There is suggestion that early time windows of exposure in primary prevention may be more relevant than secondary preventive interventions targeting demented persons in older age; yet the exact age range at which nutrition mostly impacts AD neuropathology is still unknown. Longitudinal epidemiological research, with repeated measurements of nutritional exposures and AD clinical and imaging markers, will be key to better understand the dynamic relationships between trajectories of behavioral factors and AD biomarker trajectories.

Finally, mechanistic epidemiological research will need to be expanded to fully elucidate the multiple pathways by which nutrition/diet shapes the brain and prevents AD. Indeed, a better understanding of pathways will help refine the composition of effective interventions and potentially identify vulnerable populations (e.g., those carrying specific polymorphisms for certain genes). For example, extensive research has been developed on the gut microbiome as a potential mediator of the relationship between diet and brain health—the gut-brain axis being certainly one of the most exciting fields of investigation open for the next decades.

References

1. World Alzheimer report 2016 (2016) https://www.alz.co.uk/research/world-report-2016. Accessed 16 May 2017
2. Lambert JC, Ibrahim-Verbaas CA, Harold D et al (2013) Meta-analysis of 74,046 individuals identifies 11 new susceptibility loci for Alzheimer's disease. Nat Genet 45(12):1452–1458. https://doi.org/10.1038/ng.2802
3. Scheltens P, Blennow K, Breteler MM et al (2016) Alzheimer's disease. Lancet. https://doi.org/10.1016/S0140-6736(15)01124-1
4. Grasset L, Brayne C, Joly P et al (2016) Trends in dementia incidence: evolution over a 10-year period in France. Alzheimers Dement 12(3):272–280. https://doi.org/10.1016/j.jalz.2015.11.001
5. Satizabal CL, Beiser AS, Chouraki V et al (2016) Incidence of dementia over three decades in the Framingham Heart Study. N Engl J Med 374(6):523–532. https://doi.org/10.1056/NEJMoa1504327
6. Small SA, Duff K (2008) Linking abeta and tau in late-onset Alzheimer's disease: a dual pathway hypothesis. Neuron 60(4):534–542. https://doi.org/10.1016/j.neuron.2008.11.007
7. Kumar DK, Choi SH, Washicosky KJ et al (2016) Amyloid-beta peptide protects against microbial infection in mouse and worm models of Alzheimer's disease. Sci Transl Med 8(340):340ra372. https://doi.org/10.1126/scitranslmed.aaf1059
8. Zhao Z, Nelson AR, Betsholtz C et al (2015) Establishment and dysfunction of the blood-brain barrier. Cell 163(5):1064–1078. https://doi.org/10.1016/j.cell.2015.10.067
9. Stern Y (2009) Cognitive reserve. Neuropsychologia 47(10):2015–2028
10. Shah H, Albanese E, Duggan C et al (2016) Research priorities to reduce the global burden of dementia by 2025. Lancet Neurol 15(12):1285–1294. https://doi.org/10.1016/s1474-4422(16)30235-6
11. Cunnane SC, Plourde M, Pifferi F et al (2009) Fish, docosahexaenoic acid and Alzheimer's disease. Prog Lipid Res 48(5):239–256. doi:S0163-7827(09)00018-6 [pii]10.1016/j.plipres.2009.04.001
12. Su HM (2010) Mechanisms of n-3 fatty acid-mediated development and maintenance of learning memory performance. J Nutr Biochem 21(5):364–373. https://doi.org/10.1016/j.jnutbio.2009.11.003
13. Lim GP, Calon F, Morihara T et al (2005) A diet enriched with the omega-3 fatty acid docosahexaenoic acid reduces amyloid burden in an aged Alzheimer mouse model. J Neurosci 25(12):3032–3040
14. Barberger-Gateau P, Letenneur L, Deschamps V et al (2002) Fish, meat, and risk of dementia: cohort study. BMJ 325(7370):932–933
15. Morris MC, Evans DA, Tangney CC et al (2005) Fish consumption and cognitive decline with age in a large community study. Arch Neurol 62(12):1849–1853
16. Schaefer EJ, Bongard V, Beiser AS et al (2006) Plasma phosphatidylcholine docosahexaenoic acid content and risk of dementia and Alzheimer disease: the Framingham Heart Study. Arch Neurol 63(11):1545–1550
17. Samieri C, Feart C, Letenneur L et al (2008) Low plasma eicosapentaenoic acid and depressive symptomatology are independent predictors of dementia risk. Am J Clin Nutr 88:714–721
18. Lopez LB, Kritz-Silverstein D, Barrett Connor E (2011) High dietary and plasma levels of the omega-3 fatty acid docosahexaenoic acid are associated with decreased dementia risk: the Rancho Bernardo study. J Nutr Health Aging 15(1):25–31
19. Huang TL (2010) Omega-3 fatty acids, cognitive decline, and Alzheimer's disease: a critical review and evaluation of the literature. J Alzheimers Dis 21(3):673–690. https://doi.org/10.3233/JAD-2010-090934
20. Zhang Y, Chen J, Qiu J et al (2016) Intakes of fish and polyunsaturated fatty acids and mild-to-severe cognitive impairment risks: a dose-response meta-analysis of 21 cohort studies. Am J Clin Nutr 103(2):330–340. https://doi.org/10.3945/ajcn.115.124081
21. Burdge GC, Calder PC (2005) Conversion of alpha-linolenic acid to longer-chain polyunsaturated fatty acids in human adults. Reprod Nutr Dev 45(5):581–597. https://doi.org/10.1051/rnd:2005047

22. Morris MC, Brockman J, Schneider JA et al (2016) Association of seafood consumption, brain mercury level, and APOE epsilon4 status with brain neuropathology in older adults. JAMA 315(5):489–497. https://doi.org/10.1001/jama.2015.19451
23. van de Rest O, Wang Y, Barnes LL et al (2016) APOE epsilon4 and the associations of seafood and long-chain omega-3 fatty acids with cognitive decline. Neurology 86(22):2063–2070. https://doi.org/10.1212/WNL.0000000000002719
24. Yassine HN, Braskie MN, Mack WJ et al (2017) Association of docosahexaenoic acid supplementation with Alzheimer disease stage in apolipoprotein E epsilon4 carriers: a review. JAMA Neurol. https://doi.org/10.1001/jamaneurol.2016.4899
25. Laitinen MH, Ngandu T, Rovio S et al (2006) Fat intake at midlife and risk of dementia and Alzheimer's disease: a population-based study. Dement Geriatr Cogn Disord 22(1):99–107
26. Samieri C, Feart C, Proust-Lima C et al (2011) Omega-3 fatty acids and cognitive decline: modulation by ApoEepsilon4 allele and depression. Neurobiol Aging 32(12):2313–2322. https://doi.org/10.1016/j.neurobiolaging.2010.03.020
27. Huang TL, Zandi PP, Tucker KL et al (2005) Benefits of fatty fish on dementia risk are stronger for those without APOE epsilon4. Neurology 65(9):1409–1414
28. Barberger-Gateau P, Raffaitin C, Letenneur L et al (2007) Dietary patterns and risk of dementia: the Three-City cohort study. Neurology 69(20):1921–1930
29. Whalley LJ, Deary IJ, Starr JM et al (2008) n-3 Fatty acid erythrocyte membrane content, APOE {varepsilon}4, and cognitive variation: an observational follow-up study in late adulthood. Am J Clin Nutr 87(2):449–454
30. Daiello LA, Gongvatana A, Dunsiger S et al (2015) Association of fish oil supplement use with preservation of brain volume and cognitive function. Alzheimers Dement 11(2):226–235. https://doi.org/10.1016/j.jalz.2014.02.005
31. Yassine HN, Feng Q, Azizkhanian I et al (2016) Association of serum docosahexaenoic acid with cerebral amyloidosis. JAMA Neurol. https://doi.org/10.1001/jamaneurol.2016.1924
32. Grimm MO, Mett J, Hartmann T (2016) The impact of vitamin E and other fat-soluble vitamins on Alzheimer's disease. Int J Mol Sci 17(11). https://doi.org/10.3390/ijms17111785
33. Ransom J, Morgan PJ, McCaffery PJ et al (2014) The rhythm of retinoids in the brain. J Neurochem 129(3):366–376. https://doi.org/10.1111/jnc.12620
34. Geleijnse JM, Hollman P (2008) Flavonoids and cardiovascular health: which compounds, what mechanisms? Am J Clin Nutr 88(1):12–13
35. Corder R, Mullen W, Khan NQ et al (2006) Oenology: red wine procyanidins and vascular health. Nature 444(7119):566. https://doi.org/10.1038/444566a
36. Brickman AM, Khan UA, Provenzano FA et al (2014) Enhancing dentate gyrus function with dietary flavanols improves cognition in older adults. Nat Neurosci 17(12):1798–1803. https://doi.org/10.1038/nn.3850
37. Andres-Lacueva C, Shukitt-Hale B, Galli RL et al (2005) Anthocyanins in aged blueberry-fed rats are found centrally and may enhance memory. Nutr Neurosci 8(2):111–120. https://doi.org/10.1080/10284150500078117
38. Marambaud P, Zhao H, Davies P (2005) Resveratrol promotes clearance of Alzheimer's disease amyloid-beta peptides. J Biol Chem 280(45):37377–37382. https://doi.org/10.1074/jbc.M508246200
39. Ho L, Ferruzzi MG, Janle EM et al (2013) Identification of brain-targeted bioactive dietary quercetin-3-O-glucuronide as a novel intervention for Alzheimer's disease. FASEB J 27(2):769–781. https://doi.org/10.1096/fj.12-212118
40. Wang J, Ho L, Zhao Z et al (2006) Moderate consumption of Cabernet Sauvignon attenuates Abeta neuropathology in a mouse model of Alzheimer's disease. FASEB J 20(13):2313–2320. https://doi.org/10.1096/fj.06-6281com
41. Boeing H, Bechthold A, Bub A et al (2012) Critical review: vegetables and fruit in the prevention of chronic diseases. Eur J Nutr 51(6):637–663. https://doi.org/10.1007/s00394-012-0380-y
42. Xu W, Wang H, Wan Y et al (2017) Alcohol consumption and dementia risk: a dose-response meta-analysis of prospective studies. Eur J Epidemiol. https://doi.org/10.1007/s10654-017-0225-3
43. Samieri C, Feart C, Proust-Lima C et al (2011) Olive oil consumption, plasma oleic acid, and stroke incidence: the Three-City Study. Neurology 77(5):418–425. https://doi.org/10.1212/WNL.0b013e318220abeb
44. Berr C, Portet F, Carriere I et al (2009) Olive oil and cognition: results from the three-city study. Dement Geriatr Cogn Disord 28(4):357–364. https://doi.org/10.1159/000253483
45. Kim YS, Kwak SM, Myung SK (2015) Caffeine intake from coffee or tea and cognitive disorders: a meta-analysis of observational studies. Neuroepidemiology 44(1):51–63. https://doi.org/10.1159/000371710

46. Bjelakovic G, Nikolova D, Gluud LL et al (2007) Mortality in randomized trials of antioxidant supplements for primary and secondary prevention: systematic review and meta-analysis. JAMA 297(8):842–857
47. Li FJ, Shen L, Ji HF (2012) Dietary intakes of vitamin E, vitamin C, and beta-carotene and risk of Alzheimer's disease: a meta-analysis. J Alzheimers Dis 31(2):253–258. https://doi.org/10.3233/JAD-2012-120349
48. Helmer C, Peuchant E, Letenneur L et al (2003) Association between antioxidant nutritional indicators and the incidence of dementia: results from the PAQUID prospective cohort study. Eur J Clin Nutr 57(12):1555–1561
49. Mangialasche F, Kivipelto M, Mecocci P et al (2010) High plasma levels of vitamin E forms and reduced Alzheimer's disease risk in advanced age. J Alzheimers Dis 20(4):1029–1037. https://doi.org/10.3233/JAD-2010-091450
50. Cherubini A, Martin A, Andres-Lacueva C et al (2005) Vitamin E levels, cognitive impairment and dementia in older persons: the InCHIANTI study. Neurobiol Aging 26(7):987–994
51. Ravaglia G, Forti P, Lucicesare A et al (2008) Plasma tocopherols and risk of cognitive impairment in an elderly Italian cohort. Am J Clin Nutr 87(5):1306–1313
52. Mangialasche F, Solomon A, Kareholt I et al (2013) Serum levels of vitamin E forms and risk of cognitive impairment in a Finnish cohort of older adults. Exp Gerontol 48(12):1428–1435. https://doi.org/10.1016/j.exger.2013.09.006
53. Sundelof J, Kilander L, Helmersson J et al (2009) Systemic tocopherols and F2-isoprostanes and the risk of Alzheimer's disease and dementia: a prospective population-based study. J Alzheimers Dis 18(1):71–78. https://doi.org/10.3233/JAD-2009-1125
54. Feart C, Letenneur L, Helmer C et al (2015) Plasma carotenoids are inversely associated with dementia risk in an elderly French cohort. J Gerontol A Biol Sci Med Sci. https://doi.org/10.1093/gerona/glv135
55. Hu P, Bretsky P, Crimmins EM et al (2006) Association between serum beta-carotene levels and decline of cognitive function in high-functioning older persons with or without apolipoprotein E 4 alleles: MacArthur studies of successful aging. J Gerontol A Biol Sci Med Sci 61(6):616–620
56. Kang JH, Grodstein F (2008) Plasma carotenoids and tocopherols and cognitive function: a prospective study. Neurobiol Aging 29(9):1394–1403. https://doi.org/10.1016/j.neurobiolaging.2007.03.006
57. Commenges D, Scotet V, Renaud S et al (2000) Intake of flavonoids and risk of dementia. Eur J Epidemiol 16(4):357–363
58. Letenneur L, Proust-Lima C, Le Gouge A et al (2007) Flavonoid intake and cognitive decline over a 10-year period. Am J Epidemiol 165(12):1364–1371
59. Kesse-Guyot E, Fezeu L, Andreeva VA et al (2012) Total and specific polyphenol intakes in midlife are associated with cognitive function measured 13 years later. J Nutr 142(1):76–83. https://doi.org/10.3945/jn.111.144428
60. Devore EE, Kang JH, Breteler MM et al (2012) Dietary intakes of berries and flavonoids in relation to cognitive decline. Ann Neurol 72(1):135–143. https://doi.org/10.1002/ana.23594
61. Root M, Ravine E, Harper A (2015) Flavonol intake and cognitive decline in middle-aged adults. J Med Food 18(12):1327–1332. https://doi.org/10.1089/jmf.2015.0010
62. Rabassa M, Cherubini A, Zamora-Ros R et al (2015) Low levels of a urinary biomarker of dietary polyphenol are associated with substantial cognitive decline over a 3-year period in older adults: the Invecchiare in Chianti study. J Am Geriatr Soc 63(5):938–946. https://doi.org/10.1111/jgs.13379
63. Reynolds E (2006) Vitamin B12, folic acid, and the nervous system. Lancet Neurol 5(11):949–960
64. Seshadri S, Beiser A, Selhub J et al (2002) Plasma homocysteine as a risk factor for dementia and Alzheimer's disease. N Engl J Med 346(7):476–483
65. Morris MS (2003) Homocysteine and Alzheimer's disease. Lancet Neurol 2(7):425–428
66. Selhub J, Troen A, Rosenberg IH (2010) B vitamins and the aging brain. Nutr Rev 68(Suppl 2):S112–S118. https://doi.org/10.1111/j.1753-4887.2010.00346.x
67. Morris MS (2012) The role of B vitamins in preventing and treating cognitive impairment and decline. Adv Nutr 3(6):801–812. https://doi.org/10.3945/an.112.002535
68. Agnew-Blais JC, Wassertheil-Smoller S, Kang JH et al (2015) Folate, vitamin B-6, and vitamin B-12 intake and mild cognitive impairment and probable dementia in the Women's Health Initiative Memory Study. J Acad Nutr Diet 115(2):231–241. https://doi.org/10.1016/j.jand.2014.07.006
69. Corrada MM, Kawas CH, Hallfrisch J et al (2005) Reduced risk of Alzheimer's disease with high folate intake: the Baltimore

Longitudinal Study of Aging. Alzheimers Dement 1(1):11–18. https://doi.org/10.1016/j.jalz.2005.06.001
70. Luchsinger JA, Tang MX, Miller J et al (2007) Relation of higher folate intake to lower risk of Alzheimer disease in the elderly. Arch Neurol 64(1):86–92
71. Lefevre-Arbogast S, Feart C, Dartigues JF et al (2016) Dietary B vitamins and a 10-year risk of dementia in older persons. Forum Nutr 8(12). https://doi.org/10.3390/nu8120761
72. Nelson C, Wengreen HJ, Munger RG et al (2009) Dietary folate, vitamin B-12, vitamin B-6 and incident Alzheimer's disease: the cache county memory, health and aging study. J Nutr Health Aging 13(10):899–905
73. Morris MC, Evans DA, Schneider JA et al (2006) Dietary folate and vitamins B-12 and B-6 not associated with incident Alzheimer's disease. J Alzheimers Dis 9(4):435–443
74. Morris MC, Evans DA, Bienias JL et al (2005) Dietary folate and vitamin B12 intake and cognitive decline among community-dwelling older persons. Arch Neurol 62(4):641–645. https://doi.org/10.1001/archneur.62.4.641
75. Hinterberger M, Fischer P (2013) Folate and Alzheimer: when time matters. J Neural Transm 120(1):211–224. https://doi.org/10.1007/s00702-012-0822-y
76. Landel V, Annweiler C, Millet P et al (2016) Vitamin D, cognition, and Alzheimer's disease: the therapeutic benefit is in the D-tails. J Alzheimers Dis. https://doi.org/10.3233/jad-150943
77. Hilger J, Friedel A, Herr R et al (2014) A systematic review of vitamin D status in populations worldwide. Br J Nutr 111(1):23–45. https://doi.org/10.1017/S0007114513001840
78. Annweiler C, Montero-Odasso M, Llewellyn DJ et al (2013) Meta-analysis of memory and executive dysfunctions in relation to vitamin D. J Alzheimers Dis 37(1):147–171. https://doi.org/10.3233/JAD-130452
79. Sommer I, Griebler U, Kien C et al (2017) Vitamin D deficiency as a risk factor for dementia: a systematic review and meta-analysis. BMC Geriatr 17(1):16. https://doi.org/10.1186/s12877-016-0405-0
80. Feart C, Helmer C, Merle B et al (2017) Associations of lower vitamin D concentrations with cognitive decline and long-term risk of dementia and Alzheimer's disease in older adults. Alzheimers Dement 13(11):1207–1216. https://doi.org/10.1016/j.jalz.2017.03.003.
81. Karakis I, Pase MP, Beiser A et al (2016) Association of serum vitamin D with the risk of incident dementia and subclinical indices of brain aging: the Framingham Heart Study. J Alzheimers Dis 51(2):451–461. https://doi.org/10.3233/JAD-150991
82. Olsson E, Byberg L, Karlstrom B et al (2017) Vitamin D is not associated with incident dementia or cognitive impairment: an 18-y follow-up study in community-living old men. Am J Clin Nutr. https://doi.org/10.3945/ajcn.116.141531
83. Nakagawa K, Kiko T, Hatade K et al (2009) Antioxidant effect of lutein towards phospholipid hydroperoxidation in human erythrocytes. Br J Nutr 102(9):1280–1284. https://doi.org/10.1017/S0007114509990316
84. Chen SJ, Huang LY, Hu CH (2015) Antioxidative reaction of carotenes against peroxidation of fatty acids initiated by nitrogen dioxide: a theoretical study. J Phys Chem B 119(30):9640–9650. https://doi.org/10.1021/acs.jpcb.5b04142
85. Feart C, Samieri C, Barberger-Gateau P (2015) Mediterranean diet and cognitive health: an update of available knowledge. Curr Opin Clin Nutr Metab Care 18(1):51–62. https://doi.org/10.1097/MCO.0000000000000131
86. Scarmeas N, Stern Y, Tang MX et al (2006) Mediterranean diet and risk for Alzheimer's disease. Ann Neurol 59(6):912–921
87. Feart C, Samieri C, Rondeau V et al (2009) Adherence to a Mediterranean diet, cognitive decline, and risk of dementia. JAMA 302(6):638–648
88. Singh B, Parsaik AK, Mielke MM et al (2014) Association of mediterranean diet with mild cognitive impairment and Alzheimer's disease: a systematic review and meta-analysis. J Alzheimers Dis 39(2):271–282. https://doi.org/10.3233/JAD-130830
89. Samieri C, Grodstein F, Rosner BA et al (2013) Mediterranean diet and cognitive function in older age. Epidemiology 24(4):490–499. https://doi.org/10.1097/EDE.0b013e318294a065
90. Samieri C, Okereke OI, ED E et al (2013) Long-term adherence to the Mediterranean diet is associated with overall cognitive status, but not cognitive decline, in women. J Nutr 143(4):493–499. https://doi.org/10.3945/jn.112.169896
91. Gardener SL, Rainey-Smith SR, Barnes MB et al (2014) Dietary patterns and cognitive decline in an Australian study of ageing. Mol Psychiatry. https://doi.org/10.1038/mp.2014.79
92. Smyth A, Dehghan M, O'Donnell M et al (2015) Healthy eating and reduced risk of cognitive decline: a cohort from 40 coun-

tries. Neurology. https://doi.org/10.1212/WNL.0000000000001638
93. Tangney CC, Li H, Wang Y et al (2014) Relation of DASH- and Mediterranean-like dietary patterns to cognitive decline in older persons. Neurology 83(16):1410–1416. https://doi.org/10.1212/WNL.0000000000000884
94. Berendsen AA, Kang JH, van de Rest O et al (2017) The dietary approaches to stop hypertension diet, cognitive function, and cognitive decline in American older women. J Am Med Dir Assoc. https://doi.org/10.1016/j.jamda.2016.11.026
95. Morris MC, Tangney CC, Wang Y et al (2015) MIND diet slows cognitive decline with aging. Alzheimers Dement. https://doi.org/10.1016/j.jalz.2015.04.011
96. Morris MC, Tangney CC, Wang Y et al (2015) MIND diet associated with reduced incidence of Alzheimer's disease. Alzheimers Dement. https://doi.org/10.1016/j.jalz.2014.11.009
97. Gu Y, Scarmeas N (2011) Dietary patterns in Alzheimer's disease and cognitive aging. Curr Alzheimer Res 8(5):510–519
98. Alles B, Samieri C, Feart C et al (2012) Dietary patterns: a novel approach to examine the link between nutrition and cognitive function in older individuals. Nutr Res Rev 25(2):207–222. https://doi.org/10.1017/S0954422412000133
99. Akbaraly TN, Singh-Manoux A, Marmot MG et al (2009) Education attenuates the association between dietary patterns and cognition. Dement Geriatr Cogn Disord 27(2):147–154
100. Granic A, Davies K, Adamson A et al (2016) Dietary patterns high in red meat, potato, gravy, and butter are associated with poor cognitive functioning but not with rate of cognitive decline in very old adults. J Nutr. https://doi.org/10.3945/jn.115.216952
101. Shakersain B, Santoni G, Larsson SC et al (2016) Prudent diet may attenuate the adverse effects of Western diet on cognitive decline. Alzheimers Dement 12(2):100–109. https://doi.org/10.1016/j.jalz.2015.08.002
102. Gu Y, Nieves JW, Stern Y et al (2010) Food combination and Alzheimer disease risk: a protective diet. Arch Neurol 67(6):699–706. https://doi.org/10.1001/archneurol.2010.84
103. Amadieu C, Lefevre-Arbogast S, Delcourt C et al (2017) Nutrient biomarker patterns and long-term risk of dementia in older adults. Alzheimers Dement. https://doi.org/10.1016/j.jalz.2017.01.025
104. Dubois B, Feldman HH, Jacova C et al (2014) Advancing research diagnostic criteria for Alzheimer's disease: the IWG-2 criteria. Lancet Neurol 13(6):614–629. https://doi.org/10.1016/S1474-4422(14)70090-0
105. Sperling RA, Aisen PS, Beckett LA et al (2011) Toward defining the preclinical stages of Alzheimer's disease: recommendations from the National Institute on Aging-Alzheimer's Association workgroups on diagnostic guidelines for Alzheimer's disease. Alzheimers Dement 7(3):280–292. https://doi.org/10.1016/j.jalz.2011.03.003
106. Dubois B, Hampel H, Feldman HH et al (2016) Preclinical Alzheimer's disease: definition, natural history, and diagnostic criteria. Alzheimers Dement 12(3):292–323. https://doi.org/10.1016/j.jalz.2016.02.002
107. Virtanen JK, Siscovick DS, Longstreth WT Jr et al (2008) Fish consumption and risk of subclinical brain abnormalities on MRI in older adults. Neurology 71(6):439–446
108. Virtanen JK, Siscovick DS, Lemaitre RN et al (2013) Circulating omega-3 polyunsaturated fatty acids and subclinical brain abnormalities on MRI in older adults: the Cardiovascular Health Study. J Am Heart Assoc 2(5):e000305. https://doi.org/10.1161/JAHA.113.000305
109. Tan ZS, Harris WS, Beiser AS et al (2012) Red blood cell omega-3 fatty acid levels and markers of accelerated brain aging. Neurology 78(9):658–664. doi:78/9/658 [pii] 10.1212/WNL.0b013e318249f6a9
110. Pottala JV, Yaffe K, Robinson JG et al (2014) Higher RBC EPA + DHA corresponds with larger total brain and hippocampal volumes: WHIMS-MRI Study. Neurology. https://doi.org/10.1212/WNL.0000000000000080
111. Samieri C, Maillard P, Crivello F et al (2012) Plasma long-chain omega-3 fatty acids and atrophy of the medial temporal lobe. Neurology 79(7):642–650. https://doi.org/10.1212/WNL.0b013e318264e394
112. Walhovd KB, Storsve AB, Westlye LT et al (2014) Blood markers of fatty acids and vitamin D, cardiovascular measures, body mass index, and physical activity relate to longitudinal cortical thinning in normal aging. Neurobiol Aging 35(5):1055–1064. https://doi.org/10.1016/j.neurobiolaging.2013.11.011
113. Gu Y, Brickman AM, Stern Y et al (2015) Mediterranean diet and brain structure in a multiethnic elderly cohort. Neurology. https://doi.org/10.1212/wnl.0000000000002121
114. Titova OE, Ax E, Brooks SJ et al (2013) Mediterranean diet habits in older individuals: associations with cognitive functioning and brain volumes. Exp Gerontol 48(12):

1443–1448. https://doi.org/10.1016/j.exger.2013.10.002
115. Pelletier A, Barul C, Feart C et al (2015) Mediterranean diet and preserved brain structural connectivity in older subjects. Alzheimers Dement 11(9):1023–1031. https://doi.org/10.1016/j.jalz.2015.06.1888
116. Luciano M, Corley J, Cox SR et al (2017) Mediterranean-type diet and brain structural change from 73 to 76 years in a Scottish cohort. Neurology. https://doi.org/10.1212/WNL.0000000000003559
117. Mosconi L, Murray J, Tsui WH et al (2014) Mediterranean diet and magnetic resonance imaging-assessed brain atrophy in cognitively normal individuals at risk for Alzheimer's disease. J Prev Alzheimers Dis 1(1):23–32
118. Staubo SC, Aakre JA, Vemuri P et al (2017) Mediterranean diet, micronutrients and macronutrients, and MRI measures of cortical thickness. Alzheimers Dement 13(2):168–177. https://doi.org/10.1016/j.jalz.2016.06.2359
119. Gu Y, Vorburger RS, Gazes Y et al (2016) White matter integrity as a mediator in the relationship between dietary nutrients and cognition in the elderly. Ann Neurol 79(6):1014–1025. https://doi.org/10.1002/ana.24674
120. Witte AV, Kerti L, Hermannstadter HM et al (2014) Long-chain omega-3 fatty acids improve brain function and structure in older adults. Cereb Cortex 24(11):3059–3068. https://doi.org/10.1093/cercor/bht163
121. Mosconi L, Murray J, Davies M et al (2014) Nutrient intake and brain biomarkers of Alzheimer's disease in at-risk cognitively normal individuals: a cross-sectional neuroimaging pilot study. BMJ Open 4(6):e004850. https://doi.org/10.1136/bmjopen-2014-004850
122. Berti V, Murray J, Davies M et al (2015) Nutrient patterns and brain biomarkers of Alzheimer's disease in cognitively normal individuals. J Nutr Health Aging 19(4):413–423. https://doi.org/10.1007/s12603-014-0534-0
123. Pantoni L (2010) Cerebral small vessel disease: from pathogenesis and clinical characteristics to therapeutic challenges. Lancet Neurol 9(7):689–701. https://doi.org/10.1016/S1474-4422(10)70104-6
124. Debette S, Markus HS (2010) The clinical importance of white matter hyperintensities on brain magnetic resonance imaging: systematic review and meta-analysis. BMJ 341:c3666
125. Bowman GL, Silbert LC, Howieson D et al (2012) Nutrient biomarker patterns, cognitive function, and MRI measures of brain aging. Neurology 78(4):241–249. https://doi.org/10.1212/WNL.0b013e3182436598
126. Scarmeas N, Luchsinger JA, Stern Y et al (2011) Mediterranean diet and magnetic resonance imaging-assessed cerebrovascular disease. Ann Neurol 69(2):257–268. https://doi.org/10.1002/ana.22317
127. Gottesman RF, Schneider AL, Zhou Y et al (2017) Association between midlife vascular risk factors and estimated brain amyloid deposition. JAMA 317(14):1443–1450. https://doi.org/10.1001/jama.2017.3090
128. Mazereeuw G, Lanctot KL, Chau SA et al (2012) Effects of omega-3 fatty acids on cognitive performance: a meta-analysis. Neurobiol Aging 33(7):1482 e1417–1482 e1429. https://doi.org/10.1016/j.neurobiolaging.2011.12.014
129. Jiao J, Li Q, Chu J et al (2014) Effect of n-3 PUFA supplementation on cognitive function throughout the life span from infancy to old age: a systematic review and meta-analysis of randomized controlled trials. Am J Clin Nutr 100(6):1422–1436. https://doi.org/10.3945/ajcn.114.095315
130. Farina N, Isaac MG, Clark AR et al (2012) Vitamin E for Alzheimer's dementia and mild cognitive impairment. Cochrane Database Syst Rev 11:CD002854. https://doi.org/10.1002/14651858.CD002854.pub3
131. Clarke R, Bennett D, Parish S et al (2014) Effects of homocysteine lowering with B vitamins on cognitive aging: meta-analysis of 11 trials with cognitive data on 22,000 individuals. Am J Clin Nutr. https://doi.org/10.3945/ajcn.113.076349
132. Li MM, Yu JT, Wang HF et al (2014) Efficacy of vitamins B supplementation on mild cognitive impairment and Alzheimer's disease: a systematic review and meta-analysis. Curr Alzheimer Res 11(9):844–852
133. Feart C, Samieri C, Alles B et al (2013) Potential benefits of adherence to the Mediterranean diet on cognitive health. Proc Nutr Soc 72(1):140–152. https://doi.org/10.1017/S0029665112002959
134. Grodstein F, O'Brien J, Kang JH et al (2013) Long-term multivitamin supplementation and cognitive function in men: a randomized trial. Ann Intern Med 159(12):806–814
135. Scheltens P, Twisk JW, Blesa R et al (2012) Efficacy of Souvenaid in mild Alzheimer's disease: results from a randomized, controlled trial. J Alzheimers Dis 31(1):225–236. https://doi.org/10.3233/JAD-2012-121189
136. Shah RC, Kamphuis PJ, Leurgans S et al (2013) The S-Connect study: results from a randomized, controlled trial of Souvenaid

in mild-to-moderate Alzheimer's disease. Alzheimers Res Ther 5(6):59. https://doi.org/10.1186/alzrt224
137. Soininen H, et al. Lancet Neurol. 2017 Dec; 16(12):965–975. doi: 10.1016/S1474-4422(17)30332-0.
138. Valls-Pedret C, Sala-Vila A, Serra-Mir M et al (2015) Mediterranean diet and age-related cognitive decline: a randomized clinical trial. JAMA Intern Med. https://doi.org/10.1001/jamainternmed.2015.1668

Part II

Diagnostic Concepts

Chapter 3

The Dimensional Structure of Subjective Cognitive Decline

Miguel A. Fernández-Blázquez, Marina Ávila-Villanueva, and Miguel Medina

Abstract

The self-experienced persistent decline in one's cognitive abilities in comparison with a previously normal status is referred to as subjective cognitive decline (SCD). During the last decades, evidence has emerged about the close relationship between SCD and incident cognitive impairment to such an extent that SCD is currently considered as a very early marker of future dementia. Here, we first discuss the strengths and the weaknesses of this concept, and then we describe a procedure to measure SCD accurately. Our goal is to provide the reader with specific guidelines on how to assess and classify individuals in terms of SCD in order to identify individuals at high risk of developing cognitive impairment.

Key words Aging, Alzheimer's disease, Cognitive symptoms, Dementia, Mild cognitive impairment, Subjective cognitive decline

1 Introduction

Alzheimer's disease (AD) is a multifactorial neurodegenerative dementia that begins affecting the brain many years before cognitive impairment is even noticeable. The National Institute on Aging-Alzheimer Association (NIA-AA) has recently established three different stages of AD as progression occurs over time. First, there is a preclinical phase which is defined by the incipient presence of amyloid plaques, but objective cognitive function remains normal. At the end of this preclinical phase, the individual might subjectively experience some kind of cognitive decline [1]. A second stage called prodromal AD or mild cognitive impairment (MCI) that implies subtle objective cognitive deficits that are not severe enough to significantly affect everyday activities [2]. Finally, there is a third phase in which the extent of the cognitive deterioration leads to a functional impairment that defines a dementia syndrome [3]. Currently, there are no interventions capable of modifying the progression of AD. Clinical trials in both prodromal

Robert Perneczky (ed.), *Biomarkers for Preclinical Alzheimer's Disease*, Neuromethods, vol. 137,
https://doi.org/10.1007/978-1-4939-7674-4_3,

and dementia stages have not produced significantly positive results. This scenario has led to the growing consensus that therapeutic interventions are expected to be more effective when they are putting into practice as early as possible [4]. Thus, the focus of research in the field is increasingly placing greater emphasis on the search of early markers of preclinical AD since disease-modifying therapeutic approaches will most likely be developed for future use in at-risk populations [5].

The construct subjective cognitive decline (SCD) refers to a self-experienced persistent decline in cognitive abilities in comparison with a previously normal status and independently of the objective performance on neuropsychological tests. SCD has been a focus of debate within the research literature during the past two decades because of its potential clinical relevance in predicting the onset of future dementia in older adults. In fact, it has been described that subjects might experience some type of cognitive decline up to 15 years before they develop MCI and AD [6]. Furthermore, in the absence of objective cognitive impairment, evidence has been reported about the relationship between SCD and some AD biomarkers such as brain amyloid deposition and cerebral hippocampal hypometabolism [7].

Despite cognitive complaints have been traditionally treated as equivalent to memory failures, SCD is in fact heterogeneous and might affect all cognitive domains from a neuropsychological point of view. For instance, there are symptoms that may be specifically related to memory ("forgetting recent events"), while others are associated with attention ("being unable to follow the thread of a story"), language ("the tip of the tongue phenomenon"), visual perception ("failing to recognize, by sight, close friends or relatives"), or executive functions ("perseverations in old daily routines"). Although episodic memory impairment is usually the first cognitive manifestation of MCI due to AD, perception of memory failures might not be the primary complaint of cases with preclinical AD. For instance, attentional symptoms like "following the thread of a conversation" have been proved to be more associated with risk of dementia than forgetfulness itself such as "problems to recall recent information" [8]. In any event, research on SCD has almost exclusively focused on the overall construct rather than in examining the role of specific cognitive domains upon the progression of AD.

Since the expression of cognitive complaints is affected by various factors (e.g., aging, personality, mood, drug side effects, neurological disorders, etc.), SCD is not necessarily present in all AD patients. Nevertheless, both cross-sectional [9–11] and longitudinal studies [12–14] have provided strong evidence of SCD occurring at preclinical AD. Perhaps, all these evidences along with the ease and brevity of measuring complaints are the reasons why during the last decades there has been an important increase in the study of SCD as a very early sign of cognitive deterioration.

There are overwhelming epidemiological data in favor of the relationship between SCD and incident cognitive impairment. Thus, a recent meta-analysis focused on the longitudinal value of SCD for detecting later MCI and dementia has shown that, independently of the objective memory performance, 6.6% of older adults with SCD develop MCI per year [15]. In addition, the rate of progression to dementia among those individuals who report complaints about their own cognitive performance is also twofold during a 5-year follow-up period.

Despite increasing evidence indicating that SCD may represent a very early manifestation of AD [16], little is known about the clinical role of specific complaints on the transition from normal aging to cognitive impairment. Three types of cognitive concerns have demonstrated their usefulness to discriminate between cognitively healthy older adults and MCI [17]; particularly higher scores in specific complaints on recalling immediate events, executive functioning, and prospective memory are related to prodromal stages of dementia. The fact that other types of complaints such as forgetfulness of objects or spatial orientation did not show differences between controls and MCI could be explained because the first of them refers to a high prevalent oversight in the elderly population ("forgetting where you have put something" and "forgetting where things are normally kept or looking for them in the wrong place") and the second one is an idiosyncratic sign of very mild dementia ("getting lost or turning in the wrong direction on a journey, on a walk, or in a building that one really knows because he or she has been there before"). These findings emphasize that not all cognitive complaints have the same clinical significance for predicting cognitive impairment.

However, despite its emerging role as a marker of preclinical AD, the concept of SCD is not free from some limitations which are necessary to address. For instance, many terms such as *subjective memory complaints*, *subjective cognitive decline*, or *subjective memory impairment* have been used interchangeably to refer to the same concept. This lack of consensus on a single definition of SCD affects to the comparison of findings from different investigations and epidemiological studies. Moreover, there is no accepted approach among researchers about the assessment of SCD, including the mode of administration (structured interview versus questionnaires), the cognitive domains to be examined (memory versus non-memory domains), the number of items to be used (one or two questions versus scales with a large number of items), and the optimal way to respond the items (opened questions versus multiple choice). Finally, it becomes difficult to determine which complaints underlie AD because there is a close relationship between

SCD and subjective variables such as depression [18], anxiety [19], perceived health [20], personality [21], and quality of life [22]. In sum, the heterogeneity in definitions and the different approaches for measuring SCD emphasize the necessity of searching for shared terminology and common standards of evaluation.

To assess the potential usefulness of SCD for epidemiological studies and clinical trials, an international working group, the Subjective Cognitive Decline Initiative (SCD-I), agreed to a common terminology and research procedures to identify individuals with SCD at risk of preclinical AD [23]. The SCD-I aimed at knowing whether the self-experience of decline in cognition could actually represent the first manifestation of AD. In order to demonstrate that, some common specific features are required to establish a complete profile of SCD and to characterize two distinct groups of individuals in accordance with such profile.

The SCD-I recommended to collect information regarding features such as settings in which cognitive complaints are expressed, association of SCD with medical help seeking, duration and age at onset of SCD, subjective decline in memory and non-memory domains, and association of SCD with experience of impairment. In addition, the SCD-I proposed a set of particular features which could be helpful to identify individuals at risk of clinical progression. Those features include a more acute subjective memory decline than any other cognitive domain, onset of complaints within the last 5 years, age at onset over 60 years, worries about SCD, and feeling of worse performance than other people from the same age group. Fulfilling this set of features would give rise to a more severe form of SCD referred to as subjective cognitive decline plus (SCD plus). Thus, this new category could allow us to explain the transition from a nonsymptomatic stage to the first manifestation of AD.

2 Materials

In order to measure SCD as reliably as possible, some recommendations can be extracted from a recent review of the SCD-I [24]. First, it is very important to select the most appropriate measures according to the characteristics of the target population. Complaints may have different implications depending on the research context where they are collected. For instance, clinical samples are specifically recruited in medical settings, and therefore we can assume that concerns on SCD may be higher in these samples compared to population-based studies. Moreover, it would be desirable to rely on measures previously published with adequate psychometric properties for the reference population. Second, the SCD's measures must have adequate content coverage with regard to the target population. This recommendation means that all items should be understandable; they should only inquire for a unique domain; and they should be as specific as possible and related to difficulties often found in daily life. Third, measures should explore different cognitive and non-cognitive domains because the earliest symptoms of AD may include a great variety of complaints beyond memory. Fourth, it should be taken into account the response options for the measures depending on the aim of our study. When the purpose is to distinguish between groups, dichotomous items may be enough. However, measures with ordinal response options should be preferred when changing of SCD over time is pursued. Finally, it is very important to specify the reference period of time in which we want to examine the SCD. As a rule of thumb, it should be preferred to inquire on self-experience over short periods of time (no longer than 1 year) to avoid problems with retrospective recall or estimation of complaints. Nevertheless, this does not mean that we cannot ask for longer periods if we intend to analyze the longitudinal change of SCD through the lifetime.

There is a decisive debate on how to determine whether or not an individual has SCD. Indeed, the measurement of SCD may lead to different results depending on the approach followed. In practice, SCD might be assessed by means of either individual questions or structured questionnaires. Although these two methods tend to correlate, there are some differences among them. In many studies the assessment of SCD only involves a single question similar to "Do you have any memory problems?" which is typically coded as yes/no. Instead, other investigations include a set of SCD questions in the context of a more detailed clinical interview. Finally, it may be possible to use specific questionnaires to assess SCD. Basically, the questionnaires consist of a checklist of common cognitive failures that must be rated according to the frequency in which they are experienced by individuals. These questionnaires provide information on the presence or not of any specific complaint, the frequency and severity of them, and even the cognitive strategies of an individual to deal with complaints. In addition, since questionnaires use a relative large number of items, psychometrical properties such as internal consistency and validity tend to be higher than those for specific questions of SCD [25]. For all these reasons, when available, structured questionnaires are usually considered the best approach of gaining insight into older adults' SCD [20]. Individual questions seem to be particularly useful in population-based and epidemiological studies, when administering a questionnaire is very difficult because of time constraints. However, a recent study specifically comparing open-ended questions and comprehensive questionnaire has highlighted that spontaneously reported complaints may be preferred over questionnaires because questions better reflect the distress associated with SCD [26]. In Table 1 a list of structured questionnaires on SCD that are frequently used in research settings is showed.

Table 1
Structured SCD questionnaires frequently used in research

Questionnaires	Number of items	Response options
Cognitive Failures Questionnaire (CFQ) [27]	25	Ordinal scale (5 points)
Everyday Cognition (ECog) [28]	39	Ordinal scale (4 points)
Everyday Memory Questionnaire (EMQ) [29]	35	Ordinal scale (7, 5 or 3 points)
Informant Questionnaire on Cognitive Decline (IQCODE) [30]	26	Ordinal scale (5 points)
Memory complaint questionnaire (MAC-Q) [31]	6	Ordinal scale (5 points)
Memory Functioning Questionnaire (MFQ) [32]	28	Ordinal scale (7 points)
Subjective cognitive decline questionnaire (SCD-Q) [33]	27	Dichotomic scale (yes/no)
Subjective memory complaints questionnaire (SMCQ) [34]	14	Dichotomic scale (yes/no)
Subjective memory questionnaire (SMQ) [35]	18	Ordinal scale (9 points)

3 Methods

According to the evidence and the recommendations discussed above, epidemiological studies focused on examining the role of SCD as preclinical marker of AD should combine different approaches to measure SCD accurately. Thus, there would be four specific points which should be bear in mind when a longitudinal study is methodologically designed:

1. Use both open-ended questions and structured questionnaires to measure different aspects of SCD. Specific questions should inquire about clinical details of the self-experienced cognitive decline (e.g., age at onset, seeking medical help, memory performance compared to other people, etc.) as well as concerns and frequency of particular cognitive complaints (e.g., forgetting recent events, being unable to follow the thread of a story, difficulties to retrieve the adequate word, etc.). Finally, the multiple-choice approach should vary from dichotomic to ordinal Likert-type scales to grasp the dimensionality of SCD in the best way possible.

2. Collect information in different ways to ensure a greater internal consistency. As far as possible, an interesting approach would be to gather self-perceived data by means of a face-to-face interview with a health-care professional along with self-administered questionnaires. Since it is relatively frequent that an individual reports qualitatively different features of SCD after a short period of time, this procedure would allow to evaluate the stability of those complaints.
3. Inquire about SCD with regard to different time frames (e.g., last months, last years, youth, etc.). In this way the self-experienced cognitive change over time might be examined.
4. Include items to cover all cognitive and non-cognitive domains, not only episodic memory. Some cases of atypical AD or even non-AD dementias may begin with symptoms distinct from memory loss (e.g., problems to inhibit behavior may be indicative of frontotemporal dementias rather than AD). For this reason, it is important to request subjective information about the whole spectrum of cognition and other neuropsychiatric variables.

3.1 Assessment of SCD During Neurological and Neuropsychological Interviews

Since 2011 CIEN Foundation-Queen Sofía Foundation is carrying out a multidisciplinary study to identify early markers of AD, namely, the Vallecas Project [36]. In the context of this epidemiological research, we record detailed demographic and clinical information from every participant, including self-perception of cognitive decline. Precisely, SCD has grown a great relevance in our research as a key factor which could be able to detect individuals at high risk of future cognitive impairment.

Responses to every question of SCD are directly provided by the participants since family members are not available in all cases. Most importantly, to ensure the reliability and internal consistency, SCD is assessed twice and independently within the same visit. First, during the neurological examination, participants are asked the following nine questions regarding specific cognitive domains:

1. Attention: “Are you easily distracted?”
2. Spatial orientation: “Do you get lost in familiar surroundings or have trouble finding your way when driving?”
3. Episodic memory: “Do you often forget recent information or events?”
4. Autobiographical memory: “Do you often forget autobiographical information?”
5. Visual recognition: “Do you have trouble recognizing objects or faces?”
6. Speech: “Do you have word-finding difficulties for people’s names or common words?”

7. Language comprehension: "Do you understand simple verbal and written instructions?"
8. Executive functions: "Do you have difficulty driving, managing finances, or planning daily activities?"
9. Praxis: "Do you have difficulty sequencing movements (e.g., taking the necessary steps to prepare a bath)?"

It is important to note that all these previous questions are merely tentative; thus, they are opened and spontaneously reported in such a way that alternative questions within the same cognitive domain could arise during the interview. Ultimately, subjective experience of complaints for every cognitive domain is coded in a dichotomic way (yes/no) based on the global impression of the neurologist. Additionally, there are other questions regarding psychiatric and behavioral symptoms which are a complement to the nine questions on cognitive complaints.

Second, during the neuropsychological assessment, individuals also complete an ordinal scale of cognitive complaints composed of four items with four points each (ranged 0–3). This scale addresses the following questions to be responded:

1. "How do you perceive your memory in comparison with that of others of your age?"

   ("3, bad"; "2, somewhat worse"; "1, somewhat better"; "0, excellent")
2. "How do you perceive your memory today compared with your young adulthood?"

   ("0, better"; "1, equal"; "2, somewhat worse"; "3, much worse")
3. "Do you perceive your memory today is worse than compared with ten years ago?"

   ("0, no"; "1, a little worse"; "2, somewhat worse"; "3, much worse")
4. "Do you perceive your memory today is worse than compared with one year ago?"

   ("0, no"; "1, a little worse"; "2, somewhat worse"; "3, much worse")

   The sum of these items resulted in a total score of cognitive concerns ranging from 0 (no complaints at all) to 12 (maximum complaints). Furthermore, five more open-ended questions are also collected during the neuropsychological interview:
5. Age at onset of cognitive complaints: "How old were you when your cognitive performance began to decline?"
6. Years of SCD's progression: "How long do you believe you are experiencing cognitive complaints?"

7. Worries associated with self-perceived complaints: "Are you worried about your cognitive decline?"
8. Type of onset of cognitive complaints: "How did you perceive the beginning of the cognitive decline?" (e.g., suddenly, progressive, etc.)
9. Self-experienced functional impairment associated with SCD: "Do you believe your cognitive failures are impeding your daily life activities?"

3.2 Assessment of SCD by Means of Self-Administered Questionnaire

In addition to the whole information collected in both clinical interviews, individuals must accomplish a SCD scale, namely, the Everyday Memory Questionnaire (EMQ) [29]. This questionnaire is selected because it has been previously validated in our country and showed adequate psychometrical properties for older adults [37]. In our study, EMQ is self-administered by following the instructions provided in the validation study and always in the presence of a member of the research team; individuals are required to ask any doubt may arise.

The EMQ comprised 28 items about cognitive failures that occur in everyday life. The items must be responded according to the frequency in which they are experienced by a subject. All items are scored pursuant to a three-point Likert-type scale, with 0 indicating "never, rarely", 1 "occasionally, sometimes", and 2 "frequently, almost always." The total score ranges from 0 to 56, with lower scores indicating fewer SCD. Furthermore, responses may be scored in accordance to three different factors which have been proved to distinguish between cognitively healthy individuals and MCI [17]. Due to the content of their items, the three factors are related to different cognitive domains, namely, episodic memory, executive functions, and prospective memory.

3.3 Assessment of Non-cognitive Complaints During Medical Interview

As discussed earlier, non-cognitive symptoms may represent an initial manifestation of cognitive decline due to different etiologies. Therefore, open-ended questions should be included in the context of the medical interview in order to assess behavioral, functional, and psychiatric self-reported complaints. Specifically, the aspects that should be covered may be the following:

1. Depression: "Do you feel sad, lonely, and depressed most of the time?"
2. Anxiety: "Do you feel worried and anxious most of the time?"
3. Apathy: "Do you feel a lack of emotion, motivation, or interest in hobbies and previously activities enjoyed?"
4. Disinhibition: "Do you feel difficulties to control your own behavior in some situations?"
5. Irritability: "Do you feel an irritable mood most of the time?"

6. Hallucinations: "Do you feel you see or hear strange things which maybe are unreal?"
7. Sleep: "Do you feel your sleep routine has changed?"
8. Falls: "Have you had falls recently?"
9. Gait: "Do you feel you have problems to walk?"

3.4 Classification of Individuals in SCD Groups

Besides analyzing the implication of specific SCD features as early signs of AD, information about complaints should be also examined according to the guidelines proposed by the SCD-I [23]. Following these guidelines, individuals might be grouped in three different categories pursuant to the extent of SCD reported: (1) no complaints (NCg); (2) subjective cognitive decline group (SCDg), when subjects report some kind of cognitive complaint; and (3) subjective cognitive decline plus (SCD-Pg), when individuals show complaints in memory plus another cognitive domain and additionally they fulfill the rest of the criteria for SCD plus.

The classification in any of the three groups of SCD may be carried out in two steps (a full description of this procedure is shown in Fig. 1). Initially, based on the overall information gathered both in clinical interviews and in self-administered EMQ, SCD may be operationally defined as the self-rated presence of cognitive deterioration using two criteria: (1) at least a positive response to any yes/no-type question regarding complaints in any cognitive domain from the neurological interview and (2) scores above 1 on the SCD scale administered in the neuropsychological assessment and above 8 on the self-administered EMQ. To be classified as SCDg, individuals have to mandatorily accomplish both conditions. Thus, subjects who only fulfill one criterion or neither of them will be considered as no complaints.

The second step of classification is only applied for those cases categorized as SCDg. For these individuals some specific features must be considered such as age at onset of SCD beyond 60 years, turning up of complaints within the last 5 years, worry associated with SCD, and feeling of worse performance than others of the same age group. When all these conditions accompany the self-experience of decline, then an individual is classified as SCD-Pg.

3.5 Assessment of Demographic, Clinical, and Cognitive Variables

The assessment of demographic, clinical, and cognitive variables should be inseparable from SCD, whether one's goal is to determine the properties of cognitive concerns for detecting preclinical AD and individuals at high risk of later MCI. To this end, all subjects should undergo a detailed survey and assessment protocol to gather information on demographics (age, gender, level of education, marital status, living situation, socioeconomic status, occupation, etc.), lifestyle (physical activity, social support, eating and sleeping habits, etc.), quality of life (well-being, perceived health, etc.), medical history (vital signs, physical symptoms, clinical anamnesis,

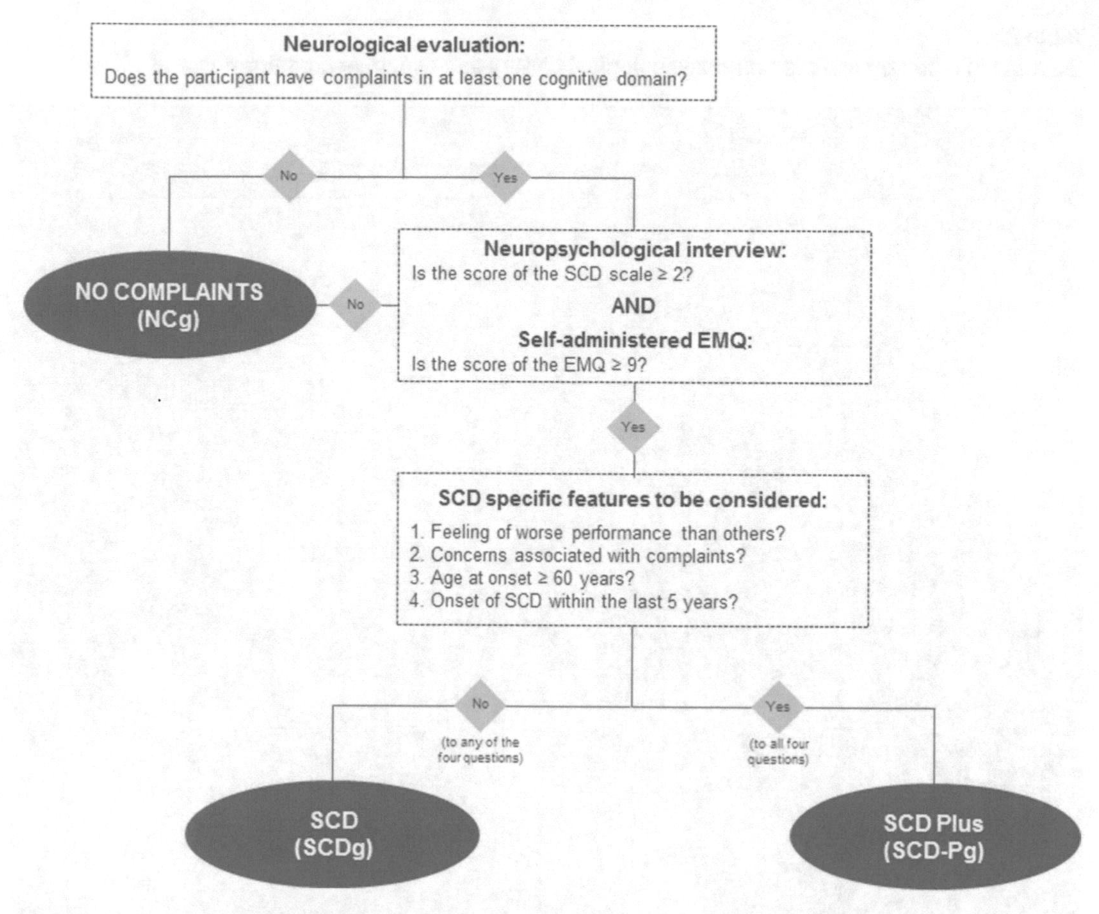

Fig. 1 Flow diagram for SCD classification. *NCg* no complaints group, *SCDg* subjective cognitive decline group, and *SCD-Pg* subjective cognitive decline plus group

medication, neurological examination, etc.), family history of dementia, and neuropsychological assessment.

Cognitive diagnosis should be always agreed between healthcare professionals at clinical consensus meetings. Every individual must be independently diagnosed according to his/her age, gender, cognitive reserve, functional information, and cognitive scores. Nevertheless, rather than psychometrically invariable cutoffs, diagnosis must be based on clinical impression. NIA-AA's criteria [2] can be applied since they are very useful to diagnose core MCI and mild dementia. Cognitively healthy subjects are given a score of 0 in the global clinical dementia rating (CDR) [38], while MCI and mild dementia must score 0.5 and 1, respectively.

The comprehensive neuropsychological battery must be administered by trained neuropsychologists. This battery should include complete information about all cognitive domains such as visual perception, attention, memory, language, praxis, and executive functions. As an example, in Table 2, a total of 12 cognitive tests which comprise the neuropsychological battery of the

Table 2
Example of a comprehensive neuropsychological battery and cognitive domains covered

	Global cognition	Visual perception	Attention	Memory	Language	Praxis	Executive functions
Mini mental state examination (MMSE) [39]	X						
Clock-drawing test	X					X	X
Rey-Osterreith complex Fig [40, 41].		X		X		X	X
Free and Cued Selective Reminding Test (FCSRT) [42]				X			
Lexical and semantic verbal fluency [43]					X		X
Forward and backward digit span [44]			X				X
Five-point test [45]							X
Rule Card Shifting Test [46]							X
Boston Naming Test (15-item version) [47]					X		
Imitation of Bilateral Postures and Symbolic Gesture [48]						X	
Digit symbol coding [44]			X				
Incomplete letters (VOSP) [49]		X					

Vallecas Project is shown. All these tests cover the whole spectrum of cognition.

In addition, scales for measuring functional activities and neuropsychiatric variables are very important. For instance, to this end, the functional activities questionnaire (FAQ) [50] can be administered to collect data with regard to instrumental activities of daily living; the Geriatric Depression Scale (GDS) [51] and the State-Trait Anxiety Inventory (STAI) [52] might be also administered as part of the neuropsychological battery to quickly estimate mood and anxiety symptoms.

4 Conclusions

Self-report of subtle cognitive complaints have been proposed to appear at the end of the preclinical phase of AD even in the absence of significant objective impairment detectable on standardized neuropsychological assessment [1]. Indeed, there are overwhelming epidemiological data that support the role of SCD, a risk factor for subsequent development of MCI and dementia in older adults [15]. That explains why SCD is gaining an increasing prominence in neurodegenerative research. Nevertheless, this construct must face up to some limitations such as the absence of an operational definition as well as the lack of harmonized criteria and assessment protocols for measuring SCD. The insight and self-experience of SCD is phenomenologically complex and may differ among individuals. Moreover, there is a great variability in SCD assessment procedures which is reflected on the different questions, response options, nature of items, and mode of administration across studies. This heterogeneity is likely affecting the outcomes obtained in each study and also determines the comparison of results. Because of this scenario, in 2014, an international working group, the SCD-I, published a position paper agreeing to a common terminology and research procedures to investigate SCD [23]. Since then, we rely on a homogeneous framework to better study the implication of cognitive complaints in dementia due to AD.

Currently, SCD could be defined as a self-experienced persistent and progressive decline in the own cognitive capacity in comparison with a previously normal performance and unrelated to any acute event [53]. This means that complaints due to AD cannot appear suddenly but gradually, and thus decline cannot be transient and be associated with any particular medical/psychological condition. In addition, from a methodologically point of view, some considerations may be made with regard to assessment procedures and protocols of SCD. First, investigations should combine different approaches to measure complaints as accurate as possible. It is highly recommended to use both open-ended questions and items belonging to structured questionnaires to measure different aspects of SCD. Specific questions should measure clinical details of the self-experienced cognitive decline (e.g., age at onset, seeking medical help, memory performance compared to other people, etc.) as well as concerns and frequency of particular cognitive complaints. Likewise, depending on the feature of SCD which is under investigation, the response option may adopt the form of dichotomic (yes/no-type question) or ordinal scales. Second, different modes of administration may be carried out to ensure a higher internal consistency of SCD. The collection of complaints data would benefit if it is made by combining face-to-face clinical interview along with self-administered questionnaires instead of one unique mode. Third, the introduction of retrospective time frames allows for the study of the course of SCD over time. It would be desirable to include items that inquire about SCD in different periods of time (e.g., from adulthood to the last year) in order to examine the role of complaints in AD. The use of questionnaires such as the EMQ is highly recommended to quantify SCD and to monitor the longitudinal progression of individuals who report those cognitive complaints. Fourth, since most instruments used to assess SCD focused on episodic memory, it would be necessary to include items that cover other cognitive and non-cognitive domains to further evaluate cases of atypical AD or cases of non-AD neurodegenerative dementia. Besides, not all cognitive complaints are effective in distinguishing healthy elderly individuals from those with MCI. Specific complaints related to episodic memory, executive functions, and prospective memory seem to discriminate between controls and cognitive impaired subjects [17].

Finally, because SCD should not be examined in isolation when one is interested in determining the effect of complaints upon AD spectrum, some other information must be obtained. Demographic variables such as age, gender, and education, as well as medical and lifestyle variables, can be gathered very easily by means of a survey. These variables have the greatest interest due to their possible implication in the expression or not of SCD. Additionally, objective cognitive performance and diagnosis are critical to establish the current stage of an individual in the continuum of AD and the relationship between SCD and risk of developing MCI and AD. Finally, neuropsychiatric variables seem to mediate between SCD and cognitive decline, and thus they should be collected as well. Symptoms such as depression, anxiety, and apathy are usually more associated with SCD than real cognitive performance, what likely indicates the mediator role of complaints between mood and cognitive status [54]. Individuals who accomplish all conditions of SCD-Pg have been proved to have a four times higher risk for developing MCI compared to those subjects without complaints (see Fig. 2); and more surprisingly, the inclusion of gender, cognitive performance, and ApoE genotyping does not seem to decrease the predictive power of the SCD-Pg [55]. Thus, particular features associated with SCD (i.e., onset of complaints within the last 5 years, age at onset over 60 years, worries associated with complaints, and feeling of worse performance than other people from the same age group) seem to help to identify individuals at high risk of fast conversion to MCI. We strongly encourage researchers to use the recommendations of SCD-I because of their implications for clinical settings. For instance, cognitive training programs should be implemented in subjects meeting all features of SCD plus proposed by SCD-I [22]. Subjects who report SCD plus might need special attention in terms of close clinical follow-up of an early cognitive or pharmacological intervention.

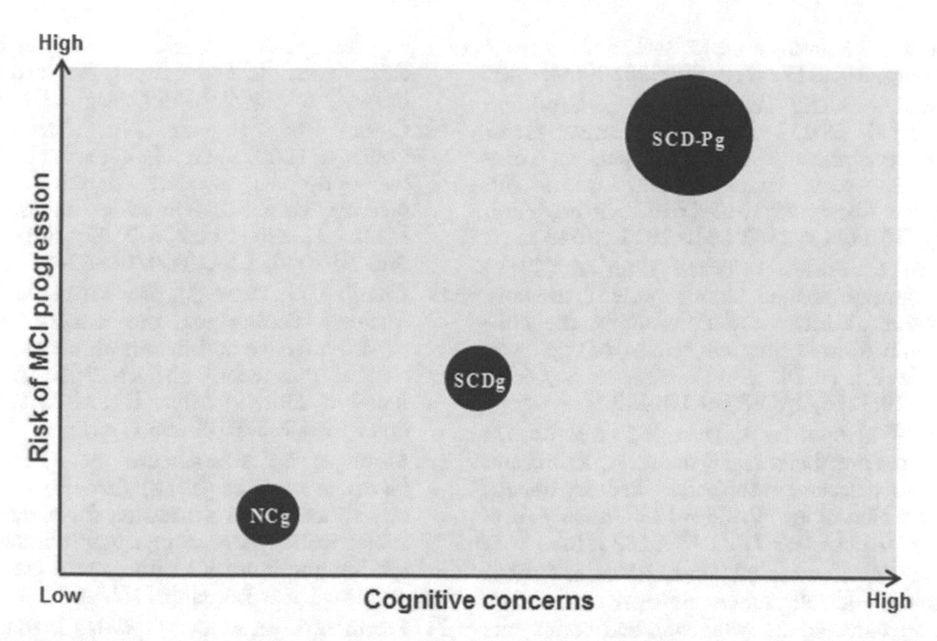

Fig. 2 Distribution of SCD groups and association with conversion rate to MCI. *NCg* no complaints group, *SCDg* subjective cognitive decline group, and *SCD-Pg* subjective cognitive decline plus group. This graph is based on the results published in a previous paper entitled "Specific Features of Subjective Cognitive Decline Predict Faster Conversion to Mild Cognitive Impairment" (Fernández-Blázquez et al. [55]). No metric multidimensional scaling was applied in order to analyze the relative position of SCD groups in terms of euclidean distance. The analysis was based on all significant variables found in nonparametric ANOVAs. Sizes of bubbles indicate the risk of annual conversion to MCI (the larger is the area, the greater is the risk). As is apparent, SCD-Pg is the group with the highest rate of conversion (18.9%) followed by SCDg (5.6%) and NCg (4.9%). In addition, the separation in perceptual map among the three groups indicates that there are three latent profiles underlying psychiatric and cognitive variables

References

1. Sperling R, Aisen PS, Beckett L et al (2011) Toward defining the preclinical stages of Alzheimer's disease: recommendations from the National Institute on Aging and the Alzheimer's Association workgroup. Alzheimers Dement 7:280–292. https://doi.org/10.1016/j.jalz.2011.03.003
2. Albert MS, DeKosky ST, Dickson D et al (2011) The diagnosis of mild cognitive impairment due to Alzheimer's disease: recommendations from the National Institute on Aging and Alzheimer's Association workgroup. Alzheimers Dement 7:270–279. https://doi.org/10.1016/j.jalz.2011.03.008
3. McKhann GM, Knopman DS, Chertkow H et al (2011) The diagnosis of dementia due to Alzheimer's disease: recommendations from the National Institute on Aging-Alzheimer's Association workgroups on diagnostic guidelines for Alzheimer's disease. Alzheimers Dement 7:263–269. https://doi.org/10.1016/j.jalz.2011.03.005
4. Selkoe DJ (2011) Resolving controversies on the path to Alzheimer's therapeutics. Nat Med 17:1060–1065. https://doi.org/10.1038/nm.2460
5. Hampel H, Lista S, Teipel SJ et al (2014) Perspective on future role of biological markers in clinical therapy trials of Alzheimer's disease: a long-range point of view beyond 2020. Biochem Pharmacol 88:426–449. https://doi.org/10.1016/j.bcp.2013.11.009
6. Reisberg B, Prichep L, Mosconi L et al (2008) The pre-mild cognitive impairment, subjective cognitive impairment stage of Alzheimer's disease. Alzheimers Dement 4:S98–S108. https://doi.org/10.1016/j.jalz.2007.11.017
7. Vannini P, Hanseeuw B, Munro CE et al (2017) Hippocampal hypometabolism in older adults with memory complaints and increased amyloid

burden. Neurology 88:1759–1767. https://doi.org/10.1212/WNL.0000000000003889

8. Amariglio RE, Townsend MK, Grodstein F et al (2011) Specific subjective memory complaints in older persons may indicate poor cognitive function. J Am Geriatr Soc 59:1612–1617. https://doi.org/10.1111/j.1532-5415.2011.03543.x
9. Rami L, Fortea J, Bosch B et al (2011) Cerebrospinal fluid biomarkers and memory present distinct associations along the continuum from healthy subjects to AD patients. J Alzheimers Dis 23:319–326. https://doi.org/10.3233/JAD-2010-101422
10. Scheef L, Spottke A, Daerr M et al (2012) Glucose metabolism, gray matter structure, and memory decline in subjective memory impairment. Neurology 79:1332–1339. https://doi.org/10.1212/WNL.0b013e31826c1a8d
11. Wang Y, Risacher SL, West JD et al (2013) Altered default mode network connectivity in older adults with cognitive complaints and amnestic mild cognitive impairment. J Alzheimers Dis 35:751–760. https://doi.org/10.3233/JAD-130080
12. Dufouil C, Fuhrer R, Alpérovitch A (2005) Subjective cognitive complaints and cognitive decline: consequence or predictor? The epidemiology of vascular aging study. J Am Geriatr Soc 53:616–621. https://doi.org/10.1111/j.1532-5415.2005.53209.x
13. Glodzik-Sobanska L, Reisberg B, De Santi S et al (2007) Subjective memory complaints: presence, severity and future outcome in normal older subjects. Dement Geriatr Cogn Disord 24:177–184. https://doi.org/10.1159/000105604
14. Reisberg B, Shulman MB, Torossian C et al (2010) Outcome over seven years of healthy adults with and without subjective cognitive impairment. Alzheimers Dement 6:11–24. https://doi.org/10.1016/j.jalz.2009.10.002
15. Mitchell AJ, Beaumont H, Ferguson D et al (2014) Risk of dementia and mild cognitive impairment in older people with subjective memory complaints: meta-analysis. Acta Psychiatr Scand 130:439–451. https://doi.org/10.1111/acps.12336
16. Jessen F (2014) Subjective and objective cognitive decline at the pre-dementia stage of Alzheimer's disease. Eur Arch Psychiatry Clin Neurosci 264:3–7. https://doi.org/10.1007/s00406-014-0539-z
17. Ávila-Villanueva M, Rebollo-Vázquez A, Ruiz-Sánchez de León JM et al (2016) Clinical relevance of specific cognitive complaints in determining mild cognitive impairment from cognitively normal states in a study of healthy elderly controls. Front Aging Neurosci 8:233. https://doi.org/10.3389/fnagi.2016.00233
18. Crane MK, Bogner HR, Brown GK, Gallo JJ (2007) The link between depressive symptoms, negative cognitive bias and memory complaints in older adults. Aging Ment Health 11:708–715. https://doi.org/10.1080/13607860701368497
19. Comijs HC, Deeg DJ, Dik MG et al (2002) Memory complaints; the association with psycho-affective and health problems and the role of personality characteristics. A 6-year follow-up study. J Affect Disord 72:157–165. doi: S0165032701004530 [pii]
20. Montejo P, Montenegro M, Fernández-blázquez M et al (2014) Association of perceived health and depression for older adults' subjective memory complaints: contrasting a specific questionnaire with general complaints questions. Eur J Ageing 11:77–87
21. Pearman A, Storandt M (2004) Predictors of subjective memory in older adults. J Gerontol B Psychol Sci Soc Sci 59:P4–P6
22. Montejo P, Montenegro M, Fernandez MA, Maestu F (2011) Subjective memory complaints in the elderly: prevalence and influence of temporal orientation, depression and quality of life in a population-based study in the city of Madrid. Aging Ment Health 15:85–96. https://doi.org/10.1080/13607863.2010.501062
23. Jessen F, Amariglio RE, van Boxtel M et al (2014) A conceptual framework for research on subjective cognitive decline in preclinical Alzheimer's disease. Alzheimers Dement 10:844–852. https://doi.org/10.1016/j.jalz.2014.01.001
24. Rabin LA, Smart CM, Crane PK et al (2015) Subjective cognitive decline in older adults: an overview of self-report measures used across 19 international research studies. J Alzheimers Dis 48(Suppl 1):S63–S86. https://doi.org/10.3233/JAD-150154
25. Embretson SE (1996) The new rules of measurement. Psychol Assess 8:341–349. https://doi.org/10.1037/1040-3590.8.4.341
26. Burmester B, Leathem J, Merrick P (2015) Assessing subjective memory complaints: a comparison of spontaneous reports and structured questionnaire methods. Int Psychogeriatr 27:61–77. https://doi.org/10.1017/S1041610214001161
27. Broadbent DE, Cooper PF, FitzGerald P, Parkes KR (1982) The Cognitive Failures Questionnaire (CFQ) and its correlates. Br J Clin Psychol 21(Pt 1):1–16. https://doi.org/10.1111/j.2044-8260.1982.tb01421.x

28. Farias ST, Mungas D, Reed BR et al (2008) The measurement of everyday cognition (ECog): scale development and psychometric properties. Neuropsychology 22:531–544. https://doi.org/10.1037/0894-4105.22.4.531
29. Sunderland A, Harris JE, Gleave J (1984) Memory failures in everyday life following severe head injury. J Clin Neuropsychol 6:127–142. https://doi.org/10.1080/01688638408401204
30. Jorm AF, Jacomb PA (1989) The Informant Questionnaire on Cognitive Decline in the Elderly (IQCODE): socio-demographic correlates, reliability, validity and some norms. Psychol Med 19:1015–1022
31. Crook TH, Feher EP, Larrabee GJ (1992) Assessment of memory complaint in age-associated memory impairment: the MAC-Q. Int Psychogeriatr 4:165–176. https://doi.org/10.1017/S1041610292000991
32. Gilewski MJ, Zelinski EM, Schaie KW (1990) The Memory Functioning Questionnaire for assessment of memory complaints in adulthood and old age. Psychol Aging 5:482–490
33. Rami L, Mollica MA, García-Sanchez C et al (2014) The Subjective Cognitive Decline Questionnaire (SCD-Q): a validation study. J Alzheimers Dis 41:453–466. https://doi.org/10.3233/JAD-132027
34. Youn JC, Kim KW, Lee DY et al (2009) Development of the Subjective Memory Complaints Questionnaire. Dement Geriatr Cogn Disord 27:310–317. https://doi.org/10.1159/000205512
35. Squire LR, Wetzel CD, Slater PC (1979) Memory complaint after electroconvulsive therapy: assessment with a new self-rating instrument. Biol Psychiatry 14:791–801
36. Olazarán J, Valentí M, Frades B et al (2015) The Vallecas Project: a cohort to identify early markers and mechanisms of Alzheimer's disease. Front Aging Neurosci 7:181. https://doi.org/10.3389/fnagi.2015.00181
37. Montejo Carrasco P, Montenegro Peña M, Suciro MJ (2012) The Memory Failures of Everyday (MFE) test: normative data in adults. Span J Psychol 15:1424–1431. https://doi.org/10.5209/rev_SJOP.2012.v15.n3.39426
38. Hughes CP, Berg L, Danziger WL (1982) A new clinical scale for the staging of dementia. Br J Psychiatry 140:566–572. https://doi.org/10.1192/bjp.140.6.566
39. Folstein MF, Folstein SE, McHugh PR (1975) "Mini-mental state": a practical method for grading the cognitive state of patients for the clinician. J Psychiatr Res 12:189–198. https://doi.org/10.1016/0022-3956(75)90026-6
40. Rey A (1941) L'examen psychologique dans les cas d'encéphalopathie traumatique. Arch Psychol (Geneve) 28:215–285. http://psycnet.apa.org/psycinfo/1943-03814-001
41. Peña-Casanova J, Gramunt-Fombuena N, Quiñones-Úbeda S et al (2009) Spanish multicenter normative studies (NEURONORMA project): norms for the rey-osterrieth complex figure (copy and memory), and free and cued selective reminding test. Arch Clin Neuropsychol 24:371–393
42. Buschke H (1984) Cued recall in amnesia. J Clin Neuropsychol 6:433–440. https://doi.org/10.1080/01688638408401233
43. Peña-Casanova J, Quiñones-Ubeda S, Gramunt-Fombuena N et al (2009) Spanish Multicenter Normative Studies (NEURONORMA Project): norms for verbal fluency tests. Arch Clin Neuropsychol 24:395–411. https://doi.org/10.1093/arclin/acp042
44. Wechsler D (1997) Wechsler Adult Intelligence Scale-III. The Psychological Corporation, San Antonio, TX
45. Lee GP, Loring DW, Newell J, McCloskey L (1994) Figural fluency on the Five-Point Test: preliminary normative and validity data. Int Neuropsychol Soc Progr Abstr, p 51
46. Wilson BA, Alderman N, Burgess PW et al (1996) The behavioural assessment of the dysexecutive syndrome. Thames Valley Company, Bury St Edmunds
47. Fernández-Blázquez MA, Ruiz-Sánchez de León JM, López-Pina JA et al (2012) A new shortened version of the Boston Naming Test for those aged over 65: an approach from item response theory. Rev Neurol 55:399–407
48. Peña-Casanova J (1990) Programa integrado de exploración neuropsicológica-Test Barcelona. Masson, Barcelona
49. Warrington EK, James M (1991) The visual object and space perception battery. Thames Valley Test Company, Bury St Edmunds
50. Pfeffer RI, Kurosaki TT, Harrah CH et al (1982) Measurement of functional activities in older adults in the community. J Gerontol 37:323–329. https://doi.org/10.1093/geronj/37.3.323
51. Yesavage JA, Brink TL, Rose TL et al (1982) Development and validation of a geriatric depression screening scale: a preliminary report. J Psychiatr Res 17:37–49. https://doi.org/10.1016/0022-3956(82)90033-4

52. Spielberger C, Gorsuch R, Leshene R (1970) Manual for the State-Trait Anxiety Inventory. Consulting Psychologists Press, Palo Alto, CA
53. Molinuevo JL, Rabin LA, Amariglio R et al (2017) Implementation of subjective cognitive decline criteria in research studies. Alzheimers Dement 13:296–311. https://doi.org/10.1016/j.jalz.2016.09.012
54. Yates JA, Clare L, Woods RT, Matthews FE (2015) Subjective memory complaints are involved in the relationship between mood and mild cognitive impairment. J Alzheimers Dis 48(Suppl 1):S115–S123. https://doi.org/10.3233/JAD-150371
55. Fernández-Blázquez MA, Ávila-Villanueva M, Maestú F, Medina M (2016) Specific features of subjective cognitive decline predict faster conversion to mild cognitive impairment. J Alzheimers Dis 52:271–281. https://doi.org/10.3233/JAD-150956

Chapter 4

Deriving and Testing the Validity of Cognitive Reserve Candidates

Yaakov Stern and Christian Habeck

Abstract

Empirical support has established cognitive reserve (CR) well as a concept since its inception three decades ago, and most brain researchers subscribe to some version of CR as a collection of subject factors that influence cognitive performance beyond simple brain structural health. We give a simple but precise analytic recipe to test requirements for any plausible cognitive reserve candidate based on brain imaging. Gradations of partial fulfillment of *some* but not all of these requirements are possible.

Key words Cognitive reserve, Brain aging, Resting-state fMRI, Cognitive aging

1 Introduction

Cognitive reserve refers to subject factors that determine cognitive performance beyond obvious indicators of brain structural health. Cognitive reserve was first formulated as concept in response to several seemingly puzzling findings: (1) autopsy data of person appearing cognitively normal at death showed varying degrees of Alzheimer-like pathology; (2) when matched for disease severity of clinical presenta tions, PET scans revealed worse metabolic deficits in parietotemporal areas for patients with higher premorbid educational and occupational attainment. Formulated differently, matching people for their brain structural and metabolic (in short, brain) health leaves unexplained incommensurate differences in cognitive performance and clinical presentations. Thus, the cognitive performance of two people with relatively similar brain health is not *necessarily* more similar than the cognitive performance of two people with appreciably different levels of brain health (even though brain health is conducive to good cognition). It became clear that a third variable was needed to adequately account for the relationship between brain structure/metabolism and cognitive performance. This variable is cognitive reserve.

There have been many empirical accounts of cognitive reserve candidates: leisure activity [1–6], lifetime mental activity [7, 8],

Robert Perneczky (ed.), *Biomarkers for Preclinical Alzheimer's Disease*, Neuromethods, vol. 137,
https://doi.org/10.1007/978-1-4939-7674-4_4, © Springer Science+Business Media, LLC 2018

and educational and occupational attainment [9–17], to name a selective and nonrepresentative sample of a much richer literature.

In the current report, we would like to present a formal framework for rigorously deriving and testing cognitive reserve candidates. First, we list three axioms regarding cognitive reserve. These axioms operate as principles in our own research program and partly originate from historical empirical findings. It is useful to make the axioms very explicit. While they may sound self-evident and somewhat trivial to some readers, others might disagree, possibly strongly. Algorithmically, these axioms are quite rigorous, but they leave open many possibilities about the implementation and mechanisms of cognitive reserve.

1.1 Cognitive Reserve Axioms

1. Cognitive reserve has well-established proxies like education, intelligence, occupational attainment, leisure activities, and other subject variables that express cognitively stimulating engagement.
2. Cognitive reserve is either an independent factor that influences cognitive performance in addition to brain structure or a moderating factor that influences the relationship between brain structure and cognitive performance; these two manifestations are not exclusive and can conveniently be incorporated in linear regression models that contain (1) brain structure and (2) the putative CR measure as direct effects and (3) an interaction term.
3. Although some of the cognitive reserve proxies might correlate with brain structural measures, cognitive reserve itself is *not* based on brain structure; this entails in particular that:
   - It is *not* a mediating influence of brain structure on cognitive performance.
   - It should be *uncorrelated* with brain structural measures.

These three axioms can be used to constrain the derivation of plausible cognitive reserve candidates. Specifically, (1) cognitive reserve should display correlations with some, or all, of the preexisting proxies and (2) should show direct effects on cognitive performance in addition to brain structure or moderate the effects of brain structure on performance while (3) being uncorrelated to brain structure. This last point is contentious with some researchers who might refer to obvious correlations between brain structure and CR proxies like intelligence or education. Recently, we have laid out how such correlations can be conceptualized as *brain maintenance* [18] which should be considered as separate from, and in correlational terms orthogonal to, cognitive reserve. CR proxies thus might correlate with cognitive reserve and brain structure, but the latter correlation strictly separates them from being a viable candidate for cognitive reserve directly. In other words, while CR proxies might indicate the state of somebody's brain structural health and how well the brain has been preserved in the course of aging, cognitive reserve strictly speaks to *how well somebody utilizes their brain regardless of how well it has been preserved.* CR proxies thus consist

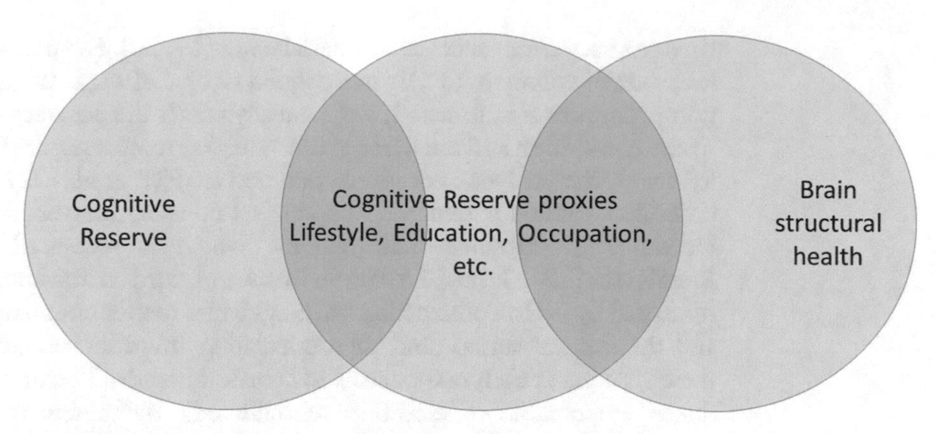

Fig. 1 Venn diagram of cognitive reserve proxies in relation to brain structure

of two parts: (1) a part that correlates with brain structure and points to possible formative influences to build and maintain a healthy brain and (2) cognitive reserve, which in our rigorous definition is independent of brain structure. The Venn diagram in Fig. 1 gives a simple schematic explanation.

2 Materials

The demonstrations of this paper were run in 451 participants, aged 20–80, using volumetric data as the indicator of brain integrity and attempting to derive a measure of CR from resting BOLD data. The methods described here could be extended to any structural and "brain activity" measures. All MR images were acquired on a 3.0T Philips Achieva magnet that has been used since the start of this study. A structural MPRAGE scan was obtained (TE/TR = 3/6.5 ms; flip angle = 8; resolution = 256 × 256; FOV = 25.6 cm × 25.6 cm; slice thickness/gap = 1/0 mm; 180 slices), as well as a resting BOLD scan (TE/TR = 20/2000 ms; flip angle = 72; resolution = 112 × 112; FOV = 22.4 cm × 22.4 cm; slice thickness/gap = 3/0 mm; 41 slices). The resting BOLD scan was either 5 minutes or 9.5 min long. In terms of demographic variables, we have education, verbal intelligence, and three measures of vocabulary, which were averaged to produce an average vocabulary domain *z*-score.

3 Methods

Using each individual's T1-weighted MPRAGE image, structural imaging measures of both global and regional brain volume were derived using the FreeSurfer software package (http://surfer.nmr.mgh.harvard.edu/). Brain volume calculations automatically assign a neuroanatomic label to each voxel; this has previously been shown to have accuracy comparable to manual labeling [19, 20]. From this labeling, a set of volumetric ROIs was defined [21]. Although the estimation procedure is automated, we manually check the accuracy of the spatial registration and the white matter and gray matter segmentations following the analytic procedures outlined in Fjell et al. [22] . The calculated volume within each region was adjusted for variations in individual's intracranial brain volume which is measured using BrainWash [23]. Using FreeSurfer software, cortical thickness was measured by first reconstructing the gray/white matter boundary [24] and the cortical surface and then calculating the distances between these surfaces at each point across the cortical mantle. The maps produced are capable of detecting submillimeter differences between groups [25]. Although the thickness estimation procedure is automated, we manually check the accuracy of the spatial registration and the white matter and gray matter segmentations following the analytic procedures outlined in Fjell et al. [22]. Using a validated automated labeling system [20], FreeSurfer divided the cortex into 33 different gyral-based areas in each hemisphere and calculates the mean thickness in each area.

Furthermore, resting-state fMRI was collected for either 5 or 9.5 min, while the participant is instructed to lie still with their eyes closed, to not think of any one thing in particular, and to not fall asleep. MCFLIRT is used to register all the volumes to a reference image [26] followed by scrubbing and motion correction as described in Power et al. [27] and incorporating the recommendation of replacing deleted volumes with zero volumes prior to filtering [28]. Using the FSL command "flsmaths –bptf t," the motion-corrected signals are passed through a band-pass filter with the cutoff frequencies of 0.01 and 0.08 Hz. After filtering, the first three volumes are discarded due to the lag of the digital filter. Finally, we residualized the motion-corrected, scrubbed, and temporally filtered volumes by regressing out the FD, RMSD, left- and right-hemisphere white matter, and lateral-ventricular signals [29]. We followed the taxonomy of functional units presented by Power et al. [30] and produced residualized and motion-corrected resting BOLD time series at 264 ROI locations. This data reduction has the advantage that it enables easy computation of global functional connectivity between all $264 \times 263/2 = 34{,}716$ node pairs.

The demonstrations that follow used omnibus measures that collapse across all ROIs and all ROI pairs. This was done for the sake of simplicity; an ecologically valid research program most likely would consider structural variables and CR candidates with a topographic specificity. In particular functional connectivity did not respect the intricate taxonomy of different resting-state networks [30] and did not break down connectivity into intra- and inter-network connectivity, which would be advisable for real basic and diagnostic research questions.

3.1 Consideration of a Connectivity-Based Measure

Deriving a candidate measure for cognitive reserve should ideally fulfill our three above postulates. Further, ideally it should be motivated with a convincing mechanistic insight or hypothesis, which suggests plausible plasticity on the basis of lifetime experience of cognitively enriched environments. This is a demanding program which, for the sake of demonstration here, we cannot do total justice.

The measure we will investigate came out of a purely empirical data exploration of the subject-level Fisher-Z matrices. This was done for didactic expediency of demonstrating the workings of the CR postulates; we are not claiming that there are not more plausible connectivity-based CR measures.

The measure we considered is the intraindividual variability of functional connectivity. This was computed on a subject level by taking the standard deviation across all 34,716 ROI-pair Fisher-Z values. Large values thus imply a lot of regional variation in functional connectivity, while lower values suggest relatively more uniformity in functional connectivity across the brain.

3.2 Test of Postulates for Cognitive Reserve

Let us now turn to our three postulates, which represent the linchpin of this report. The variability measure is denoted as *stdZ* and is a scaler value, i.e., it gives one number for each of the 451 participants.

1. The first postulate concerns a relationship with some well-known CR proxies. We checked the correlation with NART and years of education. The correlation with education years was nonsignificant ($R = -0.0376$, $p = 0.4264$), while it was modestly significant with NART ($R = -0.1076$, $p = 0.0267$). Thus, people with higher verbal intelligence had slightly less regional variability of functional connectivity in their brains.
2. The test of the second postulate is computationally somewhat more demanding and involves many more, potentially arbitrary, decisions on account of the analyst. The second postulate embodies a key requirement: Our CR candidate should account for a particular cognitive performance above and beyond brain structural independent variables and possibly moderate the structure-cognition relationship. Thus CR could be an independent factor that determines cognitive performance in addition to brain structure, or a moderator, or both. We can write down the regression equation

$$\mathbf{Dep} = \mathbf{sIV}\beta + \mathbf{CR}\gamma + \mathbf{sIV} \times \mathbf{CR}\delta + \varepsilon.$$

where ***Dep*** denotes our cognitive outcome measure, vocabulary, ***sIV*** denotes a structural independent variable, ***CR*** stands for the cognitive reserve candidate, and ***sIV* × *CR*** denotes the interaction term. Greek letters $\boldsymbol{\beta\gamma\delta}$ denote regression weights, and ε stands for the unexplained residual of ***Dep***. It is easily conceivable that there are more than just one structural independent variable and correspondingly more than one interaction term. Concretely, we will use mean cortical volume and mean cortical thickness, which results in direct effects of two structural independent variables and CR, and two interaction terms. More fundamental is the dependency of the postulate check on the choice of dependent and independent variables. The test of the second postulate for any CR candidate obviously is always relative to a particular choice of cognitive outcome and structural independent variables. Ideally, the best CR measure captures similarities with regard to different cognitive outcomes and structural variables, but this obviously does not need to be so. If, for instance, a plethora of cognitive outcomes are used, it is not guaranteed a priori that the particular CR measure in question will show consistent behavior with regard to all of them. In fact, if the cognitive outcomes that are considered do not capture one underlying construct, it is unlikely that the CR measure will display consistent behavior.

Anyways, our current report is not concerned with multiple cognitive outcomes and only considered vocabulary where three indicator variables were averaged prior to the test of the second postulate to produce one cognitive outcome measure. We ran our linear regression, and *z*-scored all three independent variables, such that they were mean-centered, to avoid collinearities between direct effects and interaction terms. (This is very important in general.) The results are displayed in Table 1.

Our CR measure, *stdZ*, thus significantly accounts for vocabulary performance with consistent directionality to the correlation with NART: higher *stdZ*, i.e., greater regional variability in functional connectivity, is associated with worse vocabulary performance. The interaction terms, however, are not significant. Of the brain structural independent variables, only cortical volume has an expected positive association with performance.

3. We now come to our last postulate. We explained in our initial remarks that this postulate is at times contentious: a suitable CR candidate should *not* display any associations with brain structure. We laid out elsewhere and only restate that amply documented correlations between lifestyle factors, enrichment variables, and brain structure should be conceptualized as brain maintenance, rather than cognitive reserve. The check

Table 1
Linear regression to explain vocabulary performance. Listed are *T*- and *p*-values for each independent variable

Independent variable	*T*	*P*
Mean cortical volume	3.0931	0.0021
Mean cortical thickness	−1.7053	0.0889
stdZ	−2.6077	0.0094
stdZ × volume	0.2702	0.7871
stdZ × thickness	0.6915	0.4896

for the third postulate is therefore quite easy and involves a correlation between *stdZ* and our brain structural variables. Both mean cortical volume and mean cortical thickness showed no association with *stdZ* ($R = -0.0394$, $p = 0.4044$ and $R = -0.0604$, $p = 0.208$, respectively). Our CR candidate is also independent of brain structure and independent of how well participants' cortical volume and thickness have been maintained throughout the course of their adult life.

All three postulates are therefore fulfilled, although postulate no. 2 did not show any moderation effects by our CR candidate.

4 Conclusions

This brief presentation describes a quantitative approach for deriving direct measures of CR from brain-based measures. More conventional research may make use of proxies for CR, but these proxies do not directly measure CR, rather they represent experiences that might impart CR. Deriving direct brain measures of CR will eventually allow us to enhance our understanding of the neural implementation of this concept.

References

1. Verghese J, Lipton RB, Katz MJ et al (2003) Leisure activities and the risk of dementia in the elderly. N Engl J Med 348(25):2508–2516
2. Hu G, Sarti C, Jousilahti P et al (2005) Leisure time, occupational, and commuting physical activity and the risk of stroke. Stroke 36(9):1994–1999
3. Rovio S, Kareholt I, Helkala EL et al (2005) Leisure-time physical activity at midlife and the risk of dementia and Alzheimer's disease. Lancet Neurol 4(11):705–711. https://doi.org/10.1016/S1474-4422(05)70198-8
4. Helzner EP, Scarmeas N, Cosentino S et al (2007) Leisure activity and cognitive decline in incident Alzheimer disease. Arch Neurol 64(12):1749–1754
5. Akbaraly TN, Portet F, Fustinoni S et al (2009) Leisure activities and the risk of dementia in the elderly: results from the Three-City Study. Neurology 73(11):854–861. https://doi.org/10.1212/WNL.0b013e3181b7849b
6. Scarmeas N, Levy G, Tang MX et al (2001) Influence of leisure activity on the incidence of Alzheimer's disease. Neurology 57(12):2236–2242
7. Valenzuela MJ, Sachdev P, Wen W et al (2008) Lifespan mental activity predicts diminished rate of hippocampal atrophy. PLoS One 3(7):e2598. https://doi.org/10.1371/journal.pone.0002598
8. Gates N, Valenzuela M (2010) Cognitive exercise and its role in cognitive function in older adults. Curr Psychiatry Rep 12(1):20–27. https://doi.org/10.1007/s11920-009-0085-y
9. Bonaiuto S, Rocca W, Lippi A (1990) Impact of education and occupation on prevalence of Alzheimer's disease (AD) and multi-infarct dementia (MID) in Appignano, Macerata Province, Italy. Neurology 40(suppl 1):346
10. Cohen CI (1994) Education, occupation, and Alzheimer's disease. JAMA 272(18):1405. Author reply 1406
11. Gun RT, Korten AE, Jorm AF et al (1997) Occupational risk factors for Alzheimer disease: a case-control study. Alzheimers Dis Assoc Disord 11(1):21–27
12. Helmer C, Letenneur L, Rouch I et al (2001) Occupation during life and risk of dementia in French elderly community residents. J Neurol Neurosurg Psychiatry 71(3):303–309
13. Anttila T, Helkala EL, Kivipelto M et al (2002) Midlife income, occupation, APOE status, and dementia: a population-based study. Neurology 59(6):887–893

14. Ravaglia G, Forti P, Maioli F et al (2002) Education, occupation, and prevalence of dementia: findings from the Conselice study. Dement Geriatr Cogn Disord 14(2):90–100
15. Finkel D, Andel R, Gatz M et al (2009) The role of occupational complexity in trajectories of cognitive aging before and after retirement. Psychol Aging 24(3):563–573. https://doi.org/10.1037/a0015511
16. Garibotto V, Borroni B, Sorbi S et al (2012) Education and occupation provide reserve in both ApoE epsilon4 carrier and noncarrier patients with probable Alzheimer's disease. Neurol Sci 33(5):1037–1042. https://doi.org/10.1007/s10072-011-0889-5
17. Stern Y, Gurland B, Tatemichi TK et al (1994) Influence of education and occupation on the incidence of Alzheimer's disease. J Am Med Assoc 271:1004–1010
18. Habeck C, Razlighi Q, Gazes Y et al (2016) Cognitive reserve and brain maintenance: orthogonal concepts in theory and practice. Cereb Cortex. https://doi.org/10.1093/cercor/bhw208
19. Fischl B, Salat DH, Busa E et al (2002) Whole brain segmentation: automated labeling of neuroanatomical structures in the human brain. Neuron 33(3):341–355
20. Fischl B, van der Kouwe A, Destrieux C et al (2004) Automatically parcellating the human cerebral cortex. Cereb Cortex 14(1):11–22. https://doi.org/10.1093/cercor/bhg087
21. Kennedy KM, Erickson KI, Rodrigue KM et al (2009) Age-related differences in regional brain volumes: a comparison of optimized voxel-based morphometry to manual volumetry. Neurobiol Aging 30(10):1657–1676. https://doi.org/10.1016/j.neurobiolaging.2007.12.020
22. Fjell AM, Westlye LT, Amlien I et al (2009) High consistency of regional cortical thinning in aging across multiple samples. Cereb Cortex 19(9):2001–2012. https://doi.org/10.1093/cercor/bhn232
23. Ardekani BA, Guckemus S, Bachman A et al (2005) Quantitative comparison of algorithms for inter-subject registration of 3D volumetric brain MRI scans. J Neurosci Methods 142(1):67–76. https://doi.org/10.1016/j.jneumeth.2004.07.014
24. Dale AM, Fischl B, Sereno MI (1999) Cortical surface-based analysis. I Segmentation and surface reconstruction. Neuroimage 9(2):179–194
25. Fischl B, Dale AM (2000) Measuring the thickness of the human cerebral cortex from magnetic resonance images. Proc Natl Acad Sci U S A 97(20):11050–11055. https://doi.org/10.1073/pnas.200033797
26. Jenkinson M, Bannister P, Brady M et al (2002) Improved optimization for the robust and accurate linear registration and motion correction of brain images. NeuroImage 17(2):825–841
27. Power JD, Barnes KA, Snyder AZ et al (2012) Spurious but systematic correlations in functional connectivity MRI networks arise from subject motion. NeuroImage 59(3):2142–2154. https://doi.org/10.1016/j.neuroimage.2011.10.018
28. Carp J (2013) Optimizing the order of operations for movement scrubbing: comment on Power et al. NeuroImage 76:436–438. https://doi.org/10.1016/j.neuroimage.2011.12.061
29. Birn RM, Diamond JB, Smith MA et al (2006) Separating respiratory-variation-related fluctuations from neuronal-activity-related fluctuations in fMRI. NeuroImage 31(4):1536–1548. https://doi.org/10.1016/j.neuroimage.2006.02.048
30. Power JD, Cohen AL, Nelson SM et al (2011) Functional network organization of the human brain. Neuron 72(4):665–678. https://doi.org/10.1016/j.neuron.2011.09.006

Check for updates

Chapter 5

Methods for Pathological Classification of Alzheimer's Disease

Johannes Attems, Kirsty E. McAleese, and Lauren Walker

Abstract

The hallmark protein aggregates required for a neuropathological diagnosis of Alzheimer's disease are intracellular neurofibrillary tangles and neuropil threads consisting of hyperphosphorylated tau, extracellular parenchymal amyloid β plaques, and neuritic plaques. In this chapter, we describe the neuropathological criteria for the neuropathological diagnosis of Alzheimer's disease, as well as the fixation, processing, and histology protocols required for the assessment of *post-mortem* human brains.

Key words Neuropathology, Alzheimer's disease, Hyperphosphorylated tau, Amyloid β, Neuritic plaques, Cerebral amyloid angiopathy, *Post-mortem* tissue, Histology

1 Introduction: Neurodegeneration (General Considerations)

The unifying feature of neurodegenerative diseases is the accumulation of insoluble aggregates of physiologically soluble proteins of the central nervous system. The native function of these proteins is lost resulting in protein misfolding, oligomerization into fibrils, and subsequent accumulation and deposition as intracellular (in neurons and/or neuroglia) or extracellular aggregates. It is generally assumed that misfolded proteins exert neurotoxicity leading to progressive neuronal dysfunction with consecutive neuronal loss of specific populations of neurons. However, the underlying mechanisms are still poorly understood. These insoluble aggregates can be visualized in human *post-mortem* brain tissue for microscopic assessment using appropriate histochemical techniques and immunohistochemical antibodies, and it is the molecular nature of the respective protein deposit and its topographical localization that is the basis of the classification of neurodegenerative disease.

Robert Perneczky (ed.), *Biomarkers for Preclinical Alzheimer's Disease*, Neuromethods, vol. 137,
https://doi.org/10.1007/978-1-4939-7674-4_5,

1.1 Alzheimer's Disease: Neuropathology

The hallmark protein aggregates required for a neuropathological diagnosis of Alzheimer's disease (AD) are intracellular neurofibrillary tangles/neuropil threads (NFTs/NTs) consisting of hyperphosphorylated tau (HP-$_{T}$), extracellular amyloid β (Aβ) plaques, and neuritic plaques (NP). In addition, cerebral amyloid angiopathy (CAA) is typically present, but its assessment is not required for stating a neuropathological diagnosis of AD according to current diagnostic criteria.

1.1.1 Macroscopic Characteristics

On gross examination, *post-mortem* brains of AD patients often appear relatively normal without striking pathological changes, and reduction in brain weight may be minimal. However, the characteristic atrophy, i.e., thinning of the gyri and deepening of the sulci, in AD involves medial temporal lobe structures including the entorhinal cortex, hippocampus, amygdala, olfactory bulb, as well as inferior temporal, superior frontal, and middle frontal gyri [1]. Cerebral atrophy leads to enlargement of the ventricles (i.e., *hydrocephalus internus e vacuo)*, in particular the inferior horn of the lateral ventricles. The substantia nigra (midbrain) is well pigmented, while the locus coeruleus (pons) might be pale, and the cerebellum is normal (Fig. 1).

1.1.2 Neurofibrillary Tangles and Neuropil Threads

Neurofibrillary tangles (NFTs) and neuropil threads (NTs) are aggregates of hyperphosphorylated microtubule-associated protein (MAP) tau (HP-$_{T}$): a microtubule-stabilizing protein that is abundant in axons and dendrites. Six isoforms of MAP tau exist in the human brain, and aggregates of 3-repeat (3R) and 4-repeat (4R) isoforms are seen in AD. Under pathological conditions, MAP tau is hyperphosphorylated resulting in the disintegration of microtubules and aggregation of insoluble filaments/fibrils composed of HP-$_{T}$ [2]. NFTs/NTs are composed of aggregates of HP-$_{T}$, while so called pre-tangles represent non-aggregated HP-$_{T}$ and are considered to be precursors of NFT/NT. Using appropriate immunohistochemical antibodies, i.e., AT8 antibody, both NFT and NT can be visualized for histological assessment. NFTs are located in neuronal cell bodies and NTs in axons and dendrites, the latter appearing in immunohistochemically stained sections as immunopositive threads in the neuropil (Fig. 2). In AD, NFT and NT primarily manifest in trans-entorhinal and entorhinal cortices and then gradually spread toward the hippocampus and isocortex [3], and this pattern is the basis for diagnostic Braak NFT stages [4] (see Sect. 3.6.1). Autopsy studies have revealed pre-tangle HP-$_{T}$ was found in the locus coeruleus in the majority of individuals under 30 years of age who were devoid of cortical HP-$_{T}$ pathology, suggesting that HP-$_{T}$ pathology begins in the locus coeruleus rather than the trans-/entorhinal cortex [5, 6], but this is controversial [7]. Importantly, the extent of neocortical NFT/NT has been shown to correlate with cognitive deficits [8].

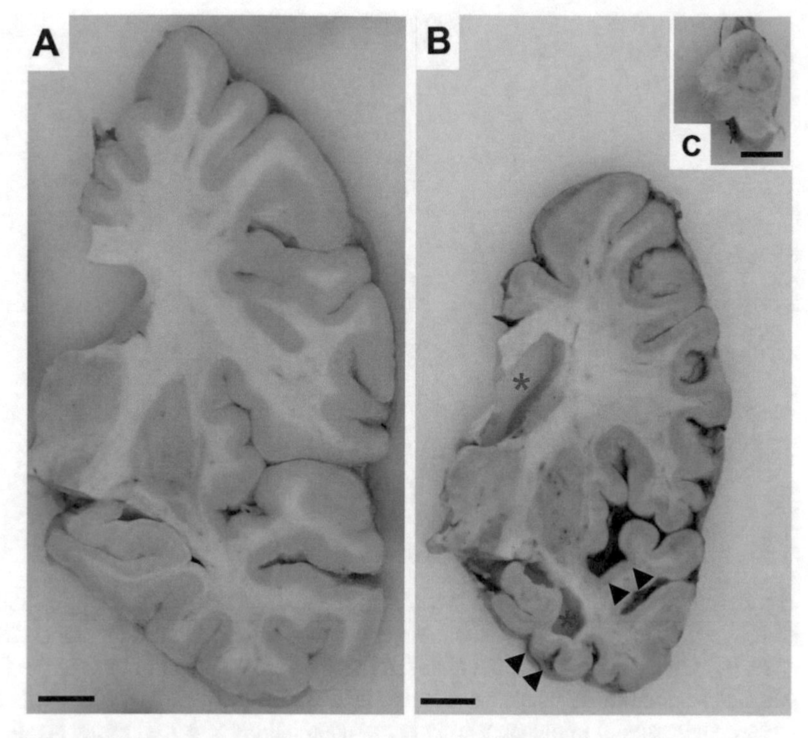

Fig. 1 Comparison between formalin-fixed tissue slices taken at the level of the anterior hippocampus of an individual without dementia (**A**) and a patient with severe AD. ((**B**) and (**C**)) There are marked enlargement of the lateral ventricle (blue asterisk), considerable atrophy of the medial temporal lobe (red asterisk), as well as thinning of the gyri and deepening of the sulci (black arrowheads). However, the substantial nigra of the midbrain remains fairly well pigmented (**c**). Photographs courtesy of Dr. Christopher Morris and the Newcastle Brain Tissue Resource. Both scale bars represent 1 cm

1.1.3 Amyloid β Plaques

Amyloid β (Aβ) peptides (4kDA) are generated by β- and γ-secretase cleavage of the transmembraneous amyloid precursor protein (APP) and comprise peptides terminating at carboxyl terminus 40 (Aβ40) and 42 (Aβ42) [9]. While Aβ40 constitutes the majority of cerebral Aβ (over 95%), Aβ42 aggregates more readily and, therefore, is believed to initiate the formation of oligomers, fibrils, and plaques [10, 11]. Aβ aggregates extracellularly, and using appropriate antibodies, i.e., 4G8 antibody, various morphological forms of Aβ aggregates can be detected. The main parenchymal Aβ deposits are diffuse or focal: diffuse Aβ deposits are usually large (up to several 100 μm) and show ill-defined borders and are often referred to as "fleecy" and "subpial band-like," while focal Aβ plaques incorporate a dense, spherical-like accumulation of Aβ peptides (Fig. 3). The topographical locations of Aβ deposits follow a distinct sequence that can be staged diagnostically using Thal Aβ phases [12]. Cerebral Aβ depositions are a hallmark of AD, but in the majority of clinicopathological correlative studies,

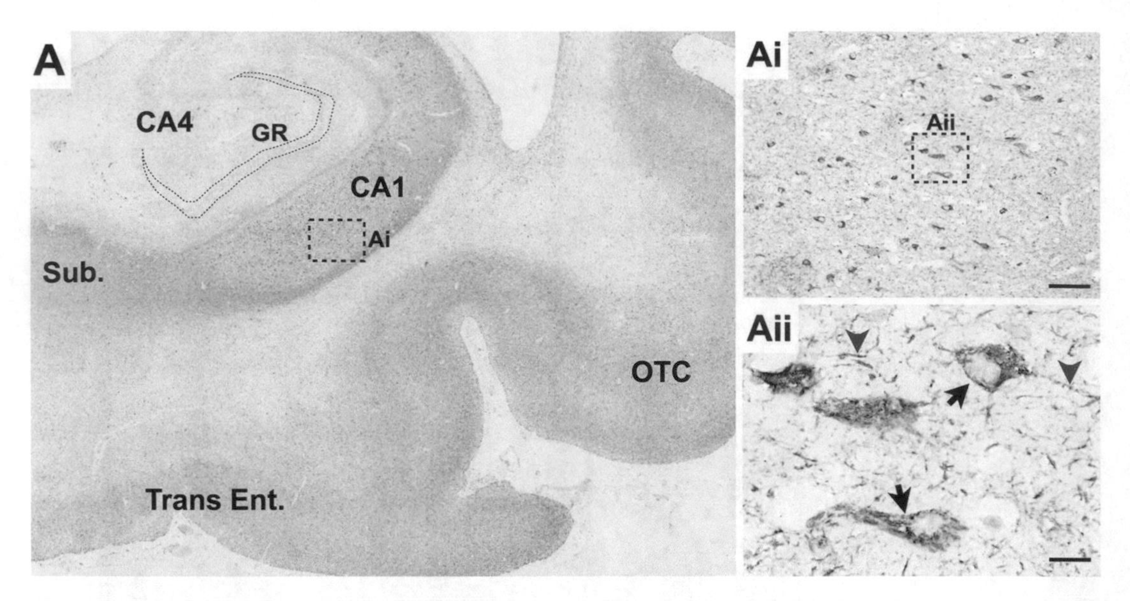

Fig. 2 Photomicrographs of neurofibrillary tangles and neuropil threads. (A) Distribution of HP-$_{T}$ immunopositivity in the hippocampus, subiculum, transentorhinal, and occipital-temporal cortex of an Alzheimer's disease case. (Ai) Magnified image of neurofibrillary tangles and neuropil threads in the CA1 region of the hippocampus. (Aii) Neurofibrillary tangles encompass immunopositivity within the neuronal soma (black arrow); neuropil thread comprises of immunopositivity in the neuronal process (red arrow head). CA1 and CA4, hippocampal cornu ammonis (Ammon's horn) sectors 1 and 4; GR, granule cell layer of the dentate gyrus; Sub, subiculum; Trans Ent, transentorhinal cortex; OTC, occipital-temporal cortex. Immunohistochemistry with antibody against hyperphosphorylated tau, AT8. Scale bars: A1, 50 μm; Aii, 10 μm

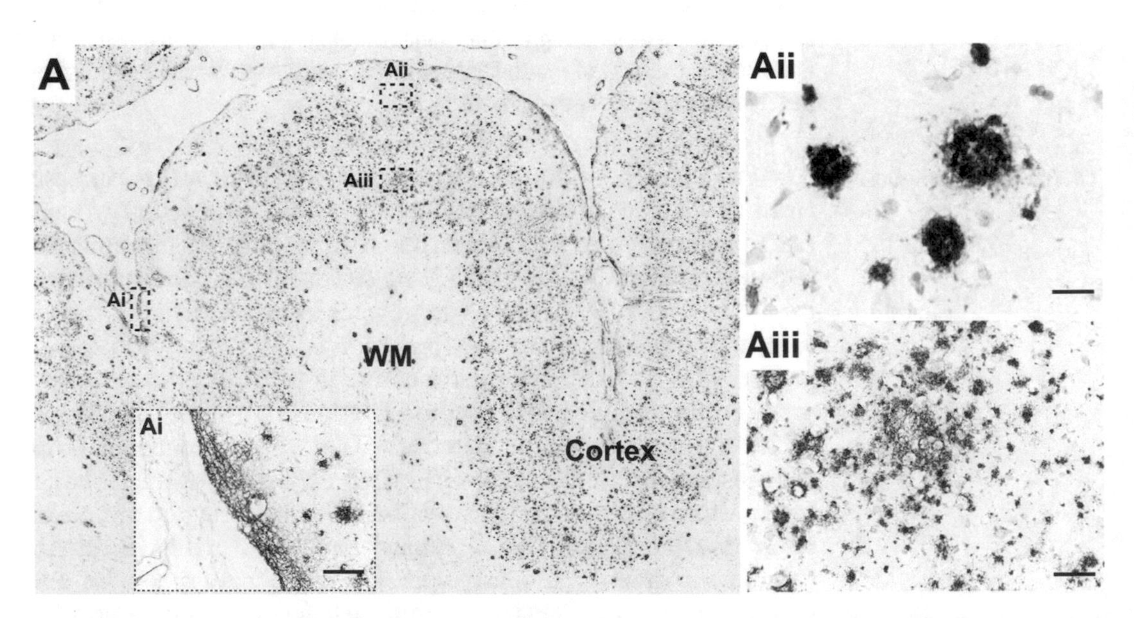

Fig. 3 Photomicrographs of Aβ plaques. (A) Distribution of Aβ immunopositivity in the prefrontal cortex of an Alzheimer's disease case. (Ai) Subpial Aβ deposits; (Aii) focal Aβ deposits; (Aiii) fleece Aβ deposits. WM, white matter; Aβ, amyloid beta. Immunohistochemistry with antibody against amyloid precursor protein, 4G8. Scale bars: Bi and Biii, 50 μm; Bii, 20 μm

cerebral Aβ load does not correlate with clinical dementia, and parenchymal Aβ deposits are frequently seen to a considerable extent in *post-mortem* brains of non-demented individuals (see Sect. 3.6.2).

1.1.4 Neuritic Plaques

Neuritic plaques (NP) consist of a focal deposit of Aβ42 [13], with an external halo (the corona) of HP-$_T$-positive processes forming dystrophic neurites (Fig. 4). NPs are strongly associated with AD, and assessment of the presence and density of NP is performed using the age-adjusted criteria of the Consortium to Establish a Registry for Alzheimer's Disease (CERAD) [14] (see Sect. 3.6.3).

1.1.5 Cerebral Amyloid Angiopathy

Cerebral amyloid angiopathy (CAA) is defined as the deposition of a congophilic material (i.e., positive staining with a Congo red dye) in cerebral leptomeningeal and intracortical arteries, arterioles, capillaries, and rarely veins. CAA is also referred to as congophilic amyloid angiopathy [15, 16] and is strongly associated with AD-related pathology, present in 80–100% of AD cases [17, 18]. CAA is characterized by the deposition of predominantly Aβ40 [9] in the tunica media of vessel walls that eventually spreads into all vessel wall layers causing vessel wall thickening, smooth muscle cell migration, and overall disruption of vessel architecture, although

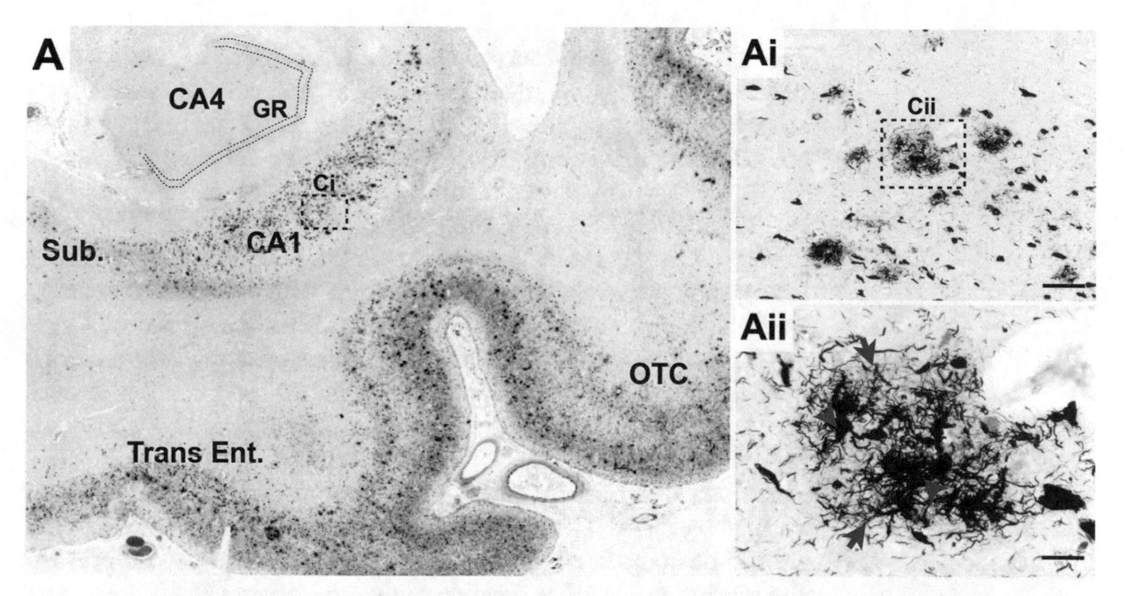

Fig. 4 Photomicrographs of neuritic plaques. (A) Distribution of neuritic plaques in the hippocampus, subiculum, transentorhinal, and occipital-temporal cortex of an Alzheimer's disease case. (Ai) Magnified image of neurotic plaques in the CA1 region of the hippocampus. (Aii) Neuritic plaque consisting of a Aβ core (red arrow head) and dystrophic neurites consisting of HPτ (blue arrow). CA1 and CA4, hippocampal cornu ammonis (Ammon's horn) sectors 1 and 4; GR, granule cell layer of the dentate gyrus; Sub, subiculum; Trans Ent, transentorhinal cortex; OTC, occipital-temporal cortex; Aβ, amyloid beta; HPτ, hyperphosphorylated tau. Histological staining with Gallyas silver stain. Scale bars: Ai, 50 μm; Aii, 10 μm

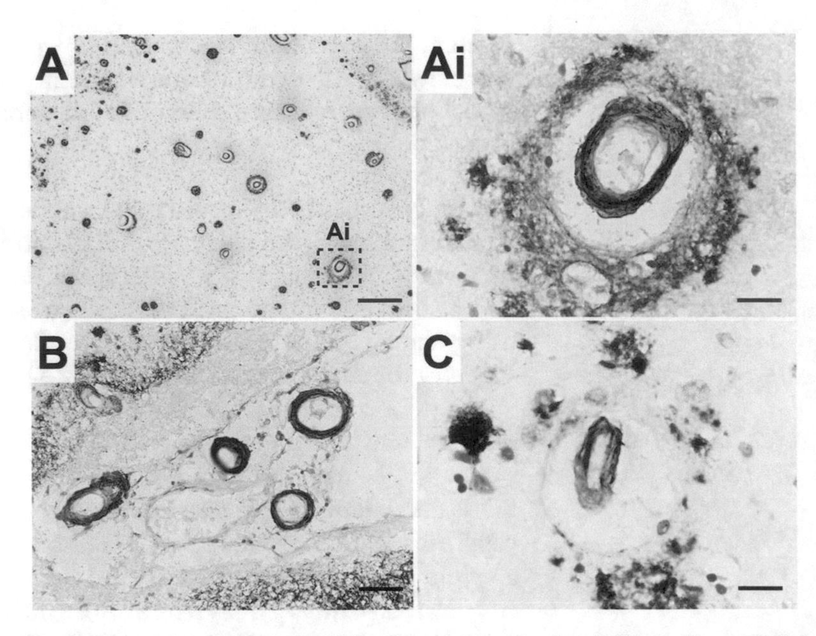

Fig. 5 Photomicrographs of cerebral amyloid antipathy (CAA) in the occipital region of an Alzheimer's disease brain (**A**) cortical CAA; (Ai) magnified image of cortical CAA; (**B**) leptomeningeal CAA; (**C**) capillary CAA. Immunohistochemistry with antibody against amyloid precursor protein, 4G8. Scale bars: **A**, 100 μm; Ai, 20 μm; **B**, 50 μm; **C**, 10 μm

endothelial cells are preserved (Fig. 5). CAA frequently has a patchy distribution and primarily affects leptomeningeal and cortical vessels of neocortical regions [19]. The occipital lobe has been reported to be the site that is both most frequently and severely affected by CAA, followed by either frontal, temporal, or parietal lobes [17, 20–22]. Aβ depositions in the walls of capillaries are referred to as capillary CAA (capCAA) and usually present as strong staining lining capillary walls in sections stained with Aβ antibodies, e.g., 4G8. The presence of capCAA distinguishes two types of CAA: CAA-type 1 is characterized by the presence of capCAA and usually also shows Aβ depositions in leptomeningeal and cortical vessels, whereas in CAA-type 2, Aβ depositions are restricted to leptomeningeal and cortical vessels without capillary involvement [23]. Histological assessment of CAA is not required for the diagnosis of AD; however, it is an important AD-associated pathology that should be considered. A neuropathological consensus criterion for the scoring of CAA has been recently established [24].

1.2 Neuropathological Criteria

Neuropathological assessment of the *post-mortem* brain is performed in a standardized manner, applying reproducible methods and following internationally recognized criteria. Neuropathological consensus criteria are available for the neuropathological hallmark lesions of AD, i.e., NFT/NT [4] Aβ plaques [12] and neuritic plaques [14], as well as the National Institute on Aging-Alzheimer's Association (NIA-AA) score for AD neuropathologic change [25] (see Sect. 3.6.4).

2 Materials

2.1 Brain Dissection (Both Fresh and Fixed)

1. Macroblade (CellPath).
2. Scalpel holder (Swann Morton).
3. Disposable scalpel blade, size 22 (Swann Morton).

2.2 Fixative and Tissue Processing

1. Formalin solution: 100 ml formaldehyde (40%); 900 ml distilled water (dH_2O) (1:10 dilution) (Genta Medical).
2. Small amount of marble chips.
3. Ethanol (100, 95, 70, 50% in dH_2O).
4. Chloroform.
5. Histology wax (melting point of 60 °C).

2.3 Tissue Preparation

1. Pretreated slides (4% 3-aminopropyltriethoxysilane (APES)) or charged slides, i.e., Superfrost Plus (Thermo Shandon).
2. Rotary microtome (HM325, Thermo Scientific).
3. Oven capable of reaching temperatures from 37 to 60 degrees.
4. Ethanol (100, 75, and 50% on dH_2O).
5. Xylene.

2.4 Gallyas Staining

1. Xylene.
2. Ethanol (100, 75, and 50% in dH_2O).
3. 4% formalin solution: 100 ml formaldehyde (40%) (Genta Medical), 900 ml dH_2O (1:10 dilution).
4. 1% potassium permanganate.
5. 5% oxalic acid.
6. 5% periodic acid.
7. *Alkaline silver solution*: 4 g NaOH in 50 ml dH_2O; add 10 g potassium iodide. Slowly add 3.5 ml silver nitrate drop by drop; make up to 100 ml dH_2O.
8. 0.25% acetic acid.
9. *Physical developer*: solution A—50 g sodium carbonate in 1 L dH_2O. Solution B—in 1 L dH_2O, add in order 2 g ammonium nitrate, 2 g silver nitrate (analar), and 10 g tungstosilicic acid (analar). Solution C—as in solution B, with the addition of 6.1 ml 40% formalin. Slowly add 30 ml solution B to 100 ml solution A while shaking vigorously, and then add 70 ml solution C (Note 1).

10. 0.2% gold chloride.
11. 5% "hypo" (sodium thiosulfate).
12. Harris hematoxylin.
13. DPX hard-set mountant (CellPath).
14. Coverslips.

2.5 Immunohistochemistry

1. Xylene.
2. Ethanol (100, 75, and 50% in dH_2O).
3. Citrate buffer: 8.4 g citric acid, 4 L dH_2O, 52 ml 2 molar (M) sodium hydroxide, pH to 6.
4. Concentrated formic acid.
5. Microwave.
6. 3% hydrogen peroxide: 10 ml hydrogen peroxide (30%) in 90 ml tap water (Note 2).
7. Tris-buffered saline (TBS): 6.95 g Tris base, 30.3 g Tris-HCL, 45 g sodium chloride made up in 5 L dH_2O, pH to 7.6.
8. Hydrophobic barrier pen.
9. Primary antibodies made up in TBS (see Table 3 for optimum dilutions).
10. TBS Tween: 4 L TBS as above, 2 ml Tween (shake well to mix).
11. MenaPath X-Cell Plus HRP Polymer Detection Kit (Menarini Diagnostics) (or alternative streptavidin-biotin system).
12. 3,3-diaminobenzidine (DAB) substrate kit (Menarini Diagnostics).
13. Harris hematoxylin.
14. DPX hard-set mountant (CellPath).
15. Coverslips.

2.6 Neuropathological Assessment Criteria for AD

1. Microscope capable of 100× magnification.
2. Ethanol for cleaning slides.

3 Methods

3.1 Fresh Brain Dissection and Fixation

The cerebrum, and spinal cord if required, is removed from the deceased patient following standard protocols. All tissue, blood, and CSF must be treated as a potentially infectious case. The pathologist and assisting personnel must wear a gown, nitrile gloves, and shoe covers, and dissection of neurologic tissues is carried out in a fume hood. The cerebrum, cerebellum, and brainstem are inspected and weighed, and the cerebrum length is measured from the frontal pole to the occipital pole. The arteries of the circle of Willis are examined for the presence of atherosclerotic changes. The dissection is carried out as follows:

1. Separation of the cerebellum and brainstem from the cerebral hemispheres at pons-midbrain junction.
2. Cerebral hemispheres are separated by a midsagittal incision through the corpus callosum; cerebellar hemispheres are separated at the point of the left dentate nucleus; and brainstem is separated down the midline.
3. The exposed "cut" surfaces are covered in 60gsm medical grade paper to prevent tissue shrinkage, and the right cerebral and right cerebellar hemispheres and the right brainstem are fixed in formalin with marble chips. The right cerebrum is suspended in formalin solution (via the junction of the middle cerebral artery and the posterior communicating artery); the left cerebral and cerebellar hemisphere and brainstem are dissected into a series of 1-cm-thick slices according to a standard protocol.
4. The serial coronal slices are placed into individual plastic envelopes, labeled, and sealed and then snap frozen on copper plates at −120 °C and stored at −80 °C.

3.2 Fixed Dissection and Embedding

Tissue is fixed for 4 weeks at room temperature. The right cerebrum is weighed, and length is measured from the frontal pole to the occipital pole. The arteries of the circle of Willis are examined for the presence of atherosclerotic changes. The midbrain is removed from the cerebrum cutting in between the mammillary body and superior colliculus. The cerebral hemisphere, cerebellum, and brainstem are then serially dissected at 7 mm intervals (Fig. 6). Regions (including specific Brodmann areas (BA)) required for

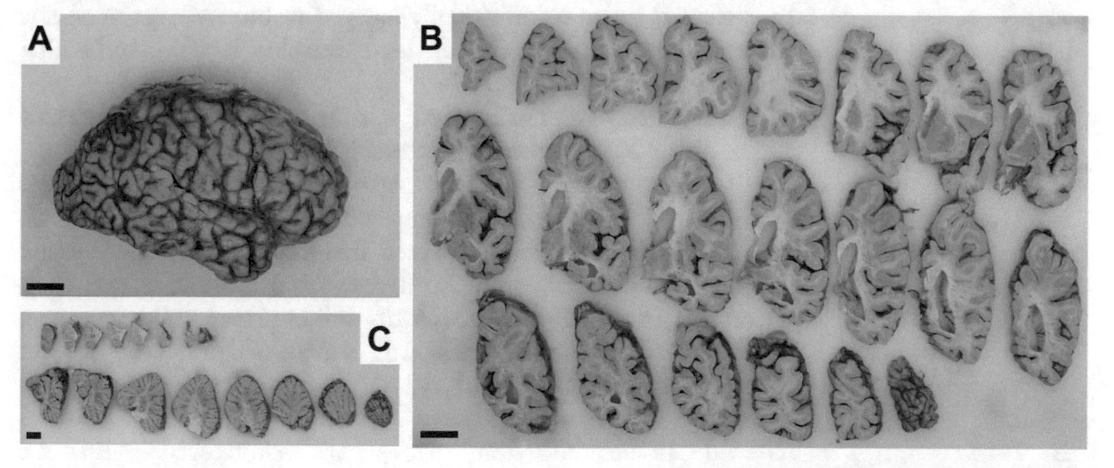

Fig. 6 Following fixation, the right hemisphere (**A**) is dissected at 7 mm intervals (**B**) as are the fixed semi-brainstem and right cerebellum (**C**). Brain regions important of the neuropathological diagnosis of AD are then sub-dissected and processed into paraffin wax. Photographs courtesy of Dr. Christopher Morris and the Newcastle Brain Tissue Resource. Scale bar in **A** and **B** represents 2 cm and 1 cm in **C**

Table 1

Brain regions (including specific Brodmann areas, BA) required for neuropathological diagnosis

1.	Prefrontal superior half (BA 9/46/32)
2.	Mid-frontal superior half (BA 9/8)
3.	Cingulate (BA 24/32)
4.	Corpus striatum and amygdala
5.	Motor (BA 4) and sensory (BA 3/1/2/5) cortex and deep white matter
6.	Trans-/entorhinal cortex (anterior: BA 27/28), occipital-temporal cortex (BA 36), and dentate fascia
7.	Hippocampus and parahippocampal gyrus (BA 27/28), occipital-temporal cortex (BA 36), and inferior temporal cortex
8.	Temporal cortex (BA 20/21/22/41/42)
9.	Thalamus
10.	Parietal cortex and deep white matter (BA 40/22)
11.	Occipital cortex inferior half (BA 17/18/19)
12.	Cerebellar hemisphere and dentate
13.	Upper midbrain
14.	Lower midbrain
15.	Upper mid pons
16.	Mid-medulla

neuropathological diagnosis are sub-dissected as required (Table 1). All dissected tissue is individually bagged in biopsy bags or plastic cassettes (depending on size of tissue sample), labeled, and sealed for tissue processing.

Tissue processing is carried out in an automatic processor (Leica ASP 300S) according to the schedule in Table 2.

Processed samples are then placed in clean molten histology wax (60 °C) in the embedding center (Leica Microsystems, EG1150H). Molten wax is poured into appropriate-sized plastic mold, and the tissue sample and case label are added and then placed on a cooling plate until solid (Leica Microsystems, EG1150C). If required, excess wax is trimmed using a hot Para Trimmer (Thermo Scientific).

3.3 Tissue Preparations and Sectioning

To aid tissue adhesion, slides are pretreated with 4% 3-aminopropyltriethoxysilane (APES), or Superfrost slides are used. To prepare APES-coated slides, clean glass slides are placed into racks and immersed in a 4% solution of APES made up in

Table 2
Three-day CNS processing schedule

Step	Time
4% formalin (no marble chips)	34 min
50% ethanol	2 h
70% ethanol	3 h 50 min
95% ethanol	5 h
100% ethanol	6 h
100% ethanol	6 h
100% ethanol	8 h
Chloroform	3 h
Chloroform	6 h
Chloroform	7 h
Histology wax	5 h 60 °C (pressure/ vacuum)
Histology wax	5 h 60 °C (pressure/ vacuum)
Histology wax	5 h 60 °C (pressure/ vacuum)

acetone for 5 min, followed by 5 min in acetone before a final 5 min immersed in 4% APES solution. Slides are drained in dH_2O, dried at 37 °C, and stored until used. The tissue is sectioned at 6 μm on a rotary microtome and dried for 2 days at 37 °C followed by 1 h at 60 °C.

3.4 Gallyas Staining

Silver staining such as Gallyas is used to visualize dystrophic neurites associated with neuritic plaques; however, amyloid deposits are not detected using this method. The protocol is as follows:

1. Sections placed in 60 °C oven to aid dewaxing.
2. Deparaffinize sections in two changes of xylene (5 min each).
3. Rehydrate through decreasing concentrations of alcohol (2 × 100%, 1 × 75%, 1 × 50%, 2 min each) to water.
4. Postfix in 4% formalin (10 min).
5. Wash in dH_2O.
6. Immerse in 1% potassium permanganate (5 min).
7. Wash in dH_2O.

8. Place in 5% oxalic acid (3 min).
9. Wash well in dH_2O.
10. Place in 5% periodic acid (5 min).
11. Wash in dH_2O.
12. Apply alkaline silver iodide solution (1 min).
13. Wash in 0.25% acetic acid (three changes, 1 min each, with gentle agitation).
14. Wash well in dH_2O.
15. Place in physical developer (look for pale gray/brown color change (5–20 min)).
16. Wash in 0.25% acetic acid (two changes, 1 minute each, with gentle agitation).
17. Wash well in water dH_2O.
18. Tone in 0.2% gold chloride (10–20 s).
19. Wash in water and fix in 5% "hypo" (sodium thiosulfate) (2 min).
20. Wash sections and counterstain with hematoxylin.
21. Dehydrate through increasing concentrations of alcohol (70%, 95%, and 2 × 100%), clear in two changes of xylene (5 min each), and mount with DPX and coverslip (Note 3).

3.5 Immunohistochemistry

Immunohistochemistry (IHC) of formalin-fixed paraffin-embedded tissues allows the visualization of pathological protein aggregates while preserving the morphological and cellular structure of the tissue. Enzyme precipitate IHC is preferred to fluorescent IHC for diagnostics as it is more cost-effective, and the chromogen will not fade with time. Formalin fixation can cross-link antigens and mask epitopes; therefore, to increase antigenicity of the required epitope, antigen retrieval is performed prior to immunohistochemistry. Information of antigen retrieval required and manufacturer's details of primary antibodies (including optimum dilutions) are provided in Table 3.

Table 3
Optimum antibody dilutions and antigen retrieval protocol

Antibody	Target	Host	Manufacturer	Working dilution (in TBS)	Antigen retrieval
AT8	Hyperphosphorylated tau Ser202/205 (HP-τ)	Mouse	Innogenetics, Ghent, Belgium	1:4000	Immersion in 0.01 m.L-1 citrate buffer (pH 6) and microwaving for 10 min
4G8	β amyloid 17-24	Mouse	Covance, Alnwick, UK	1:15,000	Immersion in concentrated formic acid for 1 h

The protocol is as follows:

1. Sections placed in 60 °C oven to aid dewaxing.
2. Deparaffinize sections in two changes of xylene (5 min each).
3. Rehydrate through decreasing concentrations of alcohol (2 × 100%, 1 × 75%, 1 × 50%, 2 min each) to water.
4. If using a peroxidase-based detection system, quench endogenous peroxidase by immersing slides in 3% hydrogen peroxide diluted in tap water (20 min).
5. Wash well in water, rinse in Tris-buffered saline (TBS), and (if using) apply hydrophobic barrier pen around the tissue.
6. Incubate in primary antibody at optimum dilution (diluted in TBS) for 1 h at room temperature.
7. Rinse in TBS, and final wash in TBS Tween.

We routinely use MenaPath X-Cell Plus HRP Polymer Detection Kit (Menarini Diagnostics), which is a sensitive biotin-free detection system capable of binding to mouse and rabbit antibodies. Following manufacturer's instructions, the primary antibody is bounded by the universal probe (1), which is then detected by the HRP-polymer (2). The increased number of enzymes bound to the polymer backbone ensures a sensitive and clean end result. Sections are rinsed well in TBS, and the chromogen is developed using 3,3-diaminobenzidine (DAB) substrate kit (also Menarini Diagnostics). Once the desired color of brown has been achieved (2–4 min), wash tissue well in tap water, and counterstain lightly in hematoxylin (Note 4). Dehydrate through increasing concentrations of alcohol, clear in xylene, and mount using DPX, which is a colorless synthetic resin mountant that dries quickly and enables long-term preservation of tissue samples (Note 5).

3.6 Neuropathological Criteria for AD

Hallmark pathological lesions associated with AD (i.e., neurofibrillary pathology, Aβ plaques, and neuritic plaques) follow a stepwise topographical pattern of deposition through the brain, which corresponds to advancing clinical symptoms [4, 12, 14]. Current consensus criteria for the degree of pathology attributed to AD (termed AD neuropathologic change) evaluate the progression of each type of pathology and yield a combined score to determine an overall level of AD-related pathology in the brain [25, 26]. For the identification of pathological lesions, stained sections are viewed under a light microscope at X100 magnification.

3.6.1 Assessment of Hyperphosphorylated Tau (NTs and NFTs)

Braak stages describe the pattern of neurofibrillary pathology (NTs and NFTs) as outlined by Braak and colleagues [4]. Subsequently, the BrainNet Europe Consortium identified four distinct anatomical regions required to accurately assign Braak stage, the anterior hippocampus at the level of the uncus, posterior hippocampus at

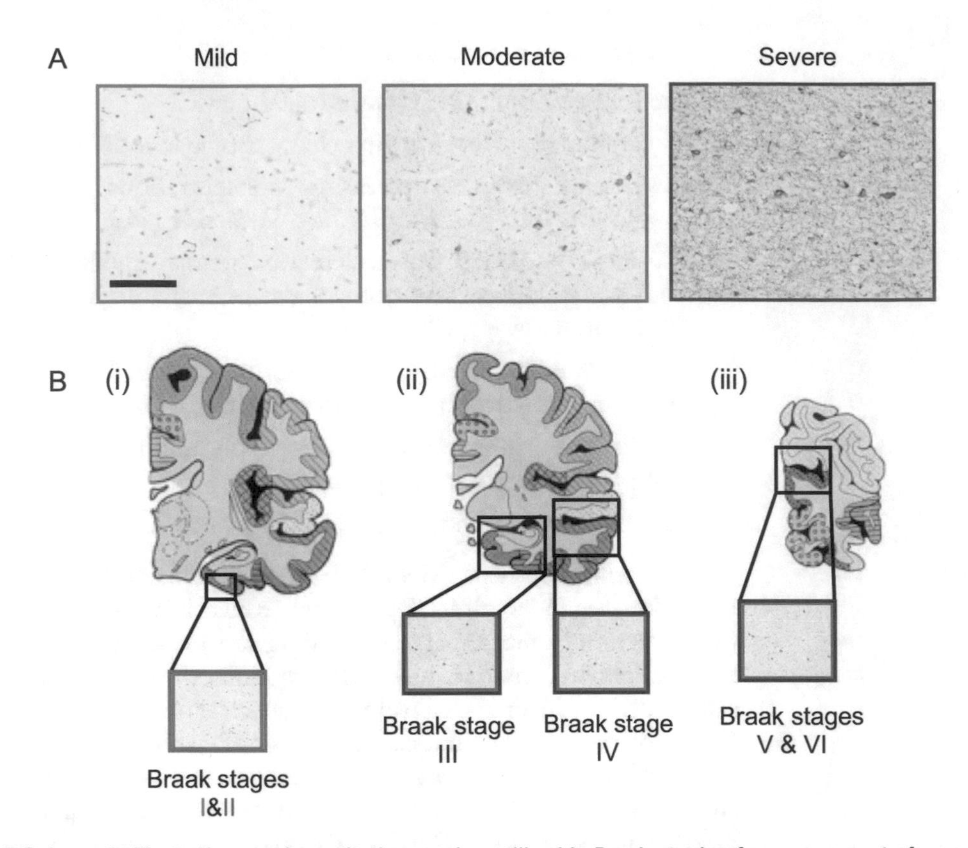

Fig. 7 Schematic illustrating semiquantitative scoring utilized in Braak staging for assessment of neurofibrillary tangles (NFTs) and neuropil threads (NTs) (**A**). Regions are classed as having mild pathology when immunoreactive structures positive for AT8 antibody are barely noted at X100 magnification. Moderate pathology is assigned when immunoreactive structures are easily visible at ×100 magnification, while the severe score is assigned when NFTs and NTs are visible even without a microscope. Specific brain regions are required to assign Braak stages (**B**) including the anterior hippocampus at the level of the uncus (i), the posterior hippocampus at the level of the lateral geniculate (ii), and the visual cortex including the calcarine fissure (iii). The semiquantitative stage is important when assessing Braak stages as Braak stage I can be assigned with mild pathology in the transentorhinal cortex (green square, Bi); however, the remaining stages require pathology of at least moderate density to qualify as the next Braak stage (orange or red squares, Bi, Bii, and Biii). For a full description of regions affected by each stage, see Sect. 3.6.1. Scale bar represents 100 μm and is valid for all photomicrographs

the level of the lateral geniculate nucleus, middle temporal gyrus, and visual cortex including calcarine fissure [27]. Slides stained for hyperphosphosphorylated tau (for protocol, see Sect. 3.5) are examined under a light microscope. In addition to the presence of NFTs and NTs in a given brain region, the amount of pathology is taken into account when assigning Braak stages, which is based on a four-tiered scale (absent, mild, moderate, and severe (Fig. 7A). Brain regions involved in each Braak stage are illustrated in Fig. 7B and described as follows:

1. Braak stage 0—All regions must be described as HP-$_T$ negative.
2. Braak stage I—Trans-entorhinal region in the anterior hippocampus at the level of uncus is immunopositive for HP-$_T$ of at least mild density.
3. Braak stage II—In the posterior hippocampus at the level of lateral geniculate nucleus, outer layers of remnants of the entorhinal region show at least moderate HP-$_T$ immunopositivity, or outer and inner layers show at least mild HP-$_T$ immunopositivity.
4. Braak stage III—At least moderate HP-$_T$ immunopositivity extends to the occipitotemporal gyrus.
5. Braak stage IV—At least moderate HP-$_T$ immunopositivity up to the middle temporal gyrus.
6. Braak stage V—At least moderate HP-$_T$ immunopositivity in the peristriate area of the occipital cortex.
7. Braak stage VI—At least moderate HP-$_T$ immunopositivity in layer V of the striate area in the occipital cortex (BA17—primary visual cortex). In Braak stages V and VI, HP-$_T$ immunopositivity is seen in other neocortical areas as well, but only the occipital cortex needs to be assessed for staging purposes.

3.6.2 Assessment of β-Amyloid Plaques

The distribution of Aβ plaques follows a hierarchical pattern throughout the brain parenchyma, and this is reflected by the Thal phases of Aβ deposition [12]. Thal phases can be assigned using Aβ-stained sections (for protocol, see Sect. 3.5, Note 6) of the specified brain regions. Of note, unlike Braak stages, Thal phases are based on a dichotomous scoring system (i.e., present or absent), and the presence of any single parenchymal Aβ deposit is sufficient to score this area positive for the respective Thal phase (Fig. 8):

1. Phase 1—Deposits are visible in the neocortex (e.g., temporal/frontal cortex).
2. Phase 2— Hippocampus (in particular cornu ammonis (CA) sector 2), entorhinal cortex, and cingulate are involved.
3. Phase 3—Subcortical structures including striatum, thalamus, and hypothalamus.
4. Phase 4—Brainstem structures including substantia nigra and medulla are affected.
5. Phase 5—Aβ deposits are visible in the cerebellum.

3.6.3 Assessment of Neuritic Plaques

The Consortium to Establish a Registry for Alzheimer's Disease (CERAD) guidelines is followed to assess neuritic plaque pathology [14]. Tissue sections are included containing middle frontal gyrus, superior/middle temporal gyri, and inferior parietal lobe and are stained with Gallyas silver stain (for protocol, see Sect. 3.4).

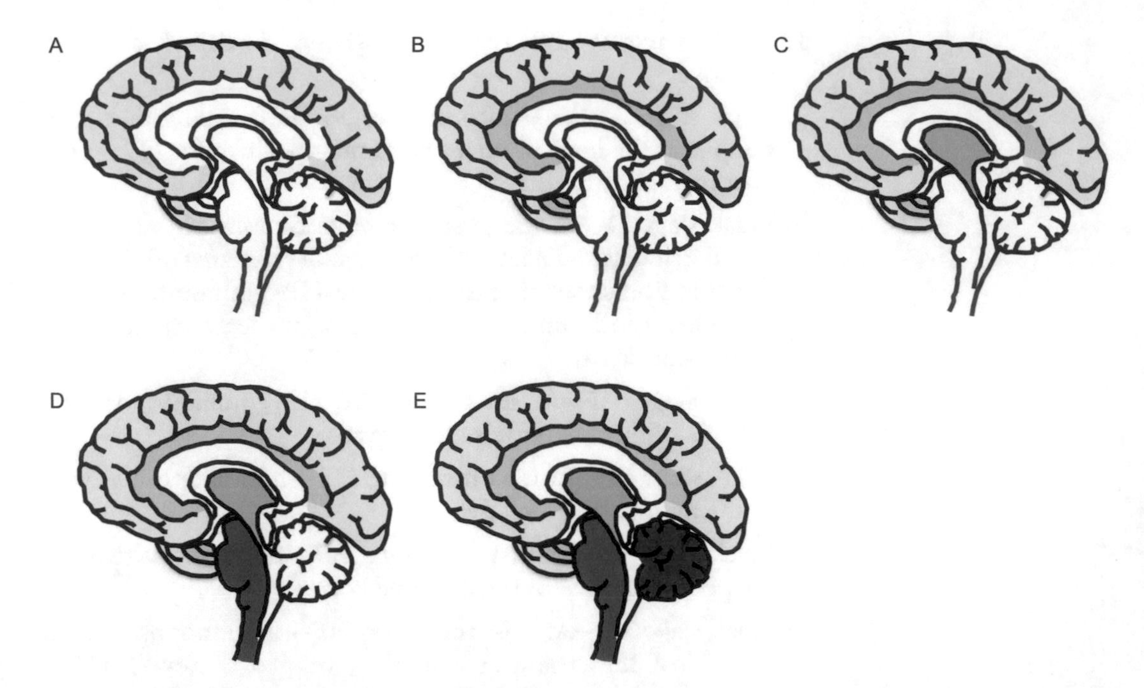

Fig. 8 Diagram illustrating progression of amyloid β (Aβ) through the brain. Aβ plaques are first seen in phase 1 in the neocortex, in particular the inferior parietal and superior temporal gyrus (**A**). In phase 2, Aβ extends to the entorhinal region, cornu ammonis (CA) sector 2 of the hippocampus and the cingulate cortex (**B**). Subcortical structures are affected in phase 3 including striatum, thalamus, hypothalamus, and claustrum (**C**). Phase 4 is characterized by Aβ deposits in the brainstem including the substantia nigra and medulla oblongata (**D**), and finally Aβ extends to the molecular layer of the cerebellum in phase 5 (**E**)

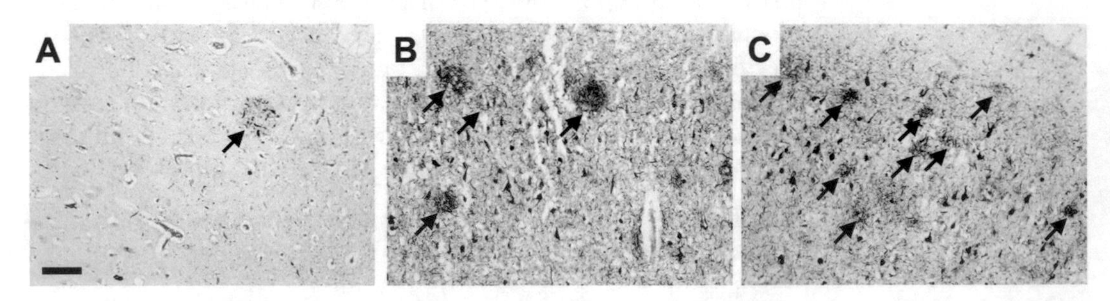

Fig. 9 Photomicrographs stained with Gallyas silver stain, demonstrating semiquantitative scoring of neuritic plaque pathology. Density of plaques per square millimeter is assessed at X100 magnification and described as mild (scored A) (**A**), moderate (scored B) (**B**), or severe (scored C) (**C**). Arrows point to individual neuritic plaques. Scale bar represents 100 μm and is also representative of **B** and **C**

Semiquantitative assessment of plaque density per square millimeter is determined and scored as sparse (A), moderate (B), or frequent (C) (Fig. 9).

3.6.4 Determining Level of AD Neuropathologic Change

Individual "scores" taken from Braak stages, Thal phases, and CERAD assessment are combined to determine the level of AD neuropathologic change using guidelines set out by the National

A score (Amyloid β Thal phase)	B score (Braak stage)			C score (CERAD)
	Braak 0-II	Braak III-IV	Braak V-VI	
0	Not	Not	Not	negative
1 or 2	Low	Low	Low	negative or A
1 or 2	Low	Intermediate	Intermediate	A or B
3	Low	Intermediate	Intermediate	negative, A,B or C
4 or 5	Low	Intermediate	Intermediate	negative or A
4 or 5	Low	Intermediate	High	B or C

Fig. 10 The level of AD neuropathologic change is determined by considering all three pathologies. *A*myloid β Thal phase (A score), *B*raak stage (B score), and *C*ERAD (C score) are combined to determine the level of AD neuropathologic change as no, low, intermediate, or high. CERAD, Consortium to Establish a Registry for Alzheimer's disease

Table 4
Protocol to assess CAA severity

Type of CAA assessed	0	1	2	3
Parenchymal CAA	Absent	Scant Aβ deposition	Some circumferential CAA	Widespread circumferential CAA
Meningeal CAA	Absent	Scant Aβ deposition	Some circumferential CAA	Widespread circumferential CAA
Capillary CAA	Absent	Present	NA	NA
Vasculopathy	Absent	Occasional vessel affected	Many vessels affected	NA

Institute on Aging-Alzheimer's Association (NIA-AA) [25, 26]. Amyloid β Thal phase, Braak stage, and CERAD score are combined to yield an "ABC" score which is reported (regardless of clinical diagnosis) to determine the level of AD neuropathologic change on a four-tiered scale that is either no, low, intermediate, or high. A summary of ABC scoring is summarized in Fig. 10.

3.7 Cerebral Amyloid Angiopathy

Although not required for the neuropathological diagnosis of AD, it is important to note the level of CAA when considering Aβ depositions in *post-mortem* tissue. A recent consensus protocol has provided a semiquantitative scoring system to assess parenchymal CAA, meningeal CAA, capillary CAA, and vasculopathy [24]. Using immunohistochemically sections for Aβ from frontal, temporal, parietal, and occipital cortices (for protocol, see Sect. 3.5), the consensus-based protocol for assigning CAA severity is outlined in Table 4.

4 Conclusions

Neuropathological *post-mortem* assessment is the gold standard in the diagnosis of neurodegenerative disease but has to be performed according to standardized methodologies. By following the published guidelines for the *post-mortem* diagnosis of AD neuropathologic change [25], an accurate neuropathological diagnosis can be made, and this is crucial to provide reliable information regarding the most frequent age-associated neurodegenerative changes (i.e., HP-$_T$ and Aβ) seen in human brains.

5 Notes

1. Physical developer solution to be made immediately before use. If precipitate forms during mixing, this will spoil the developer.
2. Hydrogen peroxide blocking solution should be prepared immediately before use.
3. The use of a solvent-based hard-set mounting media will prevent the sample from drying out and aids sample preservation.
4. DAB is a mutagenic and potential carcinogen; therefore, correct personal protective equipment must be used and safety procedures adhered to.
5. Samples to be produced in as few batches as possible to minimize variability in staining.
6. In addition to extracellular Aβ plaques, 4G8 antibody also reacts with physiological intracellular APP, and this must not be taken into consideration when assessing Aβ pathology.

Acknowledgments

We thank Lynne Ramsay and Ros Hall for their technical expertise and assistance in writing this chapter and Dr. Christopher Morris and the Newcastle Brain Tissue Resource for providing the photographic images.

References

1. Halliday GM, Double KL, Macdonald V et al (2003) Identifying severely atrophic cortical subregions in Alzheimer's disease. Neurobiol Aging 24(6):797–806
2. Ballatore C, Lee VM, Trojanowski JQ (2007) Tau-mediated neurodegeneration in Alzheimer's disease and related disorders. Nat Rev Neurosci 8(9):663–672. https://doi.org/10.1038/nrn2194
3. Braak H, Braak E (1991) Neuropathological stageing of Alzheimer-related changes. Acta Neuropathol 82(4):239–259
4. Braak H, Alafuzoff I, Arzberger T et al (2006) Staging of Alzheimer disease-associated neurofibrillary pathology using paraffin sections and immunocytochemistry. Acta Neuropathol 112(4):389–404. https://doi.org/10.1007/s00401-006-0127-z

5. Braak H, Del Tredici K (2011) The pathological process underlying Alzheimer's disease in individuals under thirty. Acta Neuropathol 121(2):171–181. https://doi.org/10.1007/s00401-010-0789-4
6. Braak H, Thal DR, Ghebremedhin E et al (2011) Stages of the pathologic process in Alzheimer disease: age categories from 1 to 100 years. J Neuropathol Exp Neurol 70(11):960–969. https://doi.org/10.1097/NEN.0b013e318232a379
7. Attems J, Thomas A, Jellinger K (2012) Correlations between cortical and subcortical tau pathology. Neuropathol Appl Neurobiol 38(6):582–590. https://doi.org/10.1111/j.1365-2990.2011.01244.x
8. Giannakopoulos P, Herrmann FR, Bussiere T et al (2003) Tangle and neuron numbers, but not amyloid load, predict cognitive status in Alzheimer's disease. Neurology 60(9):1495–1500
9. Glenner GG, Wong CW (1984) Alzheimer's disease: initial report of the purification and characterization of a novel cerebrovascular amyloid protein. Biochem Biophys Res Commun 120(3):885–890
10. Naslund J, Haroutunian V, Mohs R et al (2000) Correlation between elevated levels of amyloid beta-peptide in the brain and cognitive decline. JAMA 283(12):1571–1577
11. Younkin SG (1995) Evidence that a beta 42 is the real culprit in Alzheimer's disease. Ann Neurol 37(3):287–288. https://doi.org/10.1002/ana.410370303
12. Thal DR, Rub U, Orantes M et al (2002) Phases of a beta-deposition in the human brain and its relevance for the development of AD. Neurology 58(12):1791–1800
13. Guntert A, Dobeli H, Bohrmann B (2006) High sensitivity analysis of amyloid-beta peptide composition in amyloid deposits from human and PS2APP mouse brain. Neuroscience 143(2):461–475. https://doi.org/10.1016/j.neuroscience.2006.08.027
14. Mirra SS, Heyman A, McKeel D et al (1991) The consortium to establish a registry for Alzheimer's disease (CERAD). Part II. Standardization of the neuropathologic assessment of Alzheimer's disease. Neurology 41(4):479–486
15. Attems J, Jellinger K, Thal DR et al (2011) Review: sporadic cerebral amyloid angiopathy. Neuropathol Appl Neurobiol 37(1):75–93. https://doi.org/10.1111/j.1365-2990.2010.01137.x
16. Vinters HV (1987) Cerebral amyloid angiopathy. A critical review. Stroke 18(2):311–324
17. Attems J (2005) Sporadic cerebral amyloid angiopathy: pathology, clinical implications, and possible pathomechanisms. Acta Neuropathol 110(4):345–359. https://doi.org/10.1007/s00401-005-1074-9
18. Attems J, Jellinger KA (2004) Only cerebral capillary amyloid angiopathy correlates with Alzheimer pathology--a pilot study. Acta Neuropathol 107(2):83–90. https://doi.org/10.1007/s00401-003-0796-9
19. Thal DR, Ghebremedhin E, Orantes M et al (2003) Vascular pathology in Alzheimer disease: correlation of cerebral amyloid angiopathy and arteriosclerosis/lipohyalinosis with cognitive decline. J Neuropathol Exp Neurol 62(12):1287–1301
20. Pfeifer LA, White LR, Ross GW et al (2002) Cerebral amyloid angiopathy and cognitive function: the HAAS autopsy study. Neurology 58(11):1629–1634
21. Vinters HV, Gilbert JJ (1983) Cerebral amyloid angiopathy: incidence and complications in the aging brain. II. The distribution of amyloid vascular changes. Stroke 14(6):924–928
22. Tomonaga M (1981) Cerebral amyloid angiopathy in the elderly. J Am Geriatr Soc 29(4):151–157
23. Thal DR, Ghebremedhin E, Rub U et al (2002) Two types of sporadic cerebral amyloid angiopathy. J Neuropathol Exp Neurol 61(3):282–293
24. Love S, Chalmers K, Ince P et al (2014) Development, appraisal, validation and implementation of a consensus protocol for the assessment of cerebral amyloid angiopathy in post-mortem brain tissue. Am J Neurodegener Dis 3(1):19–32
25. Montine T, Phelps C, Beach T et al (2012) National Institute on Aging–Alzheimer's Association guidelines for the neuropathologic assessment of Alzheimer's disease: a practical approach. Acta Neuropathologica 123:1–11
26. Hyman BT, Phelps CH, Beach TG et al (2012) National Institute on Aging-Alzheimer's Association guidelines for the neuropathologic assessment of Alzheimer's disease. Alzheimer's & dementia : the journal of the Alzheimer's Association 8(1):1–13. https://doi.org/10.1016/j.jalz.2011.10.007
27. Alafuzoff I, Arzberger T, Al-Sarraj S et al (2008) Staging of neurofibrillary pathology in Alzheimer's disease: a study of the BrainNet Europe consortium. Brain pathol (Zurich, Switzerland) 18(4):484–496. https://doi.org/10.1111/j.1750-3639.2008.00147.x

Part III

Established Markers of Pathophysiology and Risk

Check for updates

Chapter 6

Cerebrospinal Fluid Biomarkers of Preclinical Alzheimer's Disease

Panagiotis Alexopoulos and Chaido Sirinian

Abstract

The recent paradigm shift toward a more biologically oriented definition of Alzheimer's disease (AD) in clinical settings increases the importance of biomarkers of AD. The established cerebrospinal fluid (CSF) biomarkers of AD, reflecting both amyloidopathy and tauopathy, are being increasingly incorporated into both diagnostic guidelines, enabling the diagnosis of AD independently of clinical symptoms, and inclusionary criteria for clinical trials. The present chapter provides an overview of the clinical utility of the three established CSF AD biomarkers and covers both clinical and methodological issues. A brief summary is given on relevant laboratory techniques to determine levels of AD CSF biomarkers; methodological and clinical challenges in the field are also discussed.

Key words Cerebrospinal fluid amyloid β 1–42, Amyloid β 1–40, Total tau, Hyperphosphorylated tau, Preclinical Alzheimer's disease, Prognosis, Diagnosis, ELISA, Luminex xMAP, MSD multi-array

1 Introduction

Alzheimer's disease, accounting for 60–70% of dementia cases, is the most common cause of dementia [1, 2]. Dementia embodies a global health challenge, the social, health care, and financial burden of which will expand in the next decades. According to the World Alzheimer Report 2016, approximately 46.8 million people worldwide suffered from major dementia in 2015, and this number is estimated to reach 131.5 million in 2050.

In recent years a change in the way AD is defined and understood has taken place [3, 4]. Traditionally the diagnosis of AD antemortem was based on dementia symptoms and on the exclusion of other potential causes of cognitive impairment, while a definite diagnosis of AD was neuropathologically established, based on brain biopsy or autopsy findings [5, 6]. The pathology-driven AD diagnosis is based on the detection of extracellular amyloid-β (Aβ) plaques, mirroring amyloidopathy; of intracellular neurofibrillary tangles, reflecting tauopathy; and of axonal degeneration, neuronal

Robert Perneczky (ed.), *Biomarkers for Preclinical Alzheimer's Disease*, Neuromethods, vol. 137,
https://doi.org/10.1007/978-1-4939-7674-4_6, © Springer Science+Business Media, LLC 2018

death, and loss of synapses [2]. However, the tremendous progress achieved in the last decades in unraveling the pathogenesis of AD has resulted in a new, more biology underpinned AD conceptualization [6, 7]. According to it, prior to its dementia phase, the disease has a preclinical and an oligosymptomatic stage, commonly termed in the current nomenclature mild cognitive impairment (MCI) [8]. The oligosymptomatic stage is characterized by declining cognitive performance, mainly memory function, and mildly affected complex activities of daily living, involving memory or complex reasoning [9]. In the preclinical stage of AD, amyloidopathy and tauopathy are present in the absence of cognitive deficits [10]. It is of note that preclinical AD, which is present one or more decades prior to the outburst of clinical symptoms and is supposed to be the stage of the disease in which future causal therapies will be most effective, can be detected in vivo through markers of AD pathology [4, 11].

Biomarkers are diagnostic tools which detect a fundamental feature of a pathobiological process, characterizing a specific disease, and increase diagnostic certainty [12]. Since cerebrospinal fluid (CSF) is in direct contact with the brain and the flow of peptides from and to it is restricted by the blood-CSF barrier, it is likely that brain pathological alterations are mirrored in CSF [13]. According to the consensus report published by the Working Group on Molecular and Biochemical Markers of Alzheimer's Disease, biomarkers of AD should have at least 80% sensitivity and specificity in the detection of neuropathologically confirmed AD cases [14]. A number of CSF peptides associated with AD pathology have been shown to fulfill the aforementioned criteria [15].

The neurochemical signature of AD consists of the 42-amino acid Aβ peptide (Aβ42; a marker of amyloid plaques); hyperphosphorylated tau (pTau), being a marker of neurofibrillary tangles; and total tau (tTau), which is a marker of axonal damage. Aβ42 is the most abundant Aβ isoform in amyloid plaques and is decreased in the CSF of patients suffering from AD to approximately 50% of control levels [16]. The marker tTau, which is not specific for AD, embraces all isoforms of tau in CSF independently of phosphorylation state, and the increase of its concentration in patients with AD is approximately 300% of the levels of controls. Protein tau has numerous phosphorylation sites. The phosphorylation at either threonine 181 or threonine 231 stands in focus of the measurement methods of pTau in CSF, which unravel a 200% increase of pTau in AD in comparison to cognitively healthy elderly individuals [7]. It is of note that pTau appears to be relative specific for AD. In a number of laboratories, the 40-amino acid Aβ peptide (Aβ40), being a common protein found in patients without the disease, is measured, in order to build the Aβ42/Aβ40 ratio, while in other centers this analysis is carried out only in the case of conflicting CSF biomarker findings such as increased tau proteins and normal concentrations of Aβ42 [17–19].

CSF AD biomarkers become abnormal years prior to the development of clinical symptoms. CSF Aβ42 levels have been shown to begin to decline 25 years before the expected onset of symptomatic AD and to plateau down to fully decreased levels at least 9 years prior to dementia onset, while tau increases approximately 15 years prior to the expected onset of symptoms [20, 21]. Longitudinal decreases in Aβ42 were observed in some individuals in early middle age, while the concentrations of tau proteins increased in mid and late middle age [22]. Nonetheless, another study points that alterations in pTau CSF levels may precede those of Aβ42 [23]. This finding is in line with postmortem evidence of the presence of neurofibrillary tangles in the entorhinal cortex and in hippocampus-related structures as the first neuropathologic event in AD [10].

The three established CSF biomarkers of AD are a useful instrument for detecting AD. CSF AD biomarker abnormality distinguishes patients suffering from dementia due to AD from controls with a sensitivity and specificity reaching 85–90% [24]. The sensitivity and specificity of these markers in detecting patients with MCI who will develop AD dementia range between 81–95% and 72–95%, respectively [25–27]. A recent meta-analysis which included more than 28,000 elderly individuals unveiled a high accuracy of established CSF AD biomarkers in differentiating AD dementia from controls (up to 2.5-fold changes between patients and controls). The differentiation between cohorts with MCI due to AD and those with stable MCI was very strong too [15]. Nonetheless, the utility of CSF AD biomarkers to differentiate AD from other dementia causes is relatively limited. Increases in tTau are commonly observed also in other diseases, for instance in traumatic and ischemic brain injury and in subtypes of frontotemporal lobar degeneration, while abnormal CSF Aβ42 levels overlap between AD and dementia with Lewy bodies [24, 28]. Aβ42/Aβ40 ratio has been shown to have added value with respect to Aβ42 alone having an accuracy of >80% in distinguishing AD dementia from dementias caused by other neurodegenerative diseases [17, 29]. Longitudinal studies in cognitively healthy individuals reported a correlation between baseline CSF biomarkers and development of oligosymptomatic AD for Aβ42 alone or in combination with tTau or pTau [10]. Of note, in individuals with abnormal Aβ42, the additional presence of abnormal tau levels is associated with a faster progression to a symptomatic stage of the disease [10]. Recently, neurochemical biomarkers have been advocated as part of the clinical diagnostic workup of memory clinics worldwide and are being increasingly incorporated into both diagnostic guidelines and inclusionary criteria for clinical trials [4, 10, 30].

2 Materials

CSF samples are obtained by lumbar puncture in the L3/L4 or L4/L5 intervertebral space (*see Notes 1* and *2*) and collected in polypropylene tubes (*see Note 3*), without fasting (*see Note 4*). They are centrifuged at 2000 × *g* for 10 min at room temperature (*see Note 5*) and then split into 0.5 mL aliquots in cryotubes and immediately frozen at −20 °C for a maximum of 2 months, until routine analysis (*see Note 6*).

3 Methods

The majority of laboratories worldwide uses enzyme-linked immunosorbent assays (ELISAs) or the bead-based xMAP platform to determine CSF AD biomarkers. Meso Scale Discovery technology is used by a smaller number of laboratories [31].

3.1 Enzyme-Linked Immunosorbent Assay (ELISA)

3.1.1 Quantitative Determination of Aβ40 and Aβ42

Levels of Aβ40 and Aβ42 can be determined using high sensitivity ELISA kits (RE59651 and RE59661) from IBL International (Hamburg, Germany) [29]. Both assays are based on microtiter plates, pre-coated with a monoclonal antibody (capture antibody) raised against the C terminus of either the Aβ40 or Aβ42 peptides. The N terminus of the Aβ40 and Aβ42 peptides is recognized by a second monoclonal antibody (detection antibody), which is conjugated to a horseradish peroxidase enzyme that is sequentially detected by the addition of the chromogenic substrate tetramethylbenzidine (TMB). The concentration of both Aβ40 and Aβ42 is proportional to the obtained optical density.

1. Add 100 μL of each Standard, Control and diluted patient sample into the respective wells of microtiter plate in duplicate. Cover plate with lid, and incubate for 120 min at room temperature (RT) (18–25 °C) on an orbital shaker (500 rpm).
2. Aspirate each well and wash all unbound antigen five times.
3. Add 100 μL of diluted enzyme conjugate in each well. Cover plate with lid, and incubate for 60 min at RT (18–25 °C) on an orbital shaker (500 rpm).
4. Aspirate each well and wash five times.
5. Pipette 100 μL of TMB substrate solution into each well. Briefly mix contents by gently shaking the plate.
6. Incubate microtiter plate for 30 min at RT (18–25 °C), and stop the substrate reaction by adding 100 μL of TMB stop solution into each well.
7. Measure optical density with a photometer at 450 nm.

Both assays present with high analytical specificity (low cross-reactivity) toward their analytes, but the Aβ42 detection kit presents with a significant higher analytical sensitivity.

3.1.2 Quantitative Determination of tTau and pTau

For quantification of tTau and Tau phosphorylated at threonine 181, two commercially available ELISA kits INNOTEST™ hTAU Ag (INX74378, Innogenetics, Gent, Belgium) and INNOTEST™ PHOSPHO-TAU$_{(181P)}$ (FRI30913, Fujirebio, Gent, Belgium) may be employed. In both assays, target proteins, Tau and pTau, are captured by a monoclonal human antibody (AT120 and HT7, respectively) which is coated on a microtiter plate. Human CSF samples are simultaneously combined with detection antibodies conjugated to biotin (conjugate 1) and loaded to the appropriate ELISA plate wells in duplicates. In the case of the tTau detection ELISA, two human monoclonal detection abs (BT2 and HT7) are employed, while for the detection of pTau, a phospho-specific Tau threonine 181 monoclonal antibody is used. The antigen-antibody complex is then recognized by peroxidase-labeled streptavidin molecules (conjugate 2). In addition, run validation controls and calibrators are used for the standardization of the quantification of both tTau and pTau proteins. The procedure is briefly described in the following.

1. Determine the number of wells (in duplicates) for controls, calibrators, and CSF samples.
2. Add conjugate working solution 1 to each well of the antibody-coated plate.
3. Add control, calibrator, and CSF samples to duplicate wells, and cover the strips with an adhesive sealer, and incubate overnight (*see Note 7*).
4. Aspirate each well and wash all unbound antigen five times.
5. Add conjugate working solution 2 to each well. Cover with a new adhesive sealer, and incubate at 25 ± 2 °C (*see Note 8*).
6. Aspirate each well and wash five times.
7. Add substrate working solution to each well for 30 min in the dark. Stop the reaction, and determine the optical density of each well immediately, using a microplate reader set to 540 nm.

3.2 Multiplexed Immunoassays

Today, single-analyte immunoassay is seeing competition from newer bead-based immunoassays that employ Luminex's xMAP and Meso Scale Discovery's Multi-Array technology. Current multiplexed immunoassays are based on multi-marker strategies, in a single reaction vessel from a relatively smaller sample volume with high efficiency.

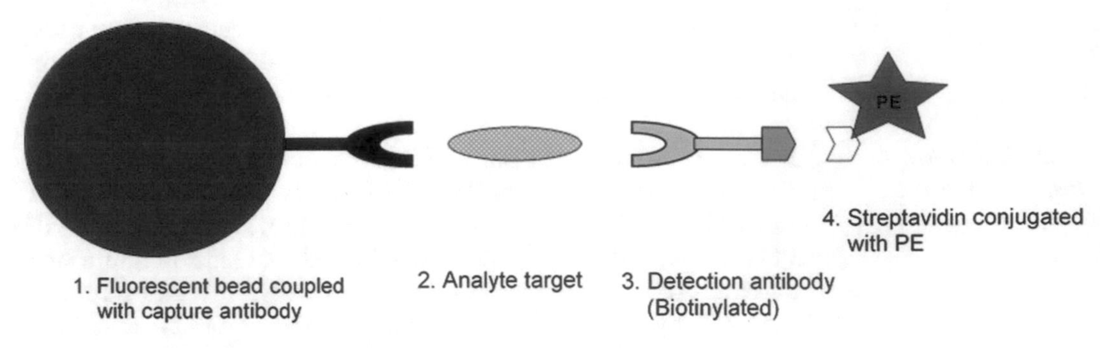

Fig. 1 General overview of the xMAP immunoassay

3.2.1 Luminex Multi-Analyte Profiling (xMAP)

The Luminex xMAP technology allows the measurement of multiple proteins simultaneously in a single sample. This technology uses 5.6-μm microsphere beads, either polystyrene or paramagnetic. The microsphere beads are internally dyed with precise amounts of two or three spectrally distinct fluorescent dyes to produce up to 100–500 different sets of microspheres, depending on the system design. Any molecule, such as proteins, antibodies, ligands, or nucleic acids, interacting with the desired target, can be coupled to the beads [32]. Initially, the coupled beads are incubated with the sample and bind to the desire target. Subsequently the bead-target complex is incubated with a mixture of biotinylated antibodies, which are specific against the target analyte, associated with a fluorescent bead. A fluorescent dye (phycoerythrin, PE) is pre-conjugated to streptavidin molecules and is used to indicate the occurrence of positive binding and quantify the amount of target analytes (Fig. 1). The multiplexing detection is performed on a Luminex analyzer, where two lasers (a 635-nm, 10-mW red diode laser and a 532-nm, 13-mW yttrium aluminum garnet laser) are used to identify and quantify fluorescence from each individual microsphere bead-target analyte-streptavidin complex. The analyzer, simultaneously, identifies the target analyte through the specific bead-fluorescence spectral address and determines its quantity (PE-fluorescence), as the sample passes through the analyzer [33].

Innogenetics manufactures one such xMAP-based test, the INNO-BIA AlzBio3, which determines the amount of pTau, tTau, and Aβ42 proteins in CSF samples. According to the test, a mix of the microsphere beads of a specific region number is captured selectively by a first monoclonal antibody [AT270 for pTau, AT120 for tTau, 4D7A3 for Aβ42)], coupled covalently. A mix of the microspheres is added in a specific volume to the filter plates. CSF samples or standards are added on the filter plate, together with a mix of biotinylated antibodies. Each biotinylated detector antibody detects one or several parameters [e.g., 3D6 for Aβ42, HT7 for detection of pTau and tTau]. The antigen-antibody complex is then detected by a phycoerythrin-labeled streptavidin conjugate. Each analyte is identified and quantified as described previously.

3.2.2 Meso Scale Discovery's Multi-Array (MSD)

In the MSD assay main steps that are followed are highly similar to ELISA assay. The plate is initially coated with a capture reagent before the addition of the samples and followed by the addition of detection reagents. However, MSD platform uses a different type of uncoated plates, which have a working electrode surface that absorbs capture reagent within the wells. The plates, used for the MSD assay, can be coated with a variety of capture molecules and reagents, including antibodies, virus-like particles, cells, peptides, and lysates. After coating, the sample and a solution containing detection antibodies conjugated with electrochemiluminescent labels are added to the wells. The analytes in the sample bind to the capture reagent immobilized on the electrode surface, and this complex recruits the detection antibodies forming a "sandwich." An MSD read buffer that provides the appropriate chemical environment for electrochemiluminescence is loaded to the plate and analyzed into an MSD imager. The imager applies a voltage to the plate electrodes, causing the SULFO-TAG in close proximity to the bottom of the plate to emit light through a series of reduction and oxidation reaction. The imager measures the intensity of emitted light to provide a quantitative measure of analytes in the sample [34].

Based on the above, at least two possible MSD immunoassay formats may be developed for amyloid peptides (Aβ40 and Aβ42) and Tau protein (tTau and pTau) identification and quantification. First, a typical format of an immunoassay that employs a specific antibody as the capture reagent (capture antibody) that coats the well of a plate can be developed. In a second format of the MSD assay, instead of using a capture antibody to coat a well-plate position, the sample (CSF sample) is directly coated in the plate well as the capture material. Then specific antibodies are used to identify and quantify the under-study analyte. According to recently published data, based on correlation studies between all the above-described immunoassays, the data obtained with the MSD platform for AD CSF biomarkers are sparse at this time compared to the ELISA and xMAP platforms [31].

3.3 Interpretation of CSF AD Biomarker Values

The interpretation of findings relies on cutoff values recommended by the manufacturers and is not affected by the patients' age [17]. It is of note that the manufacturers recommend that each laboratory establishes its own cutoff values (*see Note 9*), since the interlaboratory variability ranges between 20% and 30% even when the same assay is used [35, 36].

If all three CSF biomarkers are abnormal, the constellation is neurochemically compatible with AD. Constellations with all biomarkers normal point to neurochemical incompatibility with a diagnosis of AD. The use of "gray zones" is a common strategy for treating results that are somehow indeterminate. The recently published National Institute on Aging-Alzheimer's Association diagnostic guidelines refer to borderline biomarker results too, even though they avoid defining the range of borderline values [6]. A proposed pragmatic approach to the range of borderline CSF biomarker values is to define a gray zone as 10% increment of the cutoff value in the case of tau proteins and a 10% decrement of the cutoff value in the case of Aβ42 or Aβ42/Aβ40 ratio. The value should be at least 10% above or below the corresponding cutoff value in order to be treated as abnormal. In such a way the possibility of false-positive or false-negative interpretation of the findings is minimized. In the light of the absence of causal therapies, it is ethically acceptable to falsely negatively underdiagnose a number of individuals suffering from preclinical AD than to falsely overdiagnose a number of healthy controls [17]. In cases with conflicting biomarker findings, no diagnostic label can be given, and further diagnostic workup is necessary (for instance, Aβ40 measurement and/or brain imaging) (*see Note 10*). Additionally, the indication of follow-up and in-depth clinical and neuropsychological characterization of the patient should be considered before final conclusions are drawn. Isolated low CSF levels of Aβ42 or isolated tau pathology indicates that the individual is at risk for AD but are not sufficient for diagnosing preclinical AD [10].

3.4 Challenges

3.4.1 Clinical Challenges

The typical clinical phenotype of AD is not a straightforward consequence of amyloidopathy and tauopathy, and the presence of AD pathology does not deterministically lead to symptomatic AD. As autopsy reports underscore, a plethora of pathologies accompany AD pathological alterations in the aging brain (for instance, cerebrovascular alterations or Lewy body pathology) [31, 37]. Such concurrent pathologies can synergistically lower the threshold for the development of clinical symptoms. Such co-pathologies potentiate the clinical expression of AD-associated brain alterations which would have remained clinically silent in the absence of co-pathologies, because they are still not sufficiently advanced to become clinically recognizable [38]. The synergies of co-pathologies in the aging brain are modified by the influence of both harmful and protective environmental factors such as exposure to occupational hazards or cognitive, physical, and social activity levels which could superimpose or counterbalance the effects of the presence of brain pathologies [39]. This complex interplay factors undermine, in effect, the prognostic utility of established AD biomarkers at the individual level, and as a result biomarkers turn out to be less efficient in routine care in comparison to research settings [40]. Hence, the individualized estimation of prognosis of preclinical AD presupposes the development of novel strategies approaching the disease course as a multivariate system, so that the patients can benefit at the individual level from advances in biomarker research [3].

The limited accuracy of CSF AD biomarkers in estimating the individual prognosis and the absence of causal therapeutic strategies raise ethical issues and issues of cost-effectiveness of CSF AD biomarker analyses. The diagnosis of an asymptomatic neurodegenerative disease with uncertain prognosis and no available therapy may cause significant psychological distress. Therefore, the diagnosis of preclinical AD has to be accompanied by appropriate psychosocial counseling, so that the risk of unnecessary psychological distress is minimized [41]. Furthermore, in the light of the lacking disease-modifying strategies, health payers are in a number of countries reluctant to reimburse the full cost of CSF biomarker analyses due to their insufficient effect on patient management, despite their relatively low cost [40]. On the other hand, it should be underscored that an accurate diagnosis improves the quality of treatment and health care at the individual level. For instance, ruling out AD pathology shifts the focus of the diagnostic workup on alternative, and possible curable diseases or the diagnosis of preclinical or oligosymptomatic AD may not only foster protective behaviors (e.g., reduction of stress, changes of dietary patterns), which could positively affect the impact of known factors potentiating neurodegeneration (inflammation, oxidative stress) [3], but also enable the initiation of the necessary psychosocial interventions early enough to maximize their effect.

CSF biomarkers, especially Aβ42, have been employed in clinical trials of drug candidates with disease-modifying potential. In the light of the disappointing results of several clinical trials which have included patients at the dementia stage of AD, it is reckoned that the emphasis should be moved to oligosymptomatic or even preclinical stages of the disease. Individuals with preclinical AD would probably benefit most from interventions. In such trials CSF biomarkers can be employed as pharmacodynamic markers and/or as surrogate markers. For instance, changes in CSF Aβ42 concentrations in participants can be instructive in selecting dosage for development, and a go or no-go decision can be based on them. Nonetheless, CSF Aβ42 levels are decreased and reach a plateau already a decade prior to onset of dementia symptoms [21]. Thus, it seems that the assessment of the potential curative effect of a drug candidate cannot be based on CSF concentration changes of a biomarker that has already reached a plateau. As a consequence new surrogate and pharmacodynamic markers are urgently needed.

Despite having an invasive character, lumbar puncture is a straightforward procedure with few complications. The list of contradictions to lumbar puncture includes hypertension with bradycardia, bleeding diathesis, vertebral deformities, and infection at the site of the lumbar puncture. The frequency of complications such as postlumbar puncture headache, back pain, headache, and dizziness is low especially in the elderly [42–44], so that CSF collection by means of lumbar puncture is being increasingly implemented in diagnostic routine of memory clinics worldwide.

3.4.2 Analytical Challenges

The measured CSF biomarker concentrations differ between studies which is the consequence of a number of preanalytical, analytical, and/or assay-related factors. The source of overall variability can be attributed to within-assay run variability (between duplicate samples), intralaboratory longitudinal variability, between-laboratory variability, and within- and between-assay kit lot variability [36]. The overall variability is reported to be around 20%–30%, with lower numbers for ELISA than for xMAP and MSD measurements. Interlaboratory variability is a major contributor, while intralaboratory longitudinal variability has a smaller and the within-run variability an even smaller effect. For some of the analytes, a significant impact from between-lot-dependent variability has been observed. Hence, it is important that quality of kits is improved, in order to minimize lot-to-lot variation, being caused by matrix effects, variation in production of different kit components, etc. Furthermore, the development of detailed experimental protocols is an essential step for decreasing variability [31, 35, 36]. Interestingly, the development of a novel fully automated electrochemiluminescence assay for the quantitation of CSF Aβ42 has been recently reported [45]. It has a markedly improved precision in comparison to other available assays and possibly opens the road to overcoming the shortcoming of the high variability. In addition, several, already underway quality control initiatives aim to contribute to standardization efforts (harmonizing laboratory practices, defining procedures on CSF collection and handling, establishing reference measurement procedures) [31, 35]. The development of the first reference procedure for CSF Aβ42 was an additional important step toward standardization [46].

The differences in mean concentrations of CSF markers between the analytical techniques ELISA, xMAP, and MSD are a major limitation for establishing multicentric cooperation, gold-standard protocols, and reference values to be shared by distinct laboratories. This difference is attributable to the lack of certified reference materials (CRMs), which are also called standard reference materials, and calibrators for CSF biomarkers. Metrology institutes develop CRMs. Establishing a protein CRM is a complex task, since heterogeneities engendered by post-transnational modifications or contaminations make it difficult to establish the purity of proteins. The development of CRMs for CSF AD biomarkers embodies a major challenge. In order to achieve it, the coordinated efforts of researchers, industry, and metrology are necessary. If successful, full global traceability and comparability of biomarker results yielded by different analytical techniques and at different laboratories will be enabled [31, 36].

4 Notes

1. CSF biomarker concentrations from nonhemorrhagic samples are not influenced by fractionated sampling. Thus, there is no need to recommend that a certain fraction is to be taken when sampling CSF for AD biomarker analyses [47].
2. There is no need for obtaining CSF samples at a particular time of day, since there is no diurnal variation with AD biomarkers [48].
3. Glass or polystyrene tube should not be used, since Aβ peptides tend to bind nonspecifically to them, leading to lower measured concentrations of Aβ42. The degree of this phenomenon is related to the tube surface area, to the duration of contact with the tube surface, as well as to the temperature, sample, or tube type. Tau levels are also affected but to a more limited extent. Recent reports indicate, however, that tubes made of pure polypropylene may cause adsorption of Aβ peptides. The adsorption of Aβ onto the tube may be prevented by pretreating CSF with a nonionic detergent such as Tween-20 [47].
4. There is no evidence that food intake or glucose levels affect CSF biomarker levels [47].
5. Centrifugation is only necessary in the case of visually hemorrhagic samples, which could correspond to a red cell count between 500 and 1000/μL. In the case of traumatic lumbar puncture, it is recommended that the first 1–2 mL is discarded. The centrifugation should take place as soon as possible and not later than 2 h after lumbar puncture, in order to avoid cell lysis, influencing CSF AD biomarker levels [47].
6. It is of note that there is no statistically significant change in Aβ42, pTau, and tTau concentrations irrespective of whether the sample is frozen within 2 h or kept at room temperature for up to 5 days before freezing [49]. Moreover, since the method of freezing (– 20 °C or –80 °C) seems to influence tTau and pTau levels, it is recommended that freezing at –80 °C should be preferably carried out for storage. CSF AD biomarker concentrations seem to be very stable for an extended period, at least 10 years, at –80 °C [50]. Though CSF AD biomarkers are stable during freeze/thawing when measured with enzyme-linked immunosorbent assay, it is recommended that the number of freeze/thaw cycles is limited to one or two, so that any possible effects are minimized [47].
7. Overnight incubation temperatures differ between the two assays. For the detection of tTau protein, samples are incubated at 25 ± 2 °C, whereas for the detection of pTau, samples are incubated at 2–8 °C.

8. Incubation time for tTau protein is 30 min and for pTau is 60 min.
9. The validation for clinical routine of cutoff values designed for a particular laboratory presupposes that the chosen cutoff is the best compromise in terms of sensitivity and specificity for the laboratory's specific clinical objectives, that the accuracy of the cutoff value is periodically reassessed in larger samples recruited during clinical routine, and that any implemented preanalytical or analytical modifications within the framework of standardization strategies make a new validation of previous cutoffs necessary [17].
10. It should be borne in mind when interpreting conflicting biomarker findings that AD biological markers mirror a continuum between healthy aging and dementia with typical CSF biomarkers constellations and that they change over time within the framework of a continuous process lasting many years.

References

1. De-Paula VJ, Radanovic M, Diniz BS et al (2012) Alzheimer's disease. Subcell Biochem 65:329–352. https://doi.org/10.1007/978-94-007-5416-4_14
2. Kocahan S, Dogan Z (2017) Mechanisms of Alzheimer's disease pathogenesis and prevention: the brain, neural pathology, N-methyl-D-aspartate receptors, tau protein and other risk factors. Clin Psychopharmacol Neurosci 15(1):1–8. https://doi.org/10.9758/cpn.2017.15.1.1
3. Alexopoulos P, Kurz A (2015) The new conceptualization of Alzheimer's disease under the microscope of influential definitions of disease. Psychopathology 48(6):359–367. https://doi.org/10.1159/000441327
4. McKhann GM (2011) Changing concepts of Alzheimer disease. JAMA 305(23):2458–2459. https://doi.org/10.1001/jama.2011.810
5. McKhann G, Drachman D, Folstein M et al (1984) Clinical diagnosis of Alzheimer's disease: report of the NINCDS-ADRDA work group under the auspices of Department of Health and Human Services Task Force on Alzheimer's disease. Neurology 34(7):939–944
6. McKhann GM, Knopman DS, Chertkow H et al (2011) The diagnosis of dementia due to Alzheimer's disease: recommendations from the National Institute on Aging-Alzheimer's Association workgroups on diagnostic guidelines for Alzheimer's disease. Alzheimers Dement 7(3):263–269. https://doi.org/10.1016/j.jalz.2011.03.005
7. Forlenza OV, Radanovic M, Talib LL et al (2015) Cerebrospinal fluid biomarkers in Alzheimer's disease: diagnostic accuracy and prediction of dementia. Alzheimers Dement (Amst) 1(4):455–463. https://doi.org/10.1016/j.dadm.2015.09.003
8. Albert MS, DeKosky ST, Dickson D et al (2011) The diagnosis of mild cognitive impairment due to Alzheimer's disease: recommendations from the National Institute on Aging-Alzheimer's Association workgroups on diagnostic guidelines for Alzheimer's disease. Alzheimers Dement 7(3):270–279. https://doi.org/10.1016/j.jalz.2011.03.008
9. Perneczky R, Pohl C, Sorg C et al (2006) Complex activities of daily living in mild cognitive impairment: conceptual and diagnostic issues. Age Ageing 35(3):240–245. https://doi.org/10.1093/ageing/afj054
10. Dubois B, Hampel H, Feldman HH et al (2016) Preclinical Alzheimer's disease: definition, natural history, and diagnostic criteria. Alzheimers Dement 12(3):292–323. https://doi.org/10.1016/j.jalz.2016.02.002
11. Jack CR, Knopman DS, Jagust WJ et al (2013) Tracking pathophysiological processes in Alzheimer's disease: an updated hypothetical model of dynamic biomarkers. The Lancet Neurology 12(2):207–216. https://doi.org/10.1016/S1474-4422(12)70291-0
12. Martins-de-Souza D (2010) Is the word 'biomarker' being properly used by proteomics research in neuroscience? Eur Arch Psychiatry

Clin Neurosci 260(7):561–562. https://doi.org/10.1007/s00406-010-0105-2
13. Engelborghs S (2013) Clinical indications for analysis of Alzheimer's disease CSF biomarkers. Rev Neurol (Paris) 169(10):709–714. https://doi.org/10.1016/j.neurol.2013.07.024
14. The Ronald and Nancy Reagan Research Institute of the Alzheimer's Association and the National Institute on Aging Working Group (1998) Consensus report of the Working Group on: "Molecular and Biochemical Markers of Alzheimer's Disease". Neurobiol Aging 19(2):109–116
15. Olsson B, Lautner R, Andreasson U et al (2016) CSF and blood biomarkers for the diagnosis of Alzheimer's disease: a systematic review and meta-analysis. The Lancet Neurology 15(7):673–684. https://doi.org/10.1016/S1474-4422(16)00070-3
16. Blennow K, Zetterberg H, Fagan AM (2012) Fluid biomarkers in Alzheimer disease. Cold Spring Harb Perspect Med 2(9):a006221. https://doi.org/10.1101/cshperspect.a006221
17. Molinuevo JL, Blennow K, Dubois B et al (2014) The clinical use of cerebrospinal fluid biomarker testing for Alzheimer's disease diagnosis: a consensus paper from the Alzheimer's biomarkers standardization initiative. Alzheimers Dement 10(6):808–817. https://doi.org/10.1016/j.jalz.2014.03.003
18. Lewczuk P, Esselmann H, Otto M et al (2004) Neurochemical diagnosis of Alzheimer's dementia by CSF Abeta42, Abeta42/Abeta40 ratio and total tau. Neurobiol Aging 25(3):273–281. https://doi.org/10.1016/S0197-4580(03)00086-1
19. Bibl M, Mollenhauer B, Esselmann H et al (2006) CSF diagnosis of Alzheimer's disease and dementia with Lewy bodies. J Neural Transm (Vienna) 113(11):1771–1778. https://doi.org/10.1007/s00702-006-0537-z
20. Bateman RJ, Xiong C, Benzinger TLS et al (2012) Clinical and biomarker changes in dominantly inherited Alzheimer's disease. N Engl J Med 367(9):795–804. https://doi.org/10.1056/NEJMoa1202753
21. Stomrud E, Minthon L, Zetterberg H et al (2015) Longitudinal cerebrospinal fluid biomarker measurements in preclinical sporadic Alzheimer's disease: a prospective 9-year study. Alzheimers Dement (Amst) 1(4):403–411. https://doi.org/10.1016/j.dadm.2015.09.002
22. Sutphen CL, Jasielec MS, Shah AR et al (2015) Longitudinal cerebrospinal fluid biomarker changes in preclinical Alzheimer disease during middle age. JAMA Neurol 72(9):1029–1042. https://doi.org/10.1001/jamaneurol.2015.1285
23. Braak H, Zetterberg H, Del Tredici K et al (2013) Intraneuronal tau aggregation precedes diffuse plaque deposition, but amyloid-beta changes occur before increases of tau in cerebrospinal fluid. Acta Neuropathol 126(5):631–641. https://doi.org/10.1007/s00401-013-1139-0
24. Teunissen CE, Willemse E (2013) Cerebrospinal fluid biomarkers for Alzheimer's disease: emergence of the solution to an important unmet need. EJIFCC 24(3):97–104
25. Hansson O, Zetterberg H, Buchhave P et al (2006) Association between CSF biomarkers and incipient Alzheimer's disease in patients with mild cognitive impairment: a follow-up study. Lancet Neurol 5(3):228–234. https://doi.org/10.1016/S1474-4422(06)70355-6
26. Parnetti L, Chiasserini D, Eusebi P et al (2012) Performance of abeta1-40, abeta1-42, total tau, and phosphorylated tau as predictors of dementia in a cohort of patients with mild cognitive impairment. J Alzheimers Dis 29(1):229–238. https://doi.org/10.3233/JAD-2011-111349
27. Mattsson N, Zetterberg H, Hansson O et al (2009) CSF biomarkers and incipient Alzheimer disease in patients with mild cognitive impairment. JAMA 302(4):385–393. https://doi.org/10.1001/jama.2009.1064
28. Riedl L, Mackenzie IR, Forstl H et al (2014) Frontotemporal lobar degeneration: current perspectives. Neuropsychiatr Dis Treat 10:297–310. https://doi.org/10.2147/NDT.S38706
29. Alexopoulos P, Eisele T (2014) Efficiency of the ratio of cerebrospinal fluid Aβ42/Aβ40 concentrations in detecting Alzheimer's disease: a step forwards. Alzheimers Dement 10(4):P799
30. Dubois B, Feldman HH, Jacova C et al (2014) Advancing research diagnostic criteria for Alzheimer's disease: the IWG-2 criteria. Lancet Neurol 13(6):614 629. https://doi.org/10.1016/S1474-4422(14)70090-0
31. Kang J-H, Korecka M, Toledo JB et al (2013) Clinical utility and analytical challenges in measurement of cerebrospinal fluid amyloid-β1–42 and τ proteins as Alzheimer disease biomarkers. Clin Chem 59(6):903–916. https://doi.org/10.1373/clinchem.2013.202937
32. Vignali DA (2000) Multiplexed particle-based flow cytometric assays. J Immunol Methods 243(1–2):243–255
33. Dunbar SA and Li D (2010) Introduction to Luminex® xMAP® Technology and Applications

for Biological Analysis in China. Asia Pac Biotech 14(10):26–30

34. Meso Scale Discovery®. MSD sector and quickplex plates, 18075-v2-2014 Jul. www.mesoscale.com
35. Mattsson N, Andreasson U, Persson S et al (2013) CSF biomarker variability in the Alzheimer's Association quality control program. Alzheimers Dement 9(3):251–261. https://doi.org/10.1016/j.jalz.2013.01.010
36. Mattsson N, Andreasson U, Persson S et al (2011) The Alzheimer's Association external quality control program for cerebrospinal fluid biomarkers. Alzheimers Dement 7(4):386–395.e6. https://doi.org/10.1016/j.jalz.2011.05.2243
37. Lam B, Masellis M, Freedman M et al (2013) Clinical, imaging, and pathological heterogeneity of the Alzheimer's disease syndrome. Alzheimers Res Ther 5(1):1. https://doi.org/10.1186/alzrt155
38. Alexopoulos P, Roesler J, Thierjung N et al (2015) Mapping CSF biomarker profiles onto NIA-AA guidelines for Alzheimer's disease. Eur Arch Psychiatry Clin Neurosci 266(7):587–597. https://doi.org/10.1007/s00406-015-0628-7
39. Perneczky R, Alexopoulos P, Kurz A (2014) Soluble amyloid precursor proteins and secretases as Alzheimer's disease biomarkers. Trends Mol Med 20(1):8–15
40. Maier W (2016) Searching biomarkers for mental disorders-lessons from Alzheimer's disease. Eur Arch Psychiatry Clin Neurosci 266(7):583–585. https://doi.org/10.1007/s00406-016-0732-3
41. Perneczky R, Guo L-H (2016) Plasma proteomics biomarkers in Alzheimer's disease: latest advances and challenges. Methods Mol Biol 1303:521–529. https://doi.org/10.1007/978-1-4939-2627-5_32
42. Duits FH, Martinez-Lage P, Paquet C et al (2016) Performance and complications of lumbar puncture in memory clinics: results of the multicenter lumbar puncture feasibility study. Alzheimers Dement 12(2):154–163. https://doi.org/10.1016/j.jalz.2015.08.003
43. Blennow K, Wallin A, Hager O (1993) Low frequency of post-lumbar puncture headache in demented patients. Acta Neurol Scand 88(3):221–223
44. Zetterberg H, Tullhog K, Hansson O et al (2010) Low incidence of post-lumbar puncture headache in 1,089 consecutive memory clinic patients. Eur Neurol 63(6):326–330. https://doi.org/10.1159/000311703
45. Bittner T, Zetterberg H, Teunissen CE et al (2016) Technical performance of a novel, fully automated electrochemiluminescence immunoassay for the quantitation of beta-amyloid (1-42) in human cerebrospinal fluid. Alzheimers Dement 12(5):517–526. https://doi.org/10.1016/j.jalz.2015.09.009
46. Mattsson N, Zegers I, Andreasson U et al (2012) Reference measurement procedures for Alzheimer's disease cerebrospinal fluid biomarkers: definitions and approaches with focus on amyloid beta42. Biomark Med 6(4):409–417. https://doi.org/10.2217/bmm.12.39
47. Vanderstichele H, Bibl M, Engelborghs S et al (2012) Standardization of preanalytical aspects of cerebrospinal fluid biomarker testing for Alzheimer's disease diagnosis: a consensus paper from the Alzheimer's biomarkers standardization initiative. Alzheimers Dement 8(1):65–73. https://doi.org/10.1016/j.jalz.2011.07.004
48. Cicognola C, Chiasserini D, Eusebi P et al (2016) No diurnal variation of classical and candidate biomarkers of Alzheimer's disease in CSF. Mol Neurodegener 11(1):65. https://doi.org/10.1186/s13024-016-0130-3
49. Zimmermann R, Lelental N, Ganslandt O et al (2011) Preanalytical sample handling and sample stability testing for the neurochemical dementia diagnostics. J Alzheimers Dis 25(4):739–745. https://doi.org/10.3233/JAD-2011-110212
50. Bjerke M, Portelius E, Minthon L et al (2010) Confounding factors influencing amyloid Beta concentration in cerebrospinal fluid. Int J Alzheimers Dis 2010. https://doi.org/10.4061/2010/986310

Chapter 7

Brain Structural Imaging in Alzheimer's Disease

Sven Haller, Davide Zanchi, Cristelle Rodriguez, and Panteleimon Giannakopoulos

Abstract

Central nervous system diseases are usually associated with significant modifications in brain morphometry and function. These alterations may be subtle, in particular at early stages of the disease progress, and thus not evident by visual inspection alone in magnetic resonance imaging. Group-level statistical comparisons have dominated neuroimaging studies for many years, leading to better insight into the patterns of regional vulnerability in brain neurodegenerative pathologies. However, such group-level results have no diagnostic value at the individual level. Recently, pattern recognition approaches using multivariate analyses have led to a fundamental shift in paradigm, aiming to predict the cognitive fate of each individual on the basis of MRI-based algorithms of structural parameters. We review here the state-of-the-art fundamentals of pattern recognition including feature selection, cross validation, and classification techniques and discuss limitations including interindividual variation in normal brain anatomy and neurocognitive reserve. We conclude with a special reference to future trends including multimodal pattern recognition and multicenter approaches with data sharing and cloud computing.

Key words MRI, Pattern recognition, Support vector machines (SVMs), Multivariate, Predictive modeling

1 Introduction

Most central nervous system (CNS) disorders cause systematic modifications in brain structure that can be imaged using MRI. These modifications are usually subtle in early stages of the disease and cannot be identified by visual inspection alone. Initially, whole-brain morphometry from structural MRI has been used to train models that can discriminate between healthy controls and patients, such as AD and frontotemporal dementia [1–4]. Later on, a series of structural and functional MRI parameters such as cortical thinning and arterial spin labeling have been proposed as possible predictors of cognitive decline in preclinical AD [5, 6]. While such group-level results are relevant from a research viewpoint, they do not automatically translate into diagnostic procedures at the individual level. In fact, many CNS disorders are characterized by

Robert Perneczky (ed.), *Biomarkers for Preclinical Alzheimer's Disease*, Neuromethods, vol. 137,
https://doi.org/10.1007/978-1-4939-7674-4_7,

diffuse rather than focal changes [7], increasing the number of MRI variables to consider in models of prediction. Taking into account the interindividual responsibility and number of structural imaging variables, univariate techniques are frequently insufficient in this respect. Tools from pattern recognition—commonly referred to as multivariate pattern analysis (MVPA)—have the ability to integrate information across multiple variables. Exploiting multivariate data structure can significantly improve sensitivity, in particular when only subtle alterations occur.

The interest of machine learning techniques in the processing of neuroimaging data has been first cited by Haxby and colleagues [8] where they explicitly recognized the distributed nature of activation patterns from fMRI in the visual cortex. Until then, most fMRI analyses were performed using mass univariate techniques (i.e., voxel by voxel), which did not exploit inter-voxel dependencies. While regional patterns can be identified by multivariate statistical techniques (e.g., partial least squares [9]), there was a growing interest to employ them using tools from machine learning in order to predict mental states [10, 11]. During the late 2000, several contributions demonstrated that it is possible to train data-driven models that can subsequently decode information from brain images including semantic meaning of words [12], emotional prosody pronounced by actors [13], or more recently visual imagery [14] and even dreams [15].

The following chapter will summarize the way to proceed focusing on the step-by-step approach that can be applied in structural MRI data.

2 Methods

2.1 Data Processing Pipeline

Conventional confirmatory analysis is based on a (predefined) generative model that is fitted to the data. Statistical hypothesis testing then provides forward inference on how well the model explains the data. Many group studies use such schemes to identify significant differences between populations. Pattern recognition tools reverse the direction of inference; i.e., the model is *constructed by the data* during a training phase in order to predict the explanatory variable. The model performance is then validated during a validation phase. The models depend on the type of classifier, but they are usually flexible. The task of a classifier refers to the predicted variable; e.g., healthy versus pre-MCI. The data available to the classifier, the definition of the task, and the performance of the classifiers are three essential ingredients for the correct interpretation of the results (see Fig. 1).

As the first step, feature extraction converts the raw data into the best possible form that maximizes the amount of information and minimizes the effect of confounders due to various sources of

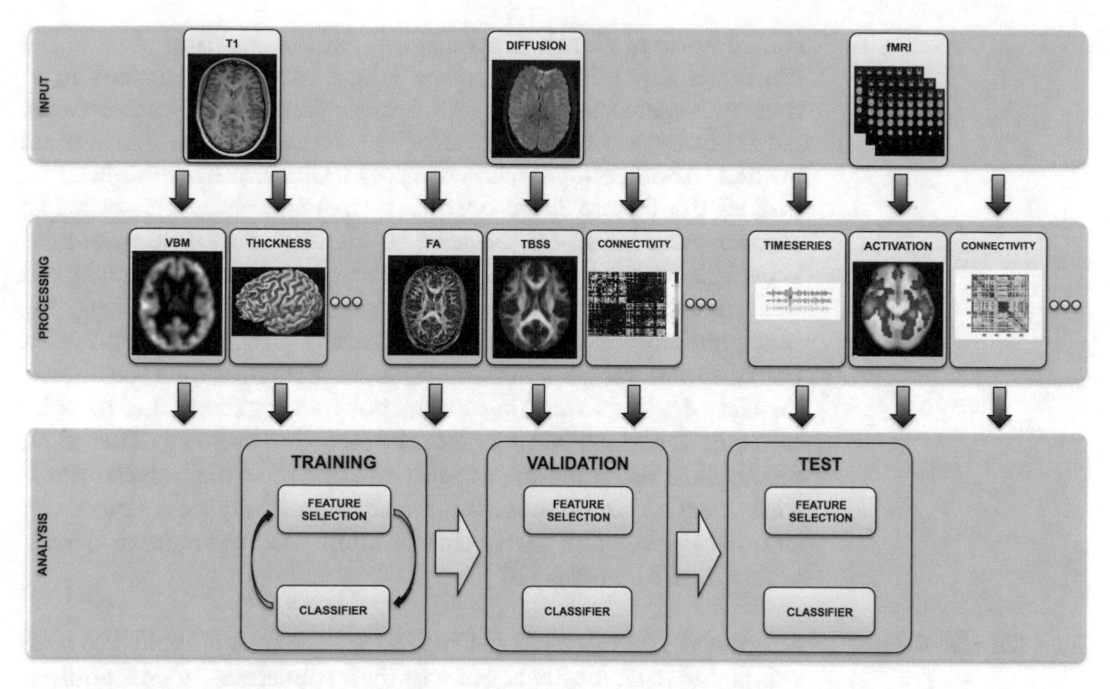

Fig. 1 Schematic illustration of the data processing pipeline. Several types of data input such as T1, diffusion-weighted imaging, or fMRI are in a first step preprocessed depending on the specific demands of spatial normalization into a standard space, including field inhomogeneity correction, spatial (and temporal) smoothing, atlasing, independent component decomposition, structural or functional connectivity analyses, and others. In the next step, the preprocessed data is used for feature selection and classification

noise. Structural features can be extracted from imaging data after a time-consuming procedure (e.g., spatial normalization to bring the data in the same "brain space"). For instance, a typical structural MRI dataset can easily contain more than 100,000 voxels. In most cases, the features are related (similarity of adjacent voxels or voxels in homologous regions), and only a limited number of brain regions (and consequently features) will carry true discriminative information. Other possibilities of features include those extracted from functional imaging (e.g., contrast of an experimental condition in a conventional activation study or functional connectivity from resting-state fMRI), diffusion-weighted imaging (e.g., measures of structural connectivity), and neuropsychological and clinical measures. These features can be combined to improve the performance of the models of prediction.

2.2 Feature Selection

In most cases, the number of features (e.g., voxels in the case of structural MRI) is much larger than the number of individuals. The selection of the best features is a long-standing problem in machine learning that can be addressed either explicitly (by a separate feature selection step) or implicitly (by regularization in the classification method). Conventional feature selection reduces the dimensionality of the feature space by ranking them according to their univariate statistical significance in case-control comparisons. An arbitrary number of features can be retained in multivariate models. Another approach is to apply multivariate techniques that project the feature space onto a (linear) subspace; e.g., principal component analysis can be used to identify those feature dimensions that explain most of the observed variance. Overall, these features are usually not optimal for prediction (i.e., since only the best predictive feature would be selected) [16]. For instance, it is better to select a whole-brain region as predictive for a given task, instead of a few voxels that might not be very stable due to noise or slight variations in brain morphometry or the data. Therefore, features are also sometimes transformed. The spatial wavelet transform leads to a representation that is more compact for piecewise smooth signals, a property that is often used to improve detectability in MRI studies [17].

2.3 Classification

Support vector machines (SVMs) have been widely applied to neuroimaging data, mostly because of their robustness against outliers, but they are by far not the only choice available [18]. Typically, SVMs are used for binary classification such as discriminating patients from healthy controls. Recently, more advanced classifiers have been developed (e.g., decision forests [19]) and should be considered for multi-class classification. However, there is some trade-off between increasing the complexity of the classifier and improving the feature extraction. Usually, good features allow for better classification with almost any classifier.

Another issue with classifiers arises when features are multimodal; i.e., combining different types of imaging data with other measures such as scores from neuropsychological tests or clinical parameters. As the model of a single classifier will not allow the necessary flexibility, "ensemble learning" is a rich field in machine learning that deals with combining multiple classifiers (and thus multiple models) to integrate the richness of the data structure. One promising approach in the future is the use of multilevel classifier algorithms to aggregate information in a hierarchical way (see Fig. 2). This approach makes it possible to take into account previously established knowledge about the relative weight of each classifier.

Finally it is also important to mention the recent breakthrough in artificial intelligence applied to neuroimaging classification: deep learning [20]. This technique has been successfully applied to multiple fields including visual recognition, mostly with full supervision. A typical deep-learning architecture for visual recognition builds upon convolutional neural network. Given large-scale training data and the power of high-performance computational infrastructure, deep learning has achieved tremendous improvement in visual recognition with thousands of categories [21].

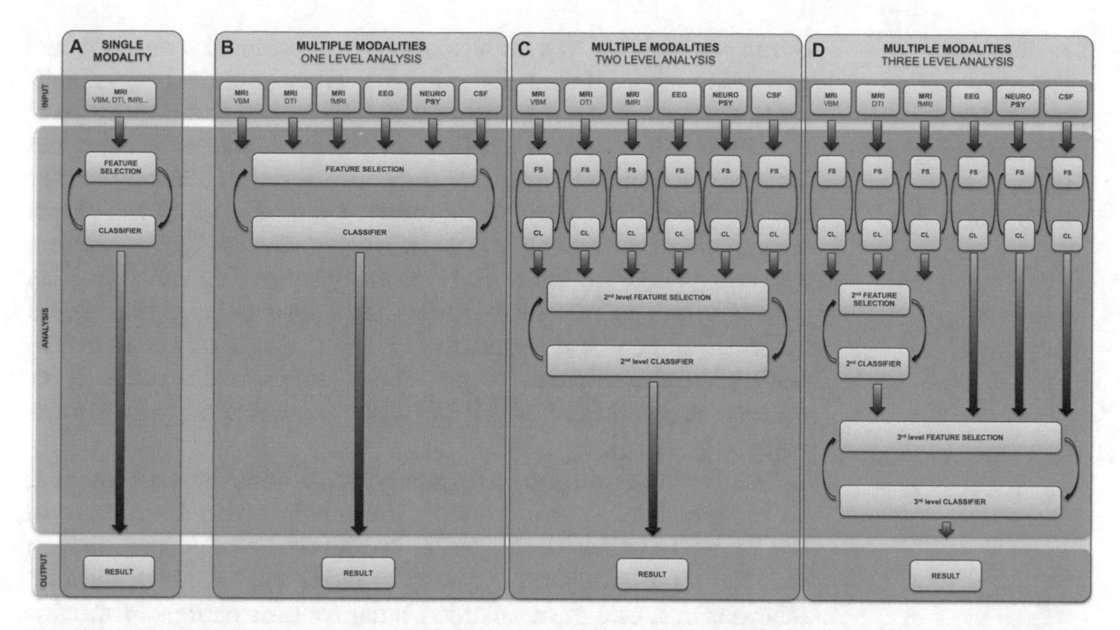

Fig. 2 Simplified processing pipeline for image classification. Processing pipeline for a single domain classification (**a**) from data input to feature selection/classification to output. In multimodal classification (**b–d**), all input data can directly enter one single feature selection/classification (**b**). Alternatively, it is possible to have one feature selection/classification per input data, followed by one single super-feature selection/classification at the second level (**c**). Moreover, it is possible to introduce an additional feature selection/classification levels, for example, regrouping all imaging modalities, followed by a super-feature selection/classification at the third level (**d**)

Its applications nowadays are moving toward feature representation for AD/MCI classification [22].

2.4 Cross Validation

A key step to consider in multimodal prediction is the distinction between training and validation of each classifier. This later phase concerns separate parts of the data according to a cross validation methodology. As already mentioned, the model of classification is built using part of the data where both the features and the predicted variable are given. Then, the classifier is evaluated on the remaining part of the data by comparing its outcome against the ground-truth prediction. The procedure is repeated after removing different parts of the data to produce an average performance across the validation folds (e.g., in terms of specificity and sensitivity). Ideally, the classifier should predict the main data structure and ignore the noise, and its model should be applicable to unseen data.

3 Challenges, Limitations, and Future Directions

We review here the fundamental challenges, potential pitfalls and limitations as well as future trends that deserve consideration, in particular with respect to clinical applications of pattern recognition analyses.

3.1 Normal Anatomical Interindividual Variability and Spatial Neighborhood Relationships

Some structural MRI variables display a substantial interindividual variability within the normal range. As an example, a 15% variability of cortical thickness has been reported for the first time 20 years ago in healthy controls [23]. At-risk mental state subjects could thus be discriminated from healthy controls only when considering the within-subject cortical asymmetry and not only the direct assessment of cortical thickness between subjects [24]. Moreover, adjacent voxels are more likely to carry similar information than distant voxels in non-related areas. It is also notable that spatial information can be integrated at several scales; e.g., across neighboring voxels, subareas of predefined anatomical structures, or even areas distributed over larger distances based on prior knowledge (e.g., [25]).

Current classification analyses typically consider each voxel as an independent feature ignoring the highly ordered structure of the brain. In other words, SVMs are invariant to permutation of the feature dimensions. To acknowledge spatial neighborhood relationships, one must include spatial transformations in the feature extraction step (e.g., spatial averaging or wavelet transforms as mentioned above). This could potentially improve classification accuracy and robustness.

3.2 Incorporating Prior Knowledge

There is a substantial body of evidences accumulated over the past 30 years on the structure/function relationship in a wide range of different diseases. This extensive prior knowledge is largely ignored in recent individual level pattern recognition analyses, although it might potentially improve classification accuracy and robustness.

Similar to the discussion about the anatomic structure of the brain above, it is likely beneficial to inject any available domain knowledge to improve the information content of the features despite the fact that feature selection and state-of-the-art classifiers are designed to deal with high-dimensional learning. From this viewpoint, well-documented case-control studies with a priori hypotheses are needed to guide and enrich the technical performance of multilevel classifiers. Purely data-driven methods could lead to biologically irrelevant observations in dementia prediction. In addition, the interpretation of the results could often become easier when features represent domain-relevant information. Modern methods have recently been proposed to exploit spatial structure; e.g., based on hierarchical clustering to regroup similar voxels and reinforce the robustness [26]. For some applications, features can also be extracted from specific regions of interest instead of the whole brain, or the "locality" of information can be probed by a "searchlight" approach where classification performance is reported in an information map for features extracted from local neighborhoods [27].

3.3 The Cognitive Reserve

Pattern recognition analyses at the individual level (as well as univariate group-level analyses) are based on the assumption that there is a direct relationship between brain pathology and symptomatology. In the example of cognitive decline, the assumption is that decreasing cognitive functions are paralleled by progressive brain atrophy. This assumption is, however, only rarely true. Due to sociodemographic factors such as education and social integration, some individuals are able to compensate clinically the structural MRI changes even for one or two decades. The example of AD is illustrative in this respect. The appearance of various biological changes (including hippocampal volume loss or tau-related cortical thinning) suggestive of AD pathology in controls has been extensively documented. They may precede clinically overt dementia by many years, or even decades, thus defining a temporally wide preclinical phase of the disease [28]. This interindividual variation in the neurocognitive reserve was described already in 1968 [29]. In the domain of pattern recognition, the same amount of structural brain alterations could be associated with clinically overt cognitive decline only if the neurocognitive reserve is exhausted. This interindividual variability in the neurocognitive reserve represents a fundamental limitation for pattern recognition approaches especially in a clinical setting.

3.4 Disbalance Between Number of Features and Number of Subjects: The Risk of Over-Fitting

As already mentioned, a typical feature set extracted from MRI data can easily contain more than 100,000 features, which clearly exceeds the number of individuals in single-center studies. In most cases, the features are related (similarity of adjacent or homologous voxels), and only a limited number will carry discriminative information.

While the cross validation methodology is essential to train and evaluate classifiers, it still has the risk of over-fitting the data as the parameters can be tuned. Consequently, several nested levels of cross validation should be used, as well as separate independent test datasets to guarantee the quality of the multivariate recognition patterns. Often the sample size for single-center studies is insufficient to estimate such real-world performance. In addition, parameter tuning for both feature selection and classifiers requires an additional inner cross validation loop that decreases the amount of available data for learning and increase the risk of over-fitting.

In summary, selecting and determining the importance of features is an essential processing step in classification analyses, yet identifying the related parameters such as optimal number of features or regularization tuning is not a trivial practice. This is still an active field of research in machine learning; e.g., stability selection, which refers to the consistency of features when subsampling the feature space, including applications to neuroimaging [30].

3.5 Variability Related to Patient Selection, Inter-Scanner Variability, and Data Preprocessing

Additional potentially confounding factors include scanner heterogeneity [31, 32], variability in data preprocessing, and patient selection.

4 Conclusions

4.1 Toward New Biomarkers

The simple classification of healthy subjects versus clinically overt AD or even MCI is of marginal importance for clinical practice. Modern AD research focuses on the prediction of cognitive decline both in MCI and preclinical AD cases. The identification of predictive biomarkers in AD research is one of the more promised and challenging domains in human neurobiology. Although there is a wide agreement about the need to identify in situ AD at the preclinical stages and if possible long before the emergence of MCI, passing from group differences to individual prediction is quite difficult in the light of the impressive interindividual variability of most biomarkers in cognitively intact cases that imply the need for including a substantial number of cases in research cohorts, the uncertainty regarding the cognitive fate of these cases, the difficulty to propose curative treatments (at the level of primary prevention) in the absence of any alert sign.

The definitions of MCI have substantially evolved and changed over the past years, which goes beyond the scope of this review. Depending on the MCI subtype, only about half of MCI subjects will progress to clinically overt dementia, whereas the other half may remain stable or evolve to other forms of dementia [33–35]. Assuming that only about half of unselected MCI individuals will progress to clinically overt dementia, the *prediction* of individuals at risk for consecutive cognitive decline is of paramount importance for early individual treatment as well as for clinical trials. In a typical placebo-controlled pharmaceutical trial, 25% of unselected MCI will remain stable despite being in the placebo group, while only 25% of individuals will progress and obtain the active compound. Therefore, *preselection* of at-risk individuals for future cognitive decline would substantially improve the design of clinical trials. As MRI is routinely performed in the clinical workup of cognitive decline, advanced data analysis techniques as those by pattern recognition tools make use of already existing data, which is thus cost effective and without additional discomfort for the patient. It is possible to *predict* future cognitive decline in MCI using baseline MRI based on gray matter voxel-based morphometry (VBM) [1, 36, 37], white matter DTI [38, 39], or iron deposition [38]. It is further possible to combine diffusion tensor imaging (DTI) and resting-state fMRI to identify MCI individuals [40] and to predict MCI to AD conversion using multimodal measures also in combination with neuropsychological scores or cerebrospinal fluid biomarkers [26, 41] or by the combination of structural MRI and FDG-PET [42–44]. Furthermore, it is possible to classify MCI subtypes, who have different risk of disease progression and who might benefit from different types of treatment based on DTI [45]. One should take into account that while Alzheimer disease lesions and vascular pathology are commonly co-occurring, they are generally considered to be separate phenomena [46]. Recently, neuroimaging findings have increasingly shown a possible interaction between pathophysiological mechanisms in AD and vascular dementia [47–49]. Along this line of research, the detection on MR images of white matter hyperintensities, which reflects vascular damage, can be used to predict future cognitive decline and diagnosis of AD [50] and represents an early and independent predictor of AD risk [46].

In summary, the application of techniques from the field of pattern recognition to neuroimaging data is an emerging field. These methods have a number of attractive features, including the use of multivariate information and the possibility to predict for previously unseen data. Ongoing research is still needed to overcome a number of limitations, including optimal feature selection that incorporates better domain knowledge and integration of multimodal measurements. In addition, future methodological developments should be increasingly based on large datasets and multicentric studies to increase both reproducibility and predictability. Recent data sharing initiatives such as ADNI [51], in combination with cloud-computing power, will provide the necessary prerequisites for these developments.

References

1. Fan Y, Batmanghelich N, Clark CM et al (2008) Spatial patterns of brain atrophy in MCI patients, identified via high-dimensional pattern classification, predict subsequent cognitive decline. Neuroimage 39:1731–1743. https://doi.org/10.1016/j.neuroimage.2007.10.031
2. Fan Y, Resnick SM, Wu X, Davatzikos C (2008) Structural and functional biomarkers of prodromal Alzheimer's disease: a high-dimensional pattern classification study. Neuroimage 41:277–285. https://doi.org/10.1016/j.neuroimage.2008.02.043
3. Klöppel S, Stonnington CM, Chu C et al (2008) Automatic classification of MR scans in Alzheimer's disease. Brain J Neurol 131:681–689. https://doi.org/10.1093/brain/awm319
4. Davatzikos C, Resnick SM, Wu X et al (2008) Individual patient diagnosis of AD and FTD via high-dimensional pattern classification of MRI. Neuroimage 41:1220–1227. https://doi.org/10.1016/j.neuroimage.2008.03.050
5. Xekardaki A, Rodriguez C, Montandon M-L et al (2015) Arterial spin labeling may contribute to the prediction of cognitive deterioration in healthy elderly individuals. Radiology 274:490–499. https://doi.org/10.1148/radiol.14140680
6. Wang L, Benzinger TL, Hassenstab J et al (2015) Spatially distinct atrophy is linked to β-amyloid and tau in preclinical Alzheimer disease. Neurology 84:1254–1260. https://doi.org/10.1212/WNL.0000000000001401
7. Seeley WW, Crawford RK, Zhou J et al (2009) Neurodegenerative diseases target large-scale human brain networks. Neuron 62:42–52. https://doi.org/10.1016/j.neuron.2009.03.024

8. Haxby JV, Gobbini MI, Furey ML et al (2001) Distributed and overlapping representations of faces and objects in ventral temporal cortex. Science 293:2425–2430. https://doi.org/10.1126/science.1063736
9. Krishnan A, Williams LJ, McIntosh AR, Abdi H (2011) Partial Least Squares (PLS) methods for neuroimaging: a tutorial and review. Neuroimage 56:455–475. https://doi.org/10.1016/j.neuroimage.2010.07.034
10. Cox DD, Savoy RL (2003) Functional magnetic resonance imaging (fMRI) "brain reading": detecting and classifying distributed patterns of fMRI activity in human visual cortex. Neuroimage 19:261–270. https://doi.org/10.1016/S1053-8119(03)00049-1
11. Pereira F, Mitchell T, Botvinick M (2009) Machine learning classifiers and fMRI: a tutorial overview. Neuroimage 45:S199–S209. https://doi.org/10.1016/j.neuroimage.2008.11.007
12. Mitchell TM, Shinkareva SV, Carlson A et al (2008) Predicting human brain activity associated with the meanings of nouns. Science 320:1191–1195. https://doi.org/10.1126/science.1152876
13. Ethofer T, Van De Ville D, Scherer K, Vuilleumier P (2009) Decoding of emotional information in voice-sensitive cortices. Curr Biol 19:1028–1033. https://doi.org/10.1016/j.cub.2009.04.054
14. Nishimoto S, Vu AT, Naselaris T et al (2011) Reconstructing visual experiences from brain activity evoked by natural movies. Curr Biol 21:1641–1646. https://doi.org/10.1016/j.cub.2011.08.031
15. Horikawa T, Tamaki M, Miyawaki Y, Kamitani Y (2013) Neural decoding of visual imagery during sleep. Science 340:639–642. https://doi.org/10.1126/science.1234330
16. Tolosi L, Lengauer T (2011) Classification with correlated features: unreliability of feature ranking and solutions. Bioinformatics 27:1986–1994. https://doi.org/10.1093/bioinformatics/btr300
17. Van De Ville D, Blu T, Unser M (2006) Surfing the brain. IEEE Eng Med Biol Mag Q Mag Eng Med Biol Soc 25:65–78
18. Cristianini N, Shawe-Taylor J (2000) An introduction to support vector machines: and other kernel-based learning methods. Cambridge University Press, New York
19. Criminisi A, Shotton J (2013) Decision forests for computer vision and medical image analysis. Springer Publishing Company, Incorporated, Berlin
20. Wu J, Yu Y, Huang C, Yu K (2015) Deep multiple instance learning for image classification and auto-annotation. IEEE, pp 3460–3469
21. van der Burgh HK, Schmidt R, Westeneng H-J et al (2017) Deep learning predictions of survival based on MRI in amyotrophic lateral sclerosis. Neuroimage Clin 13:361–369. https://doi.org/10.1016/j.nicl.2016.10.008
22. Suk H-I, Shen D (2013) Deep learning-based feature representation for AD/MCI classification. Med Image Comput Comput Assist Interv 16:583–590
23. Haug H (1987) Brain sizes, surfaces, and neuronal sizes of the cortex cerebri: a stereological investigation of man and his variability and a comparison with some mammals (primates, whales, marsupials, insectivores, and one elephant). Am J Anat 180:126–142. https://doi.org/10.1002/aja.1001800203
24. Haller S, Borgwardt SJ, Schindler C et al (2009) Can cortical thickness asymmetry analysis contribute to detection of at-risk mental state and first-episode psychosis? A pilot study. Radiology 250:212–221. https://doi.org/10.1148/radiol.2501072153
25. Hackmack K, Weygandt M, Wuerfel J et al (2012) Can we overcome the "clinico-radiological paradox" in multiple sclerosis? J Neurol 259:2151–2160. https://doi.org/10.1007/s00415-012-6475-9
26. Cui Y, Sachdev PS, Lipnicki DM et al (2012) Predicting the development of mild cognitive impairment: a new use of pattern recognition. Neuroimage 60:894–901. https://doi.org/10.1016/j.neuroimage.2012.01.084
27. Kriegeskorte N, Goebel R, Bandettini P (2006) Information-based functional brain mapping. Proc Natl Acad Sci U S A 103:3863–3868. https://doi.org/10.1073/pnas.0600244103
28. Lazarczyk MJ, Hof PR, Bouras C, Giannakopoulos P (2012) Preclinical Alzheimer disease: identification of cases at risk among cognitively intact older individuals. BMC Med 10:127. https://doi.org/10.1186/1741-7015-10-127
29. Tomlinson BE, Blessed G, Roth M (1968) Observations on the brains of non-demented old people. J Neurol Sci 7:331–356
30. Langs G, Menze BH, Lashkari D, Golland P (2011) Detecting stable distributed patterns of brain activation using Gini contrast. Neuroimage 56:497–507. https://doi.org/10.1016/j.neuroimage.2010.07.074
31. Kruggel F, Turner J, Muftuler LT, Alzheimer's Disease Neuroimaging Initiative (2010) Impact of scanner hardware and imaging protocol on image quality and compartment

volume precision in the ADNI cohort. Neuroimage 49:2123–2133. https://doi.org/10.1016/j.neuroimage.2009.11.006

32. Abdulkadir A, Mortamet B, Vemuri P et al (2011) Effects of hardware heterogeneity on the performance of SVM Alzheimer's disease classifier. Neuroimage 58:785–792. https://doi.org/10.1016/j.neuroimage.2011.06.029
33. Petersen RC (2004) Mild cognitive impairment as a diagnostic entity. J Intern Med 256:183–194. https://doi.org/10.1111/j.1365-2796.2004.01388.x
34. Mariani E, Monastero R, Mecocci P (2007) Mild cognitive impairment: a systematic review. J Alzheimers Dis 12:23–35
35. Forlenza OV, Diniz BS, Nunes PV et al (2009) Diagnostic transitions in mild cognitive impairment subtypes. Int Psychogeriatr 21:1088–1095. https://doi.org/10.1017/S1041610209990792
36. Misra C, Fan Y, Davatzikos C (2009) Baseline and longitudinal patterns of brain atrophy in MCI patients, and their use in prediction of short-term conversion to AD: results from ADNI. Neuroimage 44:1415–1422. https://doi.org/10.1016/j.neuroimage.2008.10.031
37. Plant C, Teipel SJ, Oswald A et al (2010) Automated detection of brain atrophy patterns based on MRI for the prediction of Alzheimer's disease. Neuroimage 50:162–174. https://doi.org/10.1016/j.neuroimage.2009.11.046
38. Haller S, Bartsch A, Nguyen D et al (2010) Cerebral microhemorrhage and iron deposition in mild cognitive impairment: susceptibility-weighted MR imaging assessment. Radiology 257:764–773. https://doi.org/10.1148/radiol.10100612
39. O'Dwyer L, Lamberton F, Bokde ALW et al (2012) Using support vector machines with multiple indices of diffusion for automated classification of mild cognitive impairment. PLoS One 7:e32441. https://doi.org/10.1371/journal.pone.0032441
40. Wee C-Y, Yap P-T, Zhang D et al (2012) Identification of MCI individuals using structural and functional connectivity networks. Neuroimage 59:2045–2056. https://doi.org/10.1016/j.neuroimage.2011.10.015
41. Cui Y, Liu B, Luo S et al (2011) Identification of conversion from mild cognitive impairment to Alzheimer's disease using multivariate predictors. PLoS One 6:e21896. https://doi.org/10.1371/journal.pone.0021896
42. Zhang D, Wang Y, Zhou L et al (2011) Multimodal classification of Alzheimer's disease and mild cognitive impairment. Neuroimage 55:856–867. https://doi.org/10.1016/j.neuroimage.2011.01.008
43. Zhang D, Shen D, Alzheimer's Disease Neuroimaging Initiative (2012) Multi-modal multi-task learning for joint prediction of multiple regression and classification variables in Alzheimer's disease. Neuroimage 59:895–907. https://doi.org/10.1016/j.neuroimage.2011.09.069
44. Zhang D, Shen D, Initiative ADN (2012) Predicting future clinical changes of MCI patients using longitudinal and multimodal biomarkers. PLoS One 7:e33182. https://doi.org/10.1371/journal.pone.0033182
45. Haller S, Missonnier P, Herrmann FR et al (2013) Individual classification of mild cognitive impairment subtypes by support vector machine analysis of white matter DTI. AJNR Am J Neuroradiol 34:283–291. https://doi.org/10.3174/ajnr.A3223
46. Haller S, Barkhof F (2017) Interaction of vascular damage and Alzheimer dementia: focal damage and disconnection. Radiology 282:311–313. https://doi.org/10.1148/radiol.2016161564
47. Saito S, Ihara M (2016) Interaction between cerebrovascular disease and Alzheimer pathology. Curr Opin Psychiatry 29:168–173. https://doi.org/10.1097/YCO.0000000000000239
48. Kapasi A, Schneider JA (2016) Vascular contributions to cognitive impairment, clinical Alzheimer's disease, and dementia in older persons. Biochim Biophys Acta 1862:878–886. https://doi.org/10.1016/j.bbadis.2015.12.023
49. Madigan JB, Wilcock DM, Hainsworth AH (2016) Vascular contributions to cognitive impairment and dementia: topical review of animal models. Stroke 47:1953–1959. https://doi.org/10.1161/STROKEAHA.116.012066
50. Brickman AM (2013) Contemplating Alzheimer's disease and the contribution of white matter hyperintensities. Curr Neurol Neurosci Rep 13:415. https://doi.org/10.1007/s11910-013-0415-7
51. Mueller SG, Weiner MW, Thal LJ et al (2005) Ways toward an early diagnosis in Alzheimer's disease: the Alzheimer's Disease Neuroimaging Initiative (ADNI). Alzheimers Dement 1:55–66. https://doi.org/10.1016/j.jalz.2005.06.003

Chapter 8

Brain Functional Imaging in Preclinical Alzheimer's Disease

Peter Häussermann, Thorsten Bartsch, and Oliver Granert

Abstract

In this chapter, we summarize the most important methods and techniques of functional brain imaging (FBI) in the context of biomarkers in preclinical Alzheimer's disease (pAD). We analyze putative models of pAD, covering both genetic and clinical aspects of pAD. Studies on cognitively healthy individuals carrying the APOE-ε4 allele and patients with mild cognitive impairment have been included. To address methodologies, studies on metabolism and cerebral perfusion are analyzed. Although perfusion is not part of the criteria for pAD, there is increasing evidence that perfusion imaging techniques, including PET/SPECT methods, and magnetic resonance imaging (MRI) techniques, such as functional MRI (fMRI) and arterial spin labeling (ASL), can be useful for elucidating the underlying pathophysiological changes in pAD. However, metabolic imaging with glucose PET still remains the most established FBI biomarker of pAD.

Key words Functional brain imaging, Neurodegeneration, Age-associated cognitive decline, Subjective memory dysfunction, Mild cognitive impairment, Alzheimer's disease, PET, MRI, Default mode network

1 Introduction

1.1 Definition and Potential Models of Preclinical AD

Actually, there is only sparse knowledge about changes occurring in the brain during the years preceding the cognitive and functional impairment associated with Alzheimer's disease (AD). Normal cognitive status in the elderly (i.e., incidental forgetfulness of aging) can be described as aging-associated cognitive decline without relevant memory complaints.

In 2011, the National Institute on Aging-Alzheimer's Association (NIA-AA) published criteria to stage preclinical Alzheimer's disease (pAD), mainly based on the amyloid cascade hypothesis of AD. These criteria [1] define three consecutive stages:

- Stage 1: Asymptomatic cerebral amyloidosis
- Stage 2: Cerebral amyloid deposition + downstream neurodegeneration with pathological CSF/PET or MRI markers

Robert Perneczky (ed.), *Biomarkers for Preclinical Alzheimer's Disease*, Neuromethods, vol. 137,
https://doi.org/10.1007/978-1-4939-7674-4_8, © Springer Science+Business Media, LLC 2018

(tau/phospho-tau; ^{18}F-FDG-PET, atrophy as visualized by structural MRI)

- Stage 3: Cerebral amyloidosis + neurodegeneration + subtle cognitive or behavioral decline, not meeting the criteria for MCI

Clinically, elderly patients reporting cognitive problems have recently been classified as suffering from subjective cognitive decline (SCD). SCD is defined as a subjectively experienced persistent continuous cognitive decline, or decline in cognitive capacity, compared with a previously normal state, not related to an acute event. Performance on standard tests used to define mild cognitive impairment (MCI) is normal. Patients with SCD expressing concern about these deficits have a threefold increased risk of later developing dementia and a sixfold increased risk of later developing AD. In models of the very early stages of AD, SCD can be considered to occur in late stages of pAD, characterized by increasing compensatory cognitive efforts and subtle cognitive changes before the threshold of MCI or prodromal AD is reached [2].

There are several human models of pAD. The clinically most important human model is the mild cognitive impairment (MCI) stage of cognitive decline in the elderly. In MCI, patients are partially symptomatic with substantiated cognitive impairment that does not significantly interfere with normal daily functioning. The amnestic MCI subgroup (aMCI) has the highest risk to develop AD in the future [3]. The annual conversion rate from MCI to dementia is approximately between 5–25%, depending on MCI criteria and subtype [3, 4]. MCI can, therefore, be considered as a transitional period between normal aging and dementia.

Another approach to better understand presymptomatic AD is to study cognitively healthy individuals with known genetic risk factors. This includes nondemented individuals who are at risk of later developing AD, such as subjects with recognized genetic traits. Individuals carrying the ε4 allele of the apolipoprotein E (APOE-ε4) gene have an increased risk of later developing AD. The APOE gene is thought to play a role in cerebrovascular lesions in dementia [5] and is considered to be the single most important genetic factor in late onset AD [6]. Carriage of the APOE-ε4 allele significantly increases AD risk in a gene-dose-dependent manner and lowers the age of AD onset [7, 8].

1.2 Biomarkers and Functional Imaging in Preclinical AD

Generally, a biomarker is a characteristic that is objectively measured and evaluated as an indicator of pathologic processes [9]. Effective biomarkers in AD depend on the knowledge of the underlying pathophysiology. The most prominent etiological models of AD are related to cerebral Aß deposition and tau-hyperphosphorylation, whereas cerebrovascular changes and perfusional abnormalities are increasingly being recognized as important cofactors in the pathogenesis of AD [10, 11]. To date, these models only partially mirror underlying pathophysiological changes in AD; however, the cause of AD still remains unclear.

In AD, an ideal biomarker detects important features of AD pathophysiology and neuropathology, is neuropathologically validated, and shows a high diagnostic sensitivity and specificity. Furthermore, a good biomarker should also be trustworthy, consistent, noninvasive, easy to perform, and inexpensive [12].

Currently deployed imaging biomarkers of preclinical AD (pAD) include those related to amyloid, tau, cerebral (glucose) metabolism, and brain atrophy. Alterations of cerebral blood flow (CBF) have been studied more extensively in the recent years in pAD and its models. However, perfusion changes are not (yet) included in the definition of pAD [1].

Functional imaging is currently used to explore the sequence of pathophysiological changes occurring along the continuum from normal aging to AD. Functional brain imaging (FBI) biomarkers of pAD are mainly based on imaging techniques that explore the metabolic changes of the aging brain. In our review, we have also included FBI methods exploring CBF changes as there is evidence that CBF alterations occur early in the course of pAD, even before amyloid deposition becomes detectable.

According to a vascular AD model, neurodegeneration primarily results from cerebral hypoperfusion, leading to glial and neuronal damage and producing Aß and tau deposition with cognitive decline and later AD [11]. There is recent evidence supporting an inverse relationship between cerebral perfusion and AD progression; therefore, CBF may represent a valuable biomarker of pAD [10, 13, 14]. A recent study linked the amyloid cascade and the vascular hypotheses of AD. Using two different mouse models of AD, Maier et al. [15] showed that only in the presence of cerebral amyloid angiopathy (CAA), Aß deposition is accompanied by a decline of regional CBF. Hypoperfusion as seen in AD may, therefore, be associated with CAA.

FBI methods can be subdivided into two methodological approaches, electrophysiological techniques and hemodynamic methods. Electrophysiological techniques, such as electroencephalography (EEG), event-related potentials (ERP), and magnetencephalography (MEG), directly measure neuronal activity (i.e., action potentials) as the physiological correlates of neuronal activity.

Positron emission tomography (PET), single photon emission computed tomography (SPECT), and different magnetic resonance imaging (MRI) techniques measure secondary effects of neuronal activation (e.g., changes in CBF linked to neuronal metabolism). Recently, optical imaging (OI) and near-infrared spectroscopy (NIRS) have been added to the list of functional imaging tools in neurodegenerative disorders. Furthermore, other FBI techniques, such as MR spectroscopy, can detect the regional chemical composition of the brain. MR spectroscopy, optical imaging, and NIRS are not routinely used, neither in the clinical nor in the research setting. Therefore, these techniques are not described in this chapter.

Electrophysiological tools have a high time resolution with a relatively low spatial resolution. Hemodynamic tools have a good spatial resolution with lower time resolution. Recent developments show promising results combining electrophysiological and hemodynamic techniques (e.g., EEG and fMRI).

In this chapter, we will focus on metabolic as well as hemodynamic functional imaging methods as they have the potential to mirror underlying pathophysiological changes in pAD [10, 16, 17].

2 Functional Brain Imaging Methods in Preclinical Alzheimer's Disease

2.1 Spectrum of Techniques and Methods

2.1.1 SPECT

Single photon emission computed tomography (SPECT) is a nuclear medicine FBI technique using gamma rays. SPECT uses a gamma-emitting radioisotope attached to a specific ligand to create a radioligand, which then binds to cerebral structures. This radioligand can be visualized by a gamma camera. SPECT has a lower spatial resolution (about 10 mm) than positron emission tomography (PET). In FBI, ^{99m}Tc-HMPAO (technetium-99m-hexamethyl propylene amine oxime) is usually used to measure CBF, which is coupled to regional brain metabolism and energy use. As compared to ^{18}F-FDG-PET, ^{99m}Tc-HMPAO-SPECT is relatively cheap and simple to generate. Therefore, ^{99m}Tc-HMPAO-SPECT is more widely available in the hospital setting [18]. From a technical point of view, correction for partial volume effects using high-resolution MRI is recommended to control for effects of brain atrophy on PET/SPECT (perfusion) imaging [19].

2.1.2 PET

Positron emission tomography (PET) is an in vivo molecular imaging method detecting gamma rays emitted by a positron-emitting tracer. An injected radioisotope undergoes positron emission decay, emitting a positron that interacts with an electron to produce gamma photons detected by the PET scanner. For brain metabolic imaging, ^{18}F-FDG is by far the most widely used tracer, visualizing and quantifying mainly the presynaptic regional cerebral metabolic rate of glucose uptake (rCMRglc) with a half-life of 110 min. ^{18}F-FDG-PET represents a proxy for neuronal activity and constitutes an established biomarker for neurodegeneration. There is a slight but significant decline in rCMRglc in many cortical regions as a function of age [20]. CBF can be assessed by PET using the tracer oxygen-15 which has a short half-life (about 2 min). Several PET tracers are ligands for specific neuroreceptor subtypes such as the D2/D3 receptors, opioid receptors, serotonin receptors and transporters, nicotinic acetylcholine receptors, as well as enzyme substrates. PET imaging can also be used to visualize protein aggregates (e.g., tau and amyloid) within the brain [21].

2.1.3 Functional MRI (fMRI)

Functional magnetic resonance imaging (fMRI) is a FBI method that uses the phenomenon of nuclear magnetic resonance to indirectly measure and map brain function (not anatomy) by detecting cerebral blood flow (CBF) changes following alterations in neural activity. fMRI is based on the so-called blood-oxygen-level-dependent (BOLD) contrast. This method quantifies circumscribed changes in deoxyhemoglobin level, which exhibits an interwoven dependency on CBF, cerebral metabolic rate of O_2, as well as on cerebral blood volume [22]. To understand the fMRI signal, two effects need to be considered. First, there is a difference in magnetic properties between oxygen-rich arterial (oxygenated) and oxygen-poor venous (deoxygenated) blood. Second, neural activity paradoxically leads to a much greater increase in CBF than oxygen consumption, which in turn leads to the blood being more oxygenated with increasing neural activity. An increase in neural activity in certain brain areas thereby leads to a small but detectable change in the magnetic resonance signal. The fMRI signal, therefore, depends on physiological changes associated with CBF and oxygen metabolism [22].

fMRI is widely available and relatively cheap and does not use ionizing radiation, being able to localize cerebral activity in a millimeter range. fMRI can be combined with electrophysiological techniques such as EEG and transcranial magnetic stimulation [22].

fMRI acquisition is usually coupled to a carefully designed experiment, carried out by the subject during imaging. Experimental paradigms trigger an activation of functional brain networks or regions coupled. Limitations concern the reliance on cooperation of the patient; children and dementia patients often cannot perform complex fMRI tasks. These experiments are time-consuming and require a complex experimental setup with MRI-compatible equipment. In contrast to the very controlled task-based fMRI experiments, the so-called resting-state fMRI (rs-fMRI) is much easier to perform and independent from a specific task. By analyzing correlations in the BOLD signal over time between different regions or voxels, rs-fMRI tries to assess the functional connectivity (fc) patterns of the brain during rest. The term *functional connectivity* is used to describe the specific connectivity between brain areas sharing functional properties (i.e., the temporal correlation between spatially remote neurophysiological events in distributed neuronal networks). Slow fluctuations (<0.08 Hz) of the BOLD signal with a high degree of temporal correlation between functionally connected brain regions during rest and intrinsic processing represent the neuronal baseline activity. These fluctuations correspond to inter- and intraindividually stable resting-state networks (RSNs). Resting-state fc analyses measure task-independent changes of brain function, thereby providing important insights into brain plasticity [23].

Several RSNs have been described, linking regions with functional dependencies. RSNs include brain areas involved in sensory and motor control, visual processing, executive functioning, auditory processing, memory, and the default mode network (DMN). Using rs-fMRI, brain activity in patients (e.g., children and patients suffering from AD) that would not be able to complete long experiments or perform complex fMRI tasks can be examined. Unlike task-based imaging, typically highlighting one single brain network associated with the specific paradigm, rs-fMRI allows the observation of several networks simultaneously. The simplicity and short duration (5–10 min as compared to usually 30 min or more in task-based fMRI studies) make rs-fMRI an attractive tool in the research setting [22–26].

The default mode network (DMN) represents an association of brain regions active during daydreaming, retrieving memories or planning for the future. Therefore, the DMN represents the resting state of our brain, showing a direct correspondence to structural connections. Neuroanatomically, the posterior cingulate cortex (PCC), inferior parietal, medial temporal lobe (MTL), and medial frontal gyri belong to the DMN, exhibiting a high level of correlated BOLD signal activity. These spatially separated but functionally coupled brain regions interact even if the brain is not engaged in focused mental activity. Neurophysiologically, there are slow, synchronized oscillations within the DMN, which are quite robust and even persist during sleep [27]. The DMN and other RSNs can be visualized by various methods, including fMRI, ^{18}F-FDG-PET, and EEG, but rs-fMRI provides the best spatial imaging resolution and does not utilize radioactivity. There is some evidence for altered processing of the DMN in AD [28].

2.1.4 Arterial Spin Labeling: MRI

Arterial spin labeling (ASL) is a noninvasive MRI technique, magnetically labeling arterial water and using it as an endogenous tracer to measure CBF changes associated with neuronal activity [10, 29]. The BOLD signal, which is the usual basis for the fMRI signal, is linked to a signal change between two or more conditions. In fMRI, there is no information about the absolute level of blood flow. In contrast to this, the ASL technique has the ability to quantify the absolute level of CBF by subtracting the signal of magnetically tagged and untagged arterial blood. Thereby, the static signal from all hydrogen nuclei subtracts out, ideally leaving only the signal arising from arterial blood flow [29, 30]. The ASL signal is therefore independent from changes in neuronal activity. However, the disadvantage is an increase in acquisition time, since the ASL technique requires the acquisition of two images (one with the labeled and one with non-labeled blood). ASL MRI has the potential to precisely estimate both extent and site of neural function [29]. Changes in CBF are more localized to the brain parenchyma (capillary bed), whereas the BOLD signal is more localized to cerebral veins and venules [29, 30]. As ASL MRI is a noninvasive, easily repeatable method with good reliability and reproducibility [29, 31], it is currently one of the most promising functional imaging biomarkers of pAD.

2.2 Imaging of Metabolism in Preclinical AD

2.2.1 ^{18}F-FDG-PET

Besson et al. [32] characterized a group of cognitively healthy subjects by structural MRI, ^{18}F-FDG-PET, and amyloid PET (florbetapir) and dichotomized these individuals into positive or negative for each neuroimaging marker. The ^{18}F-FDG-positive group showed an AD typical hypometabolic pattern. Furthermore, in the "neurodegeneration" group with both structural MRI and FDG-PET AD-like changes, the authors found a trend for an inverse relationship with Aß deposition (i.e., subjects with neurodegeneration exhibited less Aß and vice versa). Therefore, Besson et al. suggested to combine structural MRI and amyloid/glucose PET biomarkers. These methods offer unique insight into the time course of Aß deposition and neurodegeneration in pAD. Furthermore, neurodegenerative changes seen in this study were largely independent from cerebral Aß deposition.

Teipel et al. [33] analyzed hippocampal and posterior cingulate cortex (PCC) volume, PCC glucose metabolism, and Aß load in a cohort of 667 subjects participating in the Alzheimer's Disease Neuroimaging Initiative (ADNI). Controls and early MCI patients exhibited hypometabolism in PCC which was associated with hippocampal atrophy. In late stages of MCI, hypometabolism in PCC was associated both with PCC and hippocampal atrophy as well as PCC Aß deposition. In patients suffering from AD, PCC hypometabolism was associated with PCC atrophy.

In another study, Ewers et al. [34] showed that temporoparietal and prefrontal hypometabolism as well as decline in executive function predicted conversion of cognitively healthy elderly to MCI or dementia during a follow-up period of 3–4 years. Best predicting results for conversion to MCI/dementia were obtained when combining executive function assessment with rCMRglc in the above-mentioned areas [34]. The same group examined the effects of cognitive reserve capacity (CRC) on glucose metabolism in cognitively healthy individuals with (Aß+) and without (Aß−) CSF biomarkers of preclinical AD [35]. When using a ROI approach, higher CRC was associated with lower glucose metabolism in the PCC and angular gyrus in the Aß+ group, while higher CRC was associated with PCC hypermetabolism in the Aß− group. CRC may, therefore, play an important compensatory role in sustaining cognition in the presence of cerebral Aß deposition and altered glucose metabolism.

Another study [36] analyzed the longitudinal changes of Aß deposition, glucose metabolism, and medial temporal lobe (MTL) atrophy in three different cohorts (pAD, MCI and dementia). Patients with initial pAD stages II and III showed greater MTL volume loss and hypometabolism during follow-up than the stage I pAD group. Apparently, in this study, higher rates of MTL neurodegenerative changes occurred during follow-up in individuals with pAD who initially exhibited Aß and neurodegeneration biomarkers, as objectified with hypometabolism and MTL atrophy.

In another longitudinal study, Wirth et al. [37] showed accelerated cognitive decline in pAD with both positive Aß and neurodegeneration imaging markers in AD vulnerable regions (i.e., temporoparietal cortices and hippocampus) as compared to healthy controls. In this study, non-memory decline was best predicted by Aß and temporoparietal neurodegenerative alterations, while memory decline was best predicted by Aß and hippocampal neurodegenerative changes. These results suggest regional specificity of neurodegeneration in pAD. Ossenkoppele et al. [38] showed that initial glucose hypometabolism and amyloid deposition were related to longitudinal worsening in memory, attention and executive function in pAD, MCI and AD over a mean follow-up period of 2.2 years. Gray matter volume loss was mainly associated with longitudinal memory decline.

Knopman et al. [39] applied imaging biomarkers of neurodegeneration and Aß in three groups: one containing subjects with different stages of pAD, one with subjects that had negative imaging biomarkers, and one with subjects with positive biomarkers for neurodegeneration without Aß deposition (sNAP, suspected non-Alzheimer pathophysiology). The sNAP group did not differ from stages II and III pAD concerning clinical symptoms of parkinsonism, glucose metabolism, MR volumetry, cerebrovascular risk factors, or ischemic brain lesions, indicating that neurodegenerative changes are largely unrelated to Aß deposition in cognitively healthy individuals.

2.2.2 ^{18}F-FDG-PET in APOE-ε4 Carriers

Seo et al. [40] analyzed APOE-ε4, rCMRglc, and cerebral amyloid load (^{18}F-florbetapir PET) in cognitively healthy controls, MCI patients, and AD patients. In controls, carriage of the APOE-ε4 allele was associated with hypometabolism in the bilateral frontal, temporal, and left parietal regions. In MCI, the same comparison showed hypometabolism in the bilateral posterior parietal, temporal, and left frontal regions. In AD, APOE-ε4 carriers exhibited hypometabolism in the left hippocampus, right insular, and right temporal gyrus. After adjustment for Aß, the significant differences within temporal areas disappeared in controls and MCI, whereas in AD, all significant metabolic differences disappeared. The authors deducted that Aß-independent APOE-ε4 influence on glucose metabolism is restricted to frontal and parietal brain regions, mainly in the early stages of cognitive decline. The same group [41] also showed that temporoparietal hypometabolism disappears when controlling for Aß in APOE-ε4 carriers without cognitive symptoms as compared to healthy noncarriers [41], while medial frontal and anterior temporal hypermetabolism seen in the abovementioned comparison persists when controlling for Aß [41]. Global CMRglc declines with age, and in particular rCMRglc decline within the PCC/precuneus region and the lateral parietal region might be caused by the combined effects of aging and genetic (i.e., APOE-ε4 carriage) traits [20]. Depression and anxiety may also have an impact on rCMRglc. Therefore, these neuropsychiatric symptoms should always be determined and taken into consideration when interpreting metabolic patterns in models of pAD [42].

To better understand the influence of genetic factors on CMRglc, Didic et al. [43] explored glucose metabolism in healthy APOE-ε4 carriers and noncarriers, both at a local and a cerebral network level. Carriers showed hypometabolism in the left anterior MTL, including the entorhinal and perirhinal cortices. MTL metabolism was related to memory performance, and there was a stronger metabolic connectivity of the MTL with frontoparietal regions in carriers than in noncarriers, proposing compensatory metabolic network alterations in carriers.

Higher education as proxy for CRC mediates rCMRglc in cognitively healthy controls carrying the APOE-ε4 allele. Higher CRC was positively correlated to frontotemporal metabolism and episodic memory performance in these individuals as compared to noncarriers, independent from Aß deposition [44]. Furthermore, subjects with lower CRC carrying the APOE-ε4 allele exhibited medial temporal and prefrontal hypometabolism relative to noncarriers. Higher educated carriers were comparable to noncarriers in these areas and showed temporal lobe hypermetabolism. These results indicate that CRC helps to counteract the effects of APOE-ε4 on metabolism, independent from cerebral Aß load, thereby representing a protective factor which helps to delay cognitive deterioration in APOE-ε4 carriers [44].

Using ^{18}F-FDG-PET to characterize metabolic networks in a large sample of cognitively healthy elderly, patients with MCI and AD, Yao et al. [45] dichotomized these individuals into two groups: one with the APOE-ε4 allele carriage status and one without. The authors demonstrated increased local short distance interregional metabolic correlations and disrupted long distance interregional metabolic correlations in the APOE-ε4 carrier group. The pattern observed in APOE-ε4 allele-positive subjects resembled the pattern observed in MCI and AD.

Furthermore, there seems to be an APOE-ε4 gene-dose effect in cognitively normal, late-middle-aged subjects (ε4 homozygotes, ε4 heterozygotes, and noncarriers) on regional gray matter volume and rCMRglc. Gray matter atrophy and hypometabolism correlated with the APOE-ε4 gene dose [46].

2.2.3 ^{18}F-FDG-PET in MCI

MCI is clinically the most important model of pAD. The amnestic subtype has the highest risk of conversion to AD. The metabolic pattern of aMCI resembles that seen in early stages of AD.

PCC and precuneus hypometabolism constitute predictors for cognitive decline in MCI, whereas temporoparietal hypometabolism may either be associated with memory deficits or with conversion to dementia [47, 48]. Metabolic changes within the left lateral frontal cortex were described in another study to anticipate conversion to dementia in aMCI [49].

Coutinho et al. [50] studied glucose metabolism and CSF biomarkers of AD in non-amnestic MCI, aMCI, and healthy controls. As compared to controls, both MCI groups exhibited reduced CMRglc in the precuneus. In relation to aMCI, the non-amnestic MCI patients also showed right prefrontal hypometabolism and higher Aß CSF levels. In this study, the CSF biomarker profile in the aMCI group was more closely related to manifest AD than in the non-amnestic MCI group.

Bailly et al. [51] compared glucose metabolism and atrophy of the hippocampus, amygdala, precuneus, and ACC/PCC of cognitively healthy controls, MCI patients, and AD patients. In relation to controls, mean PCC volume was reduced in MCI and AD. The precuneus and PCC regions also exhibited hypometabolism in MCI and AD, as compared to controls.

Another study also found the hypometabolic pattern of MCI to AD converters to cover the precuneus and PCC, with further hypometabolism in temporoparietal and frontal cortices [52]. In aMCI, declarative memory storage impairments correlated with hippocampal atrophy, glucose hypometabolism, and cerebral Aß deposition. Delayed recall deficits correlate with neuroimaging biomarkers of MCI to AD converters and mirror the conversion from aMCI to AD [53].

Franzmeier et al. [54] demonstrated precuneus hypometabolism in aMCI patients with Aß deposition as compared to healthy controls, with more pronounced rCMRglc reductions in patients with higher CRC. Global left frontal cortex (LFC) connectivity was linked to higher CRC. The authors concluded that higher global LFC connectivity represents a putative substrate of CRC in aMCI patients which may help to maintain memory performance despite glucose hypometabolism in this model of pAD. Förster et al. [55] demonstrated that aMCI patients have significantly improved cognition and attenuated metabolic decline in the left anterior temporal pole and ACC after a 6-month multicomponent cognitive intervention as compared to an active control group.

Chen et al. [56] examined 1-year rCMRglc declines in AD, aMCI, and cognitively healthy controls. Both the AD and the aMCI group showed significant metabolic decline in the PCC, medial and lateral parietal, medial and lateral temporal, frontal, and occipital cortex as compared to controls within 1 year.

The MTL is of particular interest in MCI and helps to distinguish aMCI from other MCI syndromes. Patients with aMCI have decreased MTL glucose metabolism as compared to patients with non-amnestic MCI [57]. Also, cognitively deteriorating patients with amnestic MCI exhibit MTL hypometabolism as compared to controls and MCI patients that do not show cognitive decline [58].

2.2.4 A Methodological Example: Metabolic Topology in Early and Late Stages of Neurodegeneration

In a recent paper, Granert et al. [17] analyzed ^{18}F-FDG-PET images in a large cohort of patients with:

- Early stages of neurodegeneration (i.e., aMCI, early AD, and patients with Parkinson's disease)
- More advanced stages of neurodegeneration (i.e., patients with Parkinson's disease dementia and Lewy body dementia)
- Cognitively normal age-matched controls (Co)

There is abundant clinical, pathophysiological, and imaging overlap between the abovementioned diseases. Therefore, one can question the concept of distinct disease entities and suggest a somehow continuous spectrum of early neurodegenerative changes between normal aging, Parkinson's disease (PD), and AD. The same hypothesis might also hold true in more advanced stages of neurodegeneration. According to this concept, the authors [17] built a topological map based on rCMRglc to rank and localize single subjects' disease status according to parkinsonian and dementia typical pattern expression in patients clinically diagnosed to suffer from PD, Parkinson's disease dementia (PDD), Lewy body dementia (DLB), aMCI, or AD.

The authors describe a metabolic topological map which confirmed an inseparable spectrum of disease manifestation according to two different and dissociable expression patterns. The expression values, which were extracted from the *AD* versus *controls* pattern, were highly correlated with cognitive deficits, but not motor deficits. The opposite was found for the corresponding expression values of the *PD* versus *controls* pattern, which was highly correlated with motor deficits, but not cognitive deficits.

This metabolic imaging approach supports the notion that there is a continuous spectrum of neurodegeneration between neurodegenerative syndromes affecting predominantly motor skills (PD, PDD) and neurodegenerative syndromes primarily leading to cognitive deficits (aMCI and AD, DLB).

The voxelwise SPM group comparisons with the controls and PD as well as AD exhibited typical and distinct metabolic profiles for both the PD as well as the AD group (Fig. 1). The AD group showed a decline of rCMRglc bilaterally in the PCC, the lateral temporal lobe, and the inferior parietal cortex (Fig. 1a). Temporoparietal and PCC reductions of rCMRglc in the group of early AD patients are in accordance with previous imaging studies focusing on rCMRglc alterations in AD [59–63]. These rCMRglc reductions in early AD were localized within cortical areas closely linked to the DMN, a phenomenon that has been described before [64]. The PD group

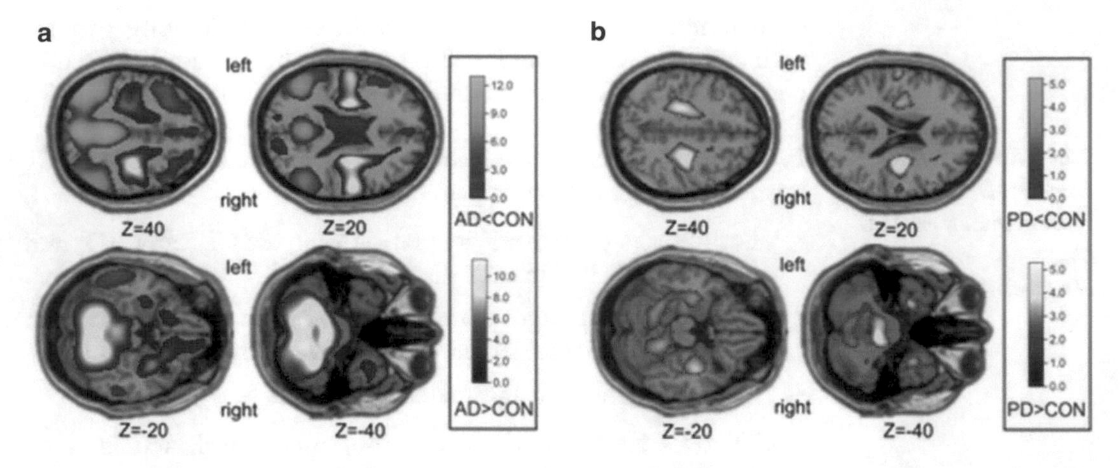

Fig. 1 Visualization of the statistical maps used to calculate the expression values corresponding to the group t-contrasts. (**Panel a**) AD versus CON. (**Panel b**) PD versus CON. Statistical maps were shown with a threshold of $p_{unc} < 0.001$ and overlaid onto the single-subject MR T1 template as delivered with the SPM software.

exhibited frontal and parieto-occipital hypometabolism as well as temporal and pontocerebellar hypermetabolism (Fig. 1b). This pattern was closely related to previous neuroimaging findings in PD [59]. Finally, there were neither clinical nor metabolic differences between the PDD and DLB groups [17].

2.3 Imaging of Perfusion in Preclinical AD

Perfusion describes the process by which oxygenated arterial blood is provided to the capillary bed in a predefined mass of brain tissue. It is measured in milliliters of blood per 100 g of tissue per min [22]. CBF values in human being approach 50 mL/100 g per min, and gray matter CBF is approximately three times higher than white matter CBF [22]. Regional CBF can be assessed using PET/SPECT methods, the more ancient ^{133}Xe inhalation technique, and MRI-based techniques such as fMRI and ASL.

In normal aging, cognitively healthy adults exhibit grey matter perfusion reductions of about 0.45% per year [31]. Hypoperfusion in aging tends to be widespread, affecting frontal, medial temporal, parietal, and subcortical regions [13, 31, 65, 66]. In the stage of clinically manifest dementia, perfusion in patients suffering from AD correlates negatively with disease severity [67]. A phenotypically normal cognitive status in the elderly despite perfusion deficits may reflect cognitive reserve capacity (CRC) of the aging brain [68].

As neither oxygen nor glucose can be stored within the brain, neuronal energy supply strongly depends on a constant delivery of oxygen and glucose by CBF. Therefore, there is a strong link between neural activity, energy metabolism, and CBF, which represents the rationale of FBI methods using perfusion, not only in pAD [10]. Several factors mediate the relationship between CBF and cognition, including CRC, age, sex, genotype, and vascular risk factors.

In this chapter, we mainly present exemplary fMRI and rs-fMRI studies in pAD and its models as there is abundant and quite heterogeneous literature on this topic, exceeding the volume of this chapter.

2.3.1 PET/SPECT Perfusion Imaging

To date, there are no recent studies applying SPECT or PET for perfusion imaging in pAD, as defined by Sperling et al. [1]. However, one study [69] investigated the relation between rCBF (as measured with ^{99m}Tc-exametazime-SPECT) and CSF biomarkers in cognitively healthy elderly. The authors found that high CSF p-tau and total tau levels are correlated with decreased rCBF in the right medial frontal lobe and that high CSF p-tau levels are correlated with increased rCBF in the left frontotemporal border zone. There were no significant associations between rCBF and CSF Aß1–42 levels. These results indicate a correlation between tau pathology rather than Aß deposition and CBF abnormalities in pAD.

Another study from the Baltimore Longitudinal Study of Aging using repeated resting-state $^{15}O\text{-}H_2O$ PET scans found rCBF differences between cognitively normal elderly with high and low cerebral amyloid load [70]. Greater longitudinal reductions of CBF in the group exhibiting high amyloid values were found in the anterior as well as middle cingulate cortex, left thalamus, right supramarginal gyrus, and midbrain. Greater longitudinal gains in rCBF in the group exhibiting high amyloid values were encountered in left medial and inferior frontal gyri, left inferior parietal gyrus, right precuneus, and left postcentral gyrus.

2.3.2 fMRI, Resting-State fMRI and Functional Connectivity in Preclinical AD

Edelman et al. [71] combined fMRI and Pittsburgh compound B-PET (PIB-PET) in a face-name memory task. In their study, Aß-positive healthy elderly displayed an increased (possibly compensatory) hippocampal activation during the task. In another study, Gordon et al. [72] combined fMRI and CSF, respectively, PET biomarkers of AD in two attentional control tasks to examine changes related to Aß and tau pathology in cognitively healthy subjects. They found that increased levels of tau and p-tau are linked to overactivations of attentional control areas, indicating that alterations of attentional networks are associated with AD neuropathology.

Sheline et al. [73] applied rs-fMRI and PIB-PET in AD patients and two groups of cognitively healthy individuals, one with evidence of brain Aß (Aß+) and one without Aß (Aß−). The Aß+ group as well as the AD group showed significant differences in functional connectivity (fc) within the DMN. The authors concluded that brain Aß deposition leads to alterations in fc of the DMN in pAD, in the same anatomical regions and in the same direction as seen in AD.

Schultz et al. [74] examined fc in several cortical association networks, including the DMN, in cognitively healthy individuals.

All subjects underwent Aß and tau PET imaging. In Aß-positive individuals, there was increased connectivity in the DMN with low tau levels, whereas elevated tau levels were associated with decreased connectivity in Aß-positive individuals. These results propose a hyperconnectivity phase followed by a hypoconnectivity phase in the course of pAD.

Brier et al. [75] used rs-fMRI in two groups of cognitively healthy subjects: one group with evidence of pAD as assessed by CSF markers of AD and the other group with normal CSF biomarkers. The authors found significant interactions between age and biomarker status in the DMN and salience and dorsal attention (DAN) resting-state networks. The presence of AD biomarkers increased the size of the aging effect within these networks. The authors observed alterations of fc in pAD with age, not existent in the group without evidence of pAD. These results suggest that underlying AD pathology in pAD may be responsible for a considerable amount of the effects of aging on fc in cognitively healthy subjects.

In another study, the same group [76] examined cognitively normal elderly with CSF biomarkers of AD and fc-fMRI. Resting-state fMRI imaging has shown that brain networks deteriorate in manifest AD. In their study, the authors found large-scale disconnections in CSF-defined pAD, similar to that seen in manifest AD. With progressive deterioration of cognition, this study identified three important hub regions (brain regions highly connected with other regions) on the medial prefrontal cortices and ACC that lose their functions gradually from normal aging to manifest AD.

The same group [77] also showed that both Aß and tau pathology affect DMN integrity before AD becomes clinically manifest. In their study, the authors found CSF biomarkers of AD to be independently associated with reduced DMN integrity. The strongest decrease in fc was seen between the PCC and MTL regions. These fc reductions could not be attributed to age or atrophy.

Drzezga et al. [78] analyzed fc in cortical hub regions in non-demented cognitively healthy individuals (CH-Aß+ and CH-Aß−) and MCI patients (MCI-Aß+ and MCI-Aß−) with and without cerebral amyloidosis. All subjects underwent amyloid PET and ^{18}F-FDG-PET imaging, as well as structural MRI and rs-fMRI. Disruptions of whole-brain connectivity were seen in the MCI-Aß+ group in the PCC/precuneus region (an important hub region), strongly overlapping with glucose hypometabolism. There were also minor connectivity disruptions and hypometabolism in the CH-Aß+ group. These findings suggest that a disruption of fc and glucose hypometabolism represent early functional consequences of amyloid burden (i.e., AD pathology), even before dementia becomes clinically apparent. The PCC/precuneus region has a particular susceptibility to amyloid deposition and glucose hypometabolism, thereby putatively reflecting a link between synaptic dysfunction and functional disconnection.

2.3.3 Arterial Spin Labeling MRI in Preclinical AD

As described above, reductions of CBF mirror the pathophysiological changes seen in pAD. Hays et al. [10] summarized ASL MRI studies in several models of pAD. In the early preclinical stage, there is evidence of hyperperfusion, whereas in later preclinical stages, there exist both hyper- and hypoperfusion. Later, in manifest AD, there is evidence for widespread reductions of CBF. Cerebral amyloid angiopathy may be an important contributing factor.

A recent study [14] found continuous CBF decreases in pAD, mainly in temporal and parietal regions, along the stages as defined by Sperling et al. [1]. Other studies have found CBF decreases related to Aß deposition in the inferior temporal and parietal cortex, In control subjects, high Aß levels were associated with greater CBF reductions [70, 79].

2.4 APOE-ε4 Carriers

2.4.1 PET/SPECT Perfusion Imaging

Generally, there have been few PET/SPECT perfusion studies focusing on cognitively normal APOE-ε4 carriers. Thambisetty et al. [80] studied longitudinal CBF differences between two groups of nondemented elderly from the Baltimore Longitudinal Study of Aging using ^{15}O-PET. One group contained APOE-ε4 carriers and the other contained noncarriers. APOE-ε4 carriers exhibited widespread and greater regional CBF decline in frontal, temporal, and parietal cortices as compared to noncarriers.

In patients with manifest AD, there was a dose-dependent effect of the APOE-ε4 allele on rCBF, with more hypoperfusion seen in frontal, temporoparietal, MTL, and occipital regions in the APOE-ε4 carrier group. AD patients homozygous for the ε4 allele also exhibit more severe MTL atrophy than noncarriers or AD patients heterozygous for the ε4 allele [19, 81–83]. Furthermore, there is a more rapid decline of CBF in the APOE-ε4 group compared to noncarriers [19].

MCI patients carrying the APOE-ε4 allele have a greater chance to convert to AD. The converters show hypoperfusion in the MTL and parietal regions [84] as well as within the postcentral region, as compared to nonconverters [85].

2.4.2 fMRI, Resting-State fMRI, and Functional Connectivity

Fleisher et al. [86] compared resting CBF (DMN signal differences) and fMRI BOLD response during an associative memory-encoding task in APOE-ε4 carriers with a positive familial history of AD versus healthy controls without these risk factors. BOLD activations during encoding did not differ between both groups. Deactivations during encoding were more pronounced in the low risk group within the parietal cortex. However, resting state DMN analysis better discriminated both groups than encoding associated fMRI. The authors conclude that resting state imaging is a more effective and promising tool in assessing AD risk as compared to activation fMRI.

Harrison et al. [87] analyzed fc of the hippocampus in cognitively normal elderly, subdivided into two groups, one consisting of APOE-ε4 carriers and one consisting of noncarriers. The paradigm used was a paired-associates memory task. During encoding, APOE-ε4 carriers showed lower fc change compared to baseline between the anterior hippocampus and right precuneus, anterior insula, and cingulate cortex. During retrieval, the bilateral supramarginal gyrus and right precuneus showed lower fc change with anterior hippocampus in the APOE-ε4 carriers. The APOE-ε4 group, in contrast to the noncarriers, showed strong negative fc changes compared to noncarriers where positive fc changes were seen. These differences may represent prodromal functional changes mediated, at least in part, by the APOE-ε4 allele. These results were considered to be consistent with the anterior-to-posterior theory of AD progression in the hippocampus. Altered fc of the hippocampus can, therefore, be regarded as an early FBI biomarker for pAD.

In another study, Sheline et al. [88] applied rs-fMRI and PIB-PET in two groups of cognitively healthy individuals, one with APOE-ε4 allele carriers (ε4+) and one group without APOE-ε4 allele carriers (ε4−). Amyloid-negative ε4+ carriers differed significantly from the amyloid-negative ε4− group in fc of the DMN. The observed alterations in the ε4+ carriers were similar to the changes described in AD. Disruptions of fc within brain regions of the DMN in the ε4+ group were seen in the absence of cerebral Aß deposition. These results suggest that genetic risk factors for AD may produce changes of resting-state networks (i.e., the DMN) even in the absence of cerebral amyloidosis.

2.4.3 Arterial Spin Labeling MRI

Michels et al. [89] assessed APOE-ε4 genotype, Aß load, and CBF in cognitively healthy (CH) individuals and aMCI patients. The global CBF was lower in the group of APOE-ε4-positive subjects as compared to noncarriers. In aMCI, the global CBF was lower compared to CH, and participants with increased Aß showed a trend for lower global CBF. Therefore, the authors concluded that the APOE-ε4 allele has an impact on CBF, which is at least partially independent from Aß.

Bangen et al. [90] used ASL MRI and fMRI to assess resting CBF as well as task-dependent CBF and BOLD response, in a memory-encoding paradigm in cognitively healthy elderly (with and without the APOE-ε4 allele) and in MCI patients (also with and without the APOE-ε4 allele). In cognitively healthy APOE-ε4 carriers, there is evidence of hyperperfusion with increased resting-state CBF in the MTL compared to ε4 noncarriers, while MCI patients have decreased resting-state CBF within MTL structures relative to controls. There were no significant group differences for task-dependent BOLD response (MCI versus controls). These findings suggest that alterations in resting-state CBF and CBF response to memory encoding may serve as early markers of brain dysfunction in different models of pAD.

Wierenga et al. [91] used ASL MRI to investigate the influence of the APOE genotype and age on CBF in two groups of cognitively healthy younger and older adults, one containing APOE-ε4 carriers and one containing noncarriers. Older adults exhibited decreased gray matter CBF, as compared to younger adults, in widespread brain areas. The ACC region exhibited a putatively compensatory increased CBF in young APOE-ε4 carriers and reduced CBF in older APOE-ε4 carriers.

In conclusion, studies on CBF in clinically nondemented APOE-ε4 carriers demonstrated both increases and decreases of CBF relative to cognitively healthy elderly [13, 80]. Studies also support the notion that effects of the APOE genotype on CBF are liaised by age, with elderly subjects displaying more areas of hypoperfusion and younger subjects exhibiting more areas of hyperperfusion [91].

In young APOE-ε4 carriers, increased CBF seems to be associated with better executive functioning, suggesting that hyperperfusion is a compensatory mechanism [91]. Elderly APOE-ε4 carriers have abnormally low CBF, which may indicate a breakdown of this initial compensatory mechanism.

Disease severity may also mediate the relationship between APOE genotype and CBF: nondemented APOE-ε4 carriers and AD patients with this genotype show widespread hypoperfusion as compared to noncarriers. In MCI, APOE-ε4 carriers exhibited regional hyperperfusion in the right parahippocampal, bilateral cingulate gyri, and the right PCC [92].

2.5 MCI

2.5.1 PET/SPECT Perfusion Imaging

Using serial ^{15}O-water PET imaging data from the Baltimore Longitudinal Study of Aging, one study [93] found alterations of rCBF in elderly individuals later converting to MCI. Compared to cognitively stable elderly, MCI converters displayed greater longitudinal rCBF decreases in temporal, parietal, and thalamic regions and rCBF increases in orbitofrontal, MTL, and ACC regions. These rCBF changes were seen within brain regions that show early accumulation of pathology of AD, suggesting a link between early neuropathological changes and rCBF alterations.

2.5.2 fMRI, Resting-State fMRI, and Functional Connectivity

Adriaanse et al. [94] analyzed fc using rs-fMRI in the DMN in cognitively healthy elderly and patients with MCI and AD. Amyloid deposition was assessed with PIB-PET. The authors found no association between fc of the DMN and Aß load in DMN areas in any of the groups or between groups. The same group described decreased fc within the DMN in AD in the precuneus and PCC compared to controls. MCI patients exhibited fc changes that were situated between patients suffering from AD and controls [95].

Liang et al. [96] examined fc in MCI patients in three different areas of the inferior parietal cortex, notably the intraparietal sulcus (IPS), the angular gyrus (AG), and the supramarginal gyrus (SG).

With these anatomical hallmarks, the authors were able to trace three different approved RSNs within this region: the DMN, the salience network (SN), and the executive control network (ECN). The AG displayed reduced fc within the DMN. The IPS also exhibited decreased fc with the right inferior frontal gyrus and increased fc with the left frontal lobes in the ECN. The SG showed decreased fc with the frontal and parietal cortex and increased fc with some subcortical areas of the salience network. Connectivity within these networks correlated with cognitive impairment in the MCI patients. The authors concluded that MCI is associated with changes of large-scale functional brain networks.

Another group [78] analyzed fc in cortical hub regions (PCC/precuneus) in MCI patients with and without cerebral amyloidosis. Disruptions of whole-brain connectivity were seen in the MCI-Aß+ group in the PCC/precuneus region, strongly overlapping with glucose hypometabolism in these regions.

2.5.3 Arterial Spin Labeling MRI

Perfusion changes seen in MCI are suggestive of neurodegeneration-induced CBF dysregulation, which may result from altered metabolic requests to preserve cognitive function [97]. Mild cognitive impairment is linked to increases and also decreases in resting-state CBF.

Kim et al. [92] used ASL MRI to study CBF in patients suffering from AD and MCI as well as in cognitively normal controls. Each group was further divided into carriers or noncarriers of the APOE-ε4 allele. In the MCI APOE-ε4 carriers, hyperperfusion was detected in the right parahippocampal gyrus, bilateral cingulate gyri, and right PCC as compared to noncarriers.

Michels et al. [89] assessed Aß, CBF, and APOE genotype in cognitively healthy individuals and aMCI patients. They found widespread CBF reductions, largely independent from Aß load, in elderly APOE-ε4 carriers. Global CBF was lower in patients with aMCI compared with cognitively healthy elderly. Interestingly, carriage of the APOE-ε4 allele had a more deleterious effect on both local and global CBF than cerebral Aß deposition.

However, another study found that cerebral amyloid pathology affected CBF across the span from cognitively healthy elderly (with cerebral amyloidosis) to clinically manifest AD. In their study, the authors [79] showed Aß pathology to be associated with hypoperfusion in pAD, whereas in later stages of neurodegeneration, atrophy becomes more prominent.

Other studies found that CBF increases within the temporal lobe, ACC, insula, putamen, left hippocampus, right amygdala, and basal ganglia. Decreases of CBF have been described in frontal and temporoparietal areas (extending to the MTL), PCC, precuneus, and left middle occipital lobe [90, 98–101].

The PCC seems to play an important role as indicator of cognitive decline, even in cognitively normal elderly [68]. This brain region, known to be involved in early AD changes, exhibits hypoperfusion in MCI and also in cognitively healthy individuals later showing cognitive decline as compared to the cognitively stable elderly.

Increases of CBF often correlate with better memory performance, not only in MCI, whereas decreases of CBF often correlate with cognitive dysfunction. Studies of functional CBF changes showed hypoperfusion in the right precuneus/cuneus in the aMCI-group as compared to age-matched controls during the control condition. During a memory encoding task condition, hypoperfusion extended to the PCC. The perfusion rates correlated with the cognitive abilities in aMCI [101]. Another study [102] demonstrated that perfusion changes induced by a memory-encoding task can also help to distinguish aMCI and cognitively healthy elderly. Task-enhanced ASL MRI may therefore represent a potential FBI biomarker of pAD.

3 Discussion

The rationale for defining pAD is that the best time for preventing neurodegeneration in elderly subjects is early in the predementia stage. Innovative pharmacological and non-pharmacological preventive interventions can best be assessed in pAD [1, 103]. Hence, there is a need for personalized, biomarker-based approaches to identify future dementia patients.

In recent years, functional brain imaging (FBI) has increasingly been recognized as having the potential to represent such biomarkers of pAD. Table 1 summarizes the most important FBI findings in pAD, MCI, and nondemented APOE-ε4 allele carriers.

Metabolic imaging (i.e., ^{18}F-FDG-PET) has significantly contributed to our understanding of pAD, showing hypometabolic changes within the PCC, precuneus, MTL, and temporoparietal regions that resemble metabolic alterations observed in AD.

Concerning perfusion imaging, more heterogeneous results have been obtained. In AD, there is evidence for widespread reductions of CBF, with a negative correlation between CBF and disease severity [67]. In pAD, both hypo- and hyperperfusion have been reported in various brain regions in different stages of cognitive decline [14]. Hays et al. [10] summarized ASL MRI studies in several models of pAD. In the preclinical phase, hyperperfusion prevails, mainly in MTL structures, while later in the subclinical phase, both hyper- and hypoperfusion coexist, with some evidence of initial hyperperfusion and later hypoperfusion.

CBF dysregulation in pAD and its models suggests the presence of vascular regulative mechanisms, putatively in response to an increased need for oxygen and glucose, or changes in cerebral metabolism to maintain cognitive function [97]. However, the direction of CBF alterations is not specified. Hyper- as well as hypoperfusion may indicate underlying pathological or compensatory changes in different predementia stages. Cerebral amyloid angiopathy may constitute an important contributing factor linking the classical amyloid cascade hypothesis with the vascular hypothesis of AD [15].

Small vessel disease and cardiovascular risk factors have an important impact on CBF; thus, CBF changes in the elderly do not necessarily mirror underlying pAD pathophysiology. Hypoperfusion, therefore, reflects the combined disease burden of microangiopathy, aging, and neurodegeneration. Normal aging processes have an impact on CBF, Aß– and tau– load. Aging is associated with a slight but continuous reduction of gray matter CBF (about 0.45% per year) [31], and hypoperfusion affects frontal, temporo-parietal, and subcortical regions [13, 31, 65, 66, 104].

Only markers of cerebral Aß deposition and later neurodegenerative changes have been included to define pAD [1]. However, there is growing evidence that perfusional alterations are also involved in the very early pathogenesis of AD, even before deposition of proteinaceous substances becomes manifest [10]. FBI has significantly contributed to this so-called vascular hypothesis of AD. Novel, easily applicable methods to quantify CBF changes in cognitively healthy individuals at risk for AD have been implemented in the clinical and research setting. ASL MRI and fMRI have largely replaced SPECT and PET in evaluating resting perfusion and task-dependent perfusion changes in pAD and AD.

There is another important point questioning the existing model of pAD [1]. Both neuroimaging studies focusing on CBF and studies focusing on metabolism in pAD have found alterations of CBF and metabolism to be at largely independent from cerebral Aß load (Table 1). Studies analyzing different biomarker profiles of neurodegeneration (brain atrophy as measured with structural MRI, ^{18}F-FDG-PET, CSF-tau levels) and amyloid deposition (PIB-/^{18}F-florbetapir PET and CSF-$Aß_{1-42}$) suggest that both, a *neurodegeneration-first* and an *amyloid-first* pathophysiological pathway, may later lead to AD [105]. It is difficult to distinguish individuals with positive biomarkers for neurodegeneration, but without Aß deposition (sNAP, suspected non-Alzheimer pathophysiology) from individuals with advanced stages of pAD, as defined by Sperling et al. [1] concerning their clinical and neuroimaging characteristics [39].

There are methodological limitations that may explain some of the apparently contradictory findings in pAD, including inconsistencies regarding increases or decreases of CBF and rCMRglc as well as regional differences in CBF/rCMRglc alterations between studies. Patient demographics with age and CRC, diagnostic criteria, comorbid depression and anxiety, disease severity, genetic traits, and differences in the cardiovascular risk profile may account for some of the disparate findings in pAD [65, 90].

Table 1

FBI findings in models of preclinical AD

	pAD[a]	MCI	APOE-ε4 carriers[b]
Imaging of metabolism			
• ^{18}F-FDG-PET	• Temporoparietal and PCC hypometabolism as typically seen in AD • Hypometabolism rather independent from Aß deposition	• Comparable metabolic changes as seen in early AD • Hypometabolism of PCC and precuneus • Temporoparietal and frontal hypometabolism	• Hypometabolism in typical AD regions • Rather independent from Aß
Imaging of perfusion			
• PET/SPECT	• rCBF decreases within the cingulate cortex • rCBF changes rather associated with tau, not Aß pathology	• rCBF alterations in brain regions exhibiting early pathological changes in AD • Compensatory hyperperfusion in different brain areas	• Frontal, temporal, and parietal rCBF reductions • Gene-dose effects on CBF
• fMRI and rs-fMRI	• Large scale disconnections as seen in manifest AD • Reduced DMN integrity • Hyperconnectivity phase followed by hypoconnectivity phase in the course of pAD	• Alterations of large-scale functional brain networks • Decreased functional connectivity within several RSNs (i.e., the DMN) • Disruptions of whole-brain connectivity within the PCC/precuneus region	• Altered functional connectivity of the hippocampus • Disruptions of fc within the DMN
• ASL MRI	• Decrease of CBF along the continuum of pAD[a] • Both hypo- and hyperperfusion may reflect pathology in different stages of pAD	• Global CBF reductions in MCI • MTL, PCC, and precuneus hypoperfusion • Relation between rCBF changes and Aß depositions remains unclear	• Alterations of rs CBF and CBF response to memory encoding in MTL • Age-dependent reductions of CBF in MTL • CBF reductions, independent from cerebral Aß

Abbreviations: *FBI* functional brain imaging, *pAD* preclinical Alzheimer's disease, *MCI* mild cognitive impairment, *Aß* amyloid-$ß_{1-42}$, *rCBF* regional cerebral blood flow, *RSNs* resting-state networks, *DMN* default mode network, *MTL* medial temporal lobe, *PCC* posterior cingulate cortex, *SPECT* single photon emission computer tomography, *^{18}F-FDG-PET* ^{18}F-fluorodesoxy-glucose positron emission tomography, *MRI* magnetic resonance imaging, *fMRI* functional magnetic resonance imaging, *rs-fMRI* resting-state fMRI, *ASL* arterial spin labeling

[a]Preclinical AD as defined by Sperling et al. [1]

[b]Only results of nondemented patients are summarized

Concerning the methods presented here, there are also some limitations for each technique. PET and SPECT have the disadvantage of being expensive and time-consuming, often inaccessible to clinicians, and associated with exposure to radiation. Studies using ASL MRI and fMRI are in line with previous PET and SPECT studies on brain perfusion, with MRI imaging being noninvasive, cheaper, and more easily accessible in clinical practice. However, ASL MRI has the disadvantage of showing low sensitivity, low temporal resolution, and also contamination from the BOLD signal. Individuals with early pAD exhibit increased MTL resting CBF, which has direct impact on the fMRI BOLD response. BOLD changes largely depend on resting-state CBF, which can be either increased or decreased in pAD. BOLD activations should therefore be interpreted with caution and do not always reflect differences in neuronal activation.

Neuroanatomically, CBF changes and metabolic alterations largely affect the similar regions in AD, thereby indicating that ^{18}F-FDG-PET might in the future be replaced by MRI techniques quantifying CBF changes [106].

Resting-state fc-fMRI imaging has shown that brain networks and their connectivity degrade during the course of AD. There is widespread loss of intranetwork as well as internetwork connectivity in AD. In pAD, studies on fc show heterogeneous results. There is some evidence for loss of long distance internetwork connectivity with increased local short distance connectivity. Resting-state fc-fMRI may also have the potential to evaluate therapeutic effects of drugs in AD as well as in pAD [107].

In AD as well as in pAD, disruptions of the DMN have been observed [76, 77, 108]. Interestingly, there is a significant degree of overlap between typical hypometabolic brain regions in AD and the DMN [17].

Concerning the models of pAD presented here, there are also several limitations for each model. As described above, the definition of pAD requires a reappraisal concerning measures of CBF and its sole dependency on the amyloid cascade hypothesis. There is some evidence that the proneness for AD in APOE-ε4 allele carriers and in pAD is mediated independently from Aß. Studies in nondemented carriers of the APOE-ε4 allele have, however, shown fc alterations that were quite similar to nondemented subjects with cerebral Aß, and both groups also resembled AD patients [88]. Furthermore, in a memory clinic setting, genetic testing is currently not recommended in the vast majority of patients, except for very early dementia cases with a clear-cut family history of dementia.

To date, there is insufficient knowledge about the role of SCD as a model of pAD. Therefore, MCI still is the most relevant clinical model. The aMCI subgroup has the highest risk to develop clinically manifest dementia, but more than 50% of these patients stay stable or even improve within 30 months [3]. Further biomarker support is required to identify MCI patients that later develop AD. In clinical practice, CSF, not FBI biomarkers, is actually state of the art to detect those subjects prone to AD.

Memory performances, mainly hippocampus-dependent declarative learning and delayed recall impairments, are hallmarks of the neuropsychological profile of AD and can also be detected in the pAD stage. These neuropsychological deficits are associated with early metabolic and CBF changes within the MTL and temporoparietal cortical areas [65]. Compensatory increases of CBF and metabolism in the MTL are associated with better memory performance in models of pAD, whereas decreases of MTL CBF and metabolism are linked to cognitive dysfunction. Normal cognitive status in the elderly despite biomarkers indicative of pAD may reflect cognitive reserve capacity (CRC) of the aging brain [68]. Years of education as a proxy for CRC should be taken into account when applying FBI biomarkers in the elderly. Strategies increasing the CRC are probably a fruitful approach to slow down cognitive decline.

In the future, fully automated and validated brain-mapping algorithms (e.g., based on machine learning techniques) will gradually replace visual reading by clinicians. With ever-improving technology, this development has yet reached FBI and also structural imaging methods. Morbelli et al. [109] described that visual analyses of ^{18}F-FDG-PET images by experts are the most accurate method; however, an automated image analyzing system also showed highly specific results, superior to moderately skilled readers.

In conclusion, there is now an urgent need to improve the diagnosis of pAD. The preexisting criteria are strongly relying on the amyloid cascade hypothesis. As described in this chapter, CBF and its measures need to be added to the existing criteria of pAD. Furthermore, there is a need to quantify fc changes, and methods to analyze fc need to be improved.

Finally, even when considering the abovementioned disadvantages (i.e., exposure to radiation), ^{18}F-FDG-PET still remains to date the most established FBI biomarker in pAD.

Clinicians may best rely on a multimodal approach combining clinical, neuropsychological, CSF, and neuroimaging-derived assessments of biomarkers to detect AD, MCI, and pAD. Although there is no causal, disease-modifying treatment for AD available to date, individuals at risk of later developing AD can be advised to control for cardiovascular risk factors, sedentary lifestyle, and exposure to contact sports and leisure activities with a relevant risk of traumatic brain injury [103].

References

1. Sperling RA, Aisen PS, Beckett LA et al (2011) Toward defining the preclinical stages of Alzheimer's disease: recommendations from the National Institute on Aging-Alzheimer's Association workgroups on diagnostic guidelines for Alzheimer's disease. Alzheimers Dement 7(3):280–292. https://doi.org/10.1016/j.jalz.2011.03.003
2. Jessen F, Amariglio RE, van Boxtel M et al (2014) A conceptual framework for research on subjective cognitive decline in preclinical Alzheimer's disease. Alzheimers Dement 10(6):844–852. https://doi.org/10.1016/j.jalz.2014.01.001
3. Fischer P, Jungwirth S, Zehetmayer S et al (2007) Conversion from subtypes of mild cognitive impairment to Alzheimer dementia. Neurology 68(4):288–291. https://doi.org/10.1212/01.wnl.0000252358.03285.9d
4. Mitchell AJ, Shiri-Feshki M (2009) Rate of progression of mild cognitive impairment to dementia—meta-analysis of 41 robust inception cohort studies. Acta Psychiatr Scand 119(4):252–265. https://doi.org/10.1111/j.1600-0447.2008.01326.x
5. Yip AG, McKee AC, Green RC et al (2005) APOE, vascular pathology, and the AD brain. Neurology 65(2):259–265. https://doi.org/10.1212/01.wnl.0000168863.49053.4d
6. Coon KD, Myers AJ, Craig DW et al (2007) A high-density whole-genome association study reveals that APOE is the major susceptibility gene for sporadic late-onset Alzheimer's disease. J Clin Psychiatry 68(4):613–618
7. Saunders AM, Strittmatter WJ, Schmechel D et al (1993) Association of apolipoprotein E allele epsilon 4 with late-onset familial and sporadic Alzheimer's disease. Neurology 43(8):1467–1472
8. Corder EH, Saunders AM, Strittmatter WJ et al (1993) Gene dose of apolipoprotein E type 4 allele and the risk of Alzheimer's disease in late onset families. Science 261(5123):921–923
9. Biomarkers Definitions Working Group (2001) Biomarkers and surrogate endpoints: preferred definitions and conceptual framework. Clin Pharmacol Ther 69(3):89–95. https://doi.org/10.1067/mcp.2001.113989
10. Hays CC, Zlatar ZZ, Wierenga CE (2016) The utility of cerebral blood flow as a biomarker of preclinical Alzheimer's disease. Cell Mol Neurobiol 36(2):167–179. https://doi.org/10.1007/s10571-015-0261-z
11. Kelleher RJ, Soiza RL (2013) Evidence of endothelial dysfunction in the development of Alzheimer's disease: is Alzheimer's a vascular disorder? Am J Cardiovasc Dis 3(4):197–226
12. Trojanowski JQ, Growdon JH (1998) A new consensus report on biomarkers for the early antemortem diagnosis of Alzheimer disease: current status, relevance to drug discovery, and recommendations for future research. J Neuropathol Exp Neurol 57(6):643–644
13. Wierenga CE, Hays CC, Zlatar ZZ (2014) Cerebral blood flow measured by arterial spin labeling MRI as a preclinical marker of Alzheimer's disease. J Alzheimers Dis 42(Suppl 4):S411–S419. https://doi.org/10.3233/JAD-141467
14. Binnewijzend MAA, Benedictus MR, Kuijer JPA et al (2016) Cerebral perfusion in the predementia stages of Alzheimer's disease. Eur Radiol 26(2):506–514. https://doi.org/10.1007/s00330-015-3834-9
15. Maier FC, Wehrl HF, Schmid AM et al (2014) Longitudinal PET-MRI reveals beta-amyloid deposition and rCBF dynamics and connects vascular amyloidosis to quantitative loss of perfusion. Nat Med 20(12):1485–1492. https://doi.org/10.1038/nm.3734
16. Kato T, Inui Y, Nakamura A et al (2016) Brain fluorodeoxyglucose (FDG) PET in dementia. Ageing Res Rev 30:73–84. https://doi.org/10.1016/j.arr.2016.02.003
17. Granert O, Drzezga AE, Boecker H et al (2015) Metabolic topology of neurodegenerative disorders: influence of cognitive and motor deficits. J Nucl Med 56(12):1916–1921. https://doi.org/10.2967/jnumed.115.156067
18. Matsuda H (2007) Role of neuroimaging in Alzheimer's disease, with emphasis on brain perfusion SPECT. J Nucl Med 48(8):1289–1300. https://doi.org/10.2967/jnumed.106.037218
19. Sakamoto S, Matsuda H, Asada T et al (2003) Apolipoprotein E genotype and early Alzheimer's disease: a longitudinal SPECT study. J Neuroimaging 13(2):113–123
20. Knopman DS, Jack CR, Wiste HJ et al (2014) 18F-fluorodeoxyglucose positron emission tomography, aging, and apolipoprotein E genotype in cognitively normal persons. Neurobiol Aging 35(9):2096–2106. https://doi.org/10.1016/j.neurobiolaging.2014.03.006
21. Nasrallah I, Dubroff J (2013) An overview of PET neuroimaging. Semin Nucl Med 43(6):449–461. https://doi.org/10.1053/j.semnuclmed.2013.06.003
22. Buxton RB (2009) Introduction to functional magnetic resonance imaging: principles and techniques, 2nd edn. Cambridge University Press, Cambridge

23. Damoiseaux JS, Rombouts SARB, Barkhof F et al (2006) Consistent resting-state networks across healthy subjects. Proc Natl Acad Sci U S A 103(37):13848–13853. https://doi.org/10.1073/pnas.0601417103
24. Guerra-Carrillo B, Mackey AP, Bunge SA (2014) Resting-state fMRI: a window into human brain plasticity. Neuroscientist 20(5):522–533. https://doi.org/10.1177/1073858414524442
25. Chuang K-H, van Gelderen P, Merkle H et al (2008) Mapping resting-state functional connectivity using perfusion MRI. Neuroimage 40(4):1595–1605. https://doi.org/10.1016/j.neuroimage.2008.01.006
26. Biswal BB, van Kylen J, Hyde JS (1997) Simultaneous assessment of flow and BOLD signals in resting-state functional connectivity maps. NMR Biomed 10(4–5):165–170
27. Watanabe T, Kan S, Koike T et al (2014) Network-dependent modulation of brain activity during sleep. Neuroimage 98:1–10. https://doi.org/10.1016/j.neuroimage.2014.04.079
28. Greicius MD, Srivastava G, Reiss AL et al (2004) Default-mode network activity distinguishes Alzheimer's disease from healthy aging: evidence from functional MRI. Proc Natl Acad Sci U S A 101(13):4637–4642. https://doi.org/10.1073/pnas.0308627101
29. Liu TT, Brown GG (2007) Measurement of cerebral perfusion with arterial spin labeling: Part 1. Methods. J Int Neuropsychol Soc 13(3):517–525. https://doi.org/10.1017/S1355617707070646
30. Lee SP, Duong TQ, Yang G et al (2001) Relative changes of cerebral arterial and venous blood volumes during increased cerebral blood flow: implications for BOLD fMRI. Magn Reson Med 45(5):791–800
31. Parkes LM, Rashid W, Chard DT et al (2004) Normal cerebral perfusion measurements using arterial spin labeling: reproducibility, stability, and age and gender effects. Magn Reson Med 51(4):736–743. https://doi.org/10.1002/mrm.20023
32. Besson FL, La Joie R, Doeuvre L et al (2015) Cognitive and brain profiles associated with current neuroimaging biomarkers of preclinical Alzheimer's disease. J Neurosci 35(29):10402–10411. https://doi.org/10.1523/JNEUROSCI.0150-15.2015
33. Teipel S, Grothe MJ (2016) Does posterior cingulate hypometabolism result from disconnection or local pathology across preclinical and clinical stages of Alzheimer's disease? Eur J Nucl Med Mol Imaging 43(3):526–536. https://doi.org/10.1007/s00259-015-3222-3
34. Ewers M, Brendel M, Rizk-Jackson A et al (2014) Reduced FDG-PET brain metabolism and executive function predict clinical progression in elderly healthy subjects. Neuroimage Clin 4:45–52. https://doi.org/10.1016/j.nicl.2013.10.018
35. Ewers M, Insel PS, Stern Y et al (2013) Cognitive reserve associated with FDG-PET in preclinical Alzheimer disease. Neurology 80(13):1194–1201. https://doi.org/10.1212/WNL.0b013e31828970c2
36. Knopman DS, Jack CR, Wiste HJ et al (2013) Selective worsening of brain injury biomarker abnormalities in cognitively normal elderly persons with β-amyloidosis. JAMA Neurol 70(8):1030–1038. https://doi.org/10.1001/jamaneurol.2013.182
37. Wirth M, Oh H, Mormino EC et al (2013) The effect of amyloid β on cognitive decline is modulated by neural integrity in cognitively normal elderly. Alzheimers Dement 9(6):687–698.e1. https://doi.org/10.1016/j.jalz.2012.10.012
38. Ossenkoppele R, van der Flier WM, Verfaillie SCJ et al (2014) Long-term effects of amyloid, hypometabolism, and atrophy on neuropsychological functions. Neurology 82(20):1768–1775. https://doi.org/10.1212/WNL.0000000000000432
39. Knopman DS, Jack CR, Wiste HJ et al (2013) Brain injury biomarkers are not dependent on β-amyloid in normal elderly. Ann Neurol 73(4):472–480. https://doi.org/10.1002/ana.23816
40. Seo EH, Kim SH, Park SH et al (2016) Topographical APOE ε4 genotype influence on cerebral metabolism in the continuum of Alzheimer's disease: amyloid burden adjusted analysis. J Alzheimers Dis 54(2):559–568. https://doi.org/10.3233/JAD-160395
41. Yi D, Lee DY, Sohn BK et al (2014) Beta-amyloid associated differential effects of APOE ε4 on brain metabolism in cognitively normal elderly. Am J Geriatr Psychiatr 22(10):961–970. https://doi.org/10.1016/j.jagp.2013.12.173
42. Krell-Roesch J, Ruider H, Lowe VJ et al (2016) FDG-PET and neuropsychiatric symptoms among cognitively normal elderly persons: the Mayo Clinic study of aging. J Alzheimers Dis 53(4):1609–1616. https://doi.org/10.3233/JAD-160326
43. Didic M, Felician O, Gour N et al (2015) Rhinal hypometabolism on FDG PET in healthy APO-E4 carriers: impact on memory function and metabolic networks. Eur J Nucl Med Mol Imaging 42(10):1512–1521. https://doi.org/10.1007/s00259-015-3057-y

44. Arenaza-Urquijo EM, Gonneaud J, Fouquet M et al (2015) Interaction between years of education and APOE ε4 status on frontal and temporal metabolism. Neurology 85(16):1392–1399. https://doi.org/10.1212/WNL.0000000000002034
45. Yao Z, Hu B, Zheng J et al (2015) A FDG-PET study of metabolic networks in apolipoprotein E ε4 allele carriers. PLoS One 10(7):e0132300. https://doi.org/10.1371/journal.pone.0132300
46. Chen K, Ayutyanont N, Langbaum JBS et al (2012) Correlations between FDG PET glucose uptake-MRI gray matter volume scores and apolipoprotein E ε4 gene dose in cognitively normal adults: a cross-validation study using voxel-based multi-modal partial least squares. Neuroimage 60(4):2316–2322. https://doi.org/10.1016/j.neuroimage.2012.02.005
47. Chételat G, Eustache F, Viader F et al (2005) FDG-PET measurement is more accurate than neuropsychological assessments to predict global cognitive deterioration in patients with mild cognitive impairment. Neurocase 11(1):14–25. https://doi.org/10.1080/13554790490896938
48. Morbelli S, Piccardo A, Villavecchia G et al (2010) Mapping brain morphological and functional conversion patterns in amnestic MCI: a voxel-based MRI and FDG-PET study. Eur J Nucl Med Mol Imaging 37(1):36–45. https://doi.org/10.1007/s00259-009-1218-6
49. Nobili F, Salmaso D, Morbelli S et al (2008) Principal component analysis of FDG PET in amnestic MCI. Eur J Nucl Med Mol Imaging 35(12):2191–2202. https://doi.org/10.1007/s00259-008-0869-z
50. Coutinho AMN, Porto FHG, Duran FLS et al (2015) Brain metabolism and cerebrospinal fluid biomarkers profile of non-amnestic mild cognitive impairment in comparison to amnestic mild cognitive impairment and normal older subjects. Alzheimers Res Ther 7(1):58. https://doi.org/10.1186/s13195-015-0143-0
51. Bailly M, Destrieux C, Hommet C et al (2015) Precuneus and cingulate cortex atrophy and hypometabolism in patients with Alzheimer's disease and mild cognitive impairment: MRI and (18)F-FDG PET quantitative analysis using FreeSurfer. Biomed Res Int 2015:583931. https://doi.org/10.1155/2015/583931
52. Pagani M, Carli F d, Morbelli S et al (2015) Volume of interest-based 18Ffluorodeoxyglucose PET discriminates MCI converting to Alzheimer's disease from healthy controls. A European Alzheimer's Disease Consortium (EADC) study. Neuroimage Clin 7:34–42. https://doi.org/10.1016/j.nicl.2014.11.007
53. Espinosa A, Alegret M, Pesini P et al (2017) Cognitive composites domain scores related to neuroimaging biomarkers within probable-amnestic mild cognitive impairment-storage subtype. J Alzheimers Dis 57(2):447–459. https://doi.org/10.3233/JAD-161223
54. Franzmeier N, Duering M, Weiner M et al (2017) Left frontal cortex connectivity underlies cognitive reserve in prodromal Alzheimer disease. Neurology 88(11):1054–1061. https://doi.org/10.1212/WNL.0000000000003711
55. Förster S, Buschert VC, Teipel SJ et al (2011) Effects of a 6-month cognitive intervention on brain metabolism in patients with amnestic MCI and mild Alzheimer's disease. J Alzheimers Dis 26(Suppl 3):337–348. https://doi.org/10.3233/JAD-2011-0025
56. Chen K, Langbaum JBS, Fleisher AS et al (2010) Twelve-month metabolic declines in probable Alzheimer's disease and amnestic mild cognitive impairment assessed using an empirically pre-defined statistical region-of-interest: findings from the Alzheimer's disease neuroimaging initiative. NeuroImage 51(2):654–664. https://doi.org/10.1016/j.neuroimage.2010.02.064
57. Clerici F, Del Sole A, Chiti A et al (2009) Differences in hippocampal metabolism between amnestic and non-amnestic MCI subjects: automated FDG-PET image analysis. Q J Nucl Med Mol Imaging 53(6):646–657
58. Pagani M, Dessi B, Morbelli S et al (2010) MCI patients declining and not-declining at mid-term follow-up: FDG-PET findings. Curr Alzheimer Res 7(4):287–294
59. Teune LK, Bartels AL, Jong BM d et al (2010) Typical cerebral metabolic patterns in neurodegenerative brain diseases. Mov Disord 25(14):2395–2404. https://doi.org/10.1002/mds.23291
60. Donnemiller E, Heilmann J, Wenning GK et al (1997) Brain perfusion scintigraphy with 99mTc-HMPAO or 99mTc-ECD and 123I-beta-CIT single-photon emission tomography in dementia of the Alzheimer-type and diffuse Lewy body disease. Eur J Nucl Med 24(3):320–325
61. Herholz K, Salmon E, Perani D et al (2002) Discrimination between Alzheimer dementia and controls by automated analysis of multicenter FDG PET. Neuroimage 17(1):302–316
62. Meltzer CC, Zubieta JK, Brandt J et al (1996) Regional hypometabolism in Alzheimer's dis-

ease as measured by positron emission tomography after correction for effects of partial volume averaging. Neurology 47(2):454–461
63. Tam CWC, Burton EJ, McKeith IG et al (2005) Temporal lobe atrophy on MRI in Parkinson disease with dementia: a comparison with Alzheimer disease and dementia with Lewy bodies. Neurology 64(5):861–865. https://doi.org/10.1212/01.WNL.0000153070.82309.D4
64. Buckner RL, Andrews-Hanna JR, Schacter DL (2008) The brain's default network: anatomy, function, and relevance to disease. Ann N Y Acad Sci 1124:1–38. https://doi.org/10.1196/annals.1440.011
65. Bangen KJ, Restom K, Liu TT et al (2009) Differential age effects on cerebral blood flow and BOLD response to encoding: associations with cognition and stroke risk. Neurobiol Aging 30(8):1276–1287. https://doi.org/10.1016/j.neurobiolaging.2007.11.012
66. Lee C, Lopez OL, Becker JT et al (2009) Imaging cerebral blood flow in the cognitively normal aging brain with arterial spin labeling: implications for imaging of neurodegenerative disease. J Neuroimaging 19(4):344–352. https://doi.org/10.1111/j.1552-6569.2008.00277.x
67. Sandson TA, O'Connor M, Sperling RA et al (1996) Noninvasive perfusion MRI in Alzheimer's disease: a preliminary report. Neurology 47(5):1339–1342
68. Xekardaki A, Rodriguez C, Montandon M-L et al (2015) Arterial spin labeling may contribute to the prediction of cognitive deterioration in healthy elderly individuals. Radiology 274(2):490–499. https://doi.org/10.1148/radiol.14140680
69. Stomrud E, Forsberg A, Hägerström D et al (2012) CSF biomarkers correlate with cerebral blood flow on SPECT in healthy elderly. Dement Geriatr Cogn Disord 33(2–3):156–163. https://doi.org/10.1159/000338185
70. Sojkova J, Beason-Held L, Zhou Y et al (2008) Longitudinal cerebral blood flow and amyloid deposition: an emerging pattern? J Nucl Med 49(9):1465–1471. https://doi.org/10.2967/jnumed.108.051946
71. Edelman K, Tudorascu D, Agudelo C et al (2017) Amyloid-beta deposition is associated with increased medial temporal lobe activation during memory encoding in the cognitively normal elderly. Am J Geriatr Psychiatry 25(5):551–560. https://doi.org/10.1016/j.jagp.2016.12.021
72. Gordon BA, Zacks JM, Blazey T et al (2015) Task-evoked fMRI changes in attention networks are associated with preclinical Alzheimer's disease biomarkers. Neurobiol Aging 36(5):1771–1779. https://doi.org/10.1016/j.neurobiolaging.2015.01.019
73. Sheline YI, Raichle ME, Snyder AZ et al (2010) Amyloid plaques disrupt resting state default mode network connectivity in cognitively normal elderly. Biol Psychiatry 67(6):584–587. https://doi.org/10.1016/j.biopsych.2009.08.024
74. Schultz AP, Chhatwal JP, Hedden T et al (2017) Phases of hyperconnectivity and hypoconnectivity in the default mode and salience networks track with amyloid and tau in clinically normal individuals. J Neurosci 37(16):4323–4331. https://doi.org/10.1523/JNEUROSCI.3263-16.2017
75. Brier MR, Thomas JB, Snyder AZ et al (2014) Unrecognized preclinical Alzheimer disease confounds rs-fcMRI studies of normal aging. Neurology 83(18):1613–1619. https://doi.org/10.1212/WNL.0000000000000939
76. Brier MR, Thomas JB, Fagan AM et al (2014) Functional connectivity and graph theory in preclinical Alzheimer's disease. Neurobiol Aging 35(4):757–768. https://doi.org/10.1016/j.neurobiolaging.2013.10.081
77. Wang L, Brier MR, Snyder AZ et al (2013) Cerebrospinal fluid Aβ42, phosphorylated Tau181, and resting-state functional connectivity. JAMA Neurol 70(10):1242–1248. https://doi.org/10.1001/jamaneurol.2013.3253
78. Drzezga A, Becker JA, van Dijk KRA et al (2011) Neuronal dysfunction and disconnection of cortical hubs in non-demented subjects with elevated amyloid burden. Brain 134(Pt 6):1635–1646. https://doi.org/10.1093/brain/awr066
79. Mattsson N, Tosun D, Insel PS et al (2014) Association of brain amyloid-beta with cerebral perfusion and structure in Alzheimer's disease and mild cognitive impairment. Brain 137(Pt 5):1550–1561. https://doi.org/10.1093/brain/awu043
80. Thambisetty M, Beason-Held L, An Y et al (2010) APOE epsilon4 genotype and longitudinal changes in cerebral blood flow in normal aging. Arch Neurol 67(1):93–98. https://doi.org/10.1001/archneurol.2009.913
81. Suwa A, Nishida K, Utsunomiya K et al (2015) Neuropsychological evaluation and cerebral blood flow effects of apolipoprotein E4 in Alzheimer's disease patients after one year of treatment: an exploratory study. Dement Geriatr Cogn Dis Extra 5(3):414–423. https://doi.org/10.1159/000440714
82. Lehtovirta M, Kuikka J, Helisalmi S et al (1998) Longitudinal SPECT study in

Alzheimer's disease: relation to apolipoprotein E polymorphism. J Neurol Neurosurg Psychiatry 64(6):742–746
83. Lehtovirta M, Soininen H, Laakso MP et al (1996) SPECT and MRI analysis in Alzheimer's disease: relation to apolipoprotein E epsilon 4 allele. J Neurol Neurosurg Psychiatry 60(6):644–649
84. Devanand DP, van Heertum RL, Kegeles LS et al (2010) (99m)Tc hexamethyl-propylene-aminoxime single-photon emission computed tomography prediction of conversion from mild cognitive impairment to Alzheimer disease. Am J Geriatr Psychiatr 18(11):959–972. https://doi.org/10.1097/JGP.0b013e3181ec8696
85. Alegret M, Cuberas-Borrós G, Espinosa A et al (2014) Cognitive, genetic, and brain perfusion factors associated with four year incidence of Alzheimer's disease from mild cognitive impairment. J Alzheimers Dis 41(3):739–748. https://doi.org/10.3233/JAD-132516
86. Fleisher AS, Sherzai A, Taylor C et al (2009) Resting-state BOLD networks versus task-associated functional MRI for distinguishing Alzheimer's disease risk groups. Neuroimage 47(4):1678–1690. https://doi.org/10.1016/j.neuroimage.2009.06.021
87. Harrison TM, Burggren AC, Small GW et al (2016) Altered memory-related functional connectivity of the anterior and posterior hippocampus in older adults at increased genetic risk for Alzheimer's disease. Hum Brain Mapp 37(1):366–380. https://doi.org/10.1002/hbm.23036
88. Sheline YI, Morris JC, Snyder AZ et al (2010) APOE4 allele disrupts resting state fMRI connectivity in the absence of amyloid plaques or decreased CSF Aβ42. J Neurosci 30(50):17035–17040. https://doi.org/10.1523/JNEUROSCI.3987-10.2010
89. Michels L, Warnock G, Buck A et al (2016) Arterial spin labeling imaging reveals widespread and Abeta-independent reductions in cerebral blood flow in elderly apolipoprotein epsilon-4 carriers. J Cereb Blood Flow Metab 36(3):581–595. https://doi.org/10.1177/0271678X15605847
90. Bangen KJ, Restom K, Liu TT et al (2012) Assessment of Alzheimer's disease risk with functional magnetic resonance imaging: an arterial spin labeling study. J Alzheimers Dis 31(Suppl 3):S59–S74. https://doi.org/10.3233/JAD-2012-120292
91. Wierenga CE, Clark LR, Dev SI et al (2013) Interaction of age and APOE genotype on cerebral blood flow at rest. J Alzheimers Dis 34(4):921–935. https://doi.org/10.3233/JAD-121897
92. Kim SM, Kim MJ, Rhee HY et al (2013) Regional cerebral perfusion in patients with Alzheimer's disease and mild cognitive impairment: effect of APOE epsilon4 allele. Neuroradiology 55(1):25–34. https://doi.org/10.1007/s00234-012-1077-x
93. Beason-Held LL, Goh JO, An Y et al (2013) Changes in brain function occur years before the onset of cognitive impairment. J Neurosci 33(46):18008–18014. https://doi.org/10.1523/JNEUROSCI.1402-13.2013
94. Adriaanse SM, Sanz-Arigita EJ, Binnewijzend MAA et al (2014) Amyloid and its association with default network integrity in Alzheimer's disease. Hum Brain Mapp 35(3):779–791. https://doi.org/10.1002/hbm.22213
95. Binnewijzend MAA, Schoonheim MM, Sanz-Arigita E et al (2012) Resting-state fMRI changes in Alzheimer's disease and mild cognitive impairment. Neurobiol Aging 33(9):2018–2028. https://doi.org/10.1016/j.neurobiolaging.2011.07.003
96. Liang P, Wang Z, Yang Y et al (2012) Three subsystems of the inferior parietal cortex are differently affected in mild cognitive impairment. J Alzheimers Dis 30(3):475–487. https://doi.org/10.3233/JAD-2012-111721
97. Ostergaard L, Aamand R, Gutierrez-Jimenez E et al (2013) The capillary dysfunction hypothesis of Alzheimer's disease. Neurobiol Aging 34(4):1018–1031. https://doi.org/10.1016/j.neurobiolaging.2012.09.011
98. Alexopoulos P, Sorg C, Forschler A et al (2012) Perfusion abnormalities in mild cognitive impairment and mild dementia in Alzheimer's disease measured by pulsed arterial spin labeling MRI. Eur Arch Psychiatry Clin Neurosci 262(1):69–77. https://doi.org/10.1007/s00406-011-0226-2
99. Dai W, Lopez OL, Carmichael OT et al (2009) Mild cognitive impairment and alzheimer disease: patterns of altered cerebral blood flow at MR imaging. Radiology 250(3):856–866. https://doi.org/10.1148/radiol.2503080751
100. Johnson NA, Jahng G-H, Weiner MW et al (2005) Pattern of cerebral hypoperfusion in Alzheimer disease and mild cognitive impairment measured with arterial spin-labeling MR imaging: initial experience. Radiology 234(3):851–859. https://doi.org/10.1148/radiol.2343040197
101. Xu G, Antuono PG, Jones J et al (2007) Perfusion fMRI detects deficits in regional CBF during memory-encoding tasks in MCI subjects. Neurology 69(17):1650–1656. https://doi.org/10.1212/01.wnl.0000296941.06685.22

102. Xie L, Dolui S, Das SR et al (2016) A brain stress test: cerebral perfusion during memory encoding in mild cognitive impairment. Neuroimage Clin 11:388–397. https://doi.org/10.1016/j.nicl.2016.03.002
103. Vanitallie TB (2013) Preclinical sporadic Alzheimer's disease: target for personalized diagnosis and preventive intervention. Metab Clin Exp 62(Suppl 1):S30–S33. https://doi.org/10.1016/j.metabol.2012.08.024
104. Lee IH, Kim ST, Kim H-J et al (2010) Analysis of perfusion weighted image of CNS lymphoma. Eur J Radiol 76(1):48–51. https://doi.org/10.1016/j.ejrad.2009.05.013
105. Jack CR, Wiste HJ, Weigand SD et al (2013) Amyloid-first and neurodegeneration-first profiles characterize incident amyloid PET positivity. Neurology 81(20):1732–1740. https://doi.org/10.1212/01.wnl.0000435556.21319.e4
106. Verfaillie SCJ, Adriaanse SM, Binnewijzend MAA et al (2015) Cerebral perfusion and glucose metabolism in Alzheimer's disease and frontotemporal dementia: two sides of the same coin? Eur Radiol 25(10): 3050–3059. https://doi.org/10.1007/s00330-015-3696-1
107. Wang L, Day J, Roe CM et al (2014) The effect of APOE ε4 allele on cholinesterase inhibitors in patients with Alzheimer disease: evaluation of the feasibility of resting state functional connectivity magnetic resonance imaging. Alzheimer Dis Assoc Disord 28(2):122–127. https://doi.org/10.1097/WAD.0b013e318299d096
108. Sheline YI, Raichle ME (2013) Resting state functional connectivity in preclinical Alzheimer's disease. Biol Psychiatry 74(5):340–347. https://doi.org/10.1016/j.biopsych.2012.11.028
109. Morbelli S, Brugnolo A, Bossert I et al (2015) Visual versus semi-quantitative analysis of 18F-FDG-PET in amnestic MCI: an European Alzheimer's Disease Consortium (EADC) project. J Alzheimers Dis 44(3): 815–826. https://doi.org/10.3233/JAD-142229

Check for updates

Chapter 9

Imaging of Brain Amyloid Load in Early Alzheimer's Disease

Timo Grimmer and Jennifer Grace Perryman

Abstract

The content of this chapter deals with the usefulness and indications of amyloid PET for the diagnosis of patients with Alzheimer's disease (AD). In addition to a description of amyloid PET itself, topics such as early and differential diagnosis of AD are discussed. Measuring amyloid in CSF as an alternative method to amyloid PET and the associations between cerebral amyloid load and the apolipoprotein E genotype are also covered in this section.

Key words Alzheimer's disease, Positron-emission tomography, Cerebrospinal fluid, Biomarker, Early diagnosis, Prognosis, Apolipoprotein E

1 Introduction

As Alzheimer's disease (AD) is the primary cause of dementia [1], it seems intuitive that a method which is able to recognize, predict, and differentiate this disease should be offered to patients worldwide. The measuring of cerebral amyloid plaque load by use of positron-emission tomography (PET) can achieve these aims.

Senile plaques and neurofibrillary tangles together with a loss of neurons and synapses are the characteristic histopathological features of AD [2, 3]. Amyloid β-protein (Aβ) is the main component of these senile plaques. As known from cross-sectional histopathological studies, cerebral amyloid load slowly increases with the progression of AD, making its way from neocortical across allocortical areas all the way to the cerebellum [3, 4]. In amyloid PET, tracers like [^{11}C]PiB [5] attach themselves to the amyloid load, thereby allowing for its imaging. A more detailed description of the process will be given in the methods section below. The signal intensity measured by amyloid PET closely corresponds to the extent of amyloid plaques as measured by histopathology [6];

Robert Perneczky (ed.), *Biomarkers for Preclinical Alzheimer's Disease*, Neuromethods, vol. 137, https://doi.org/10.1007/978-1-4939-7674-4_9, © Springer Science+Business Media, LLC 2018

hence, amyloid PET appears to be a valid method of measuring amyloid plaque load in vivo.

In addition to supporting the diagnosis of AD, amyloid PET could also be useful in three other major clinical scenarios: (1) to differentiate between causes of dementia; (2) to diagnose AD at pre-dementia stages, i.e., at early symptomatic or even asymptomatic stages; and (3) to rule out AD as an underlying cause in patients and in persons with subjective and/or objective memory impairment [7].

2 Implications and Utility of Amyloid Depiction via Positron-Emission Tomography

2.1 Differential Diagnosis

The value of amyloid PET for differential diagnosis varies between the particular causes of dementia, depending on the occurrence of cerebral amyloid load [7]. The first study that successfully used amyloid PET to differentiate between semantic dementia, a subtype of the frontotemporal lobar degenerations (FTLD) [8] usually caused by non-amyloid pathologies, and AD yielded a perfect group discrimination [9]. A similar result was achieved differentiating AD from the behavioral variant of frontotemporal dementia (bvFTD) [10], another FTLD subtype. However, distinguishing AD from dementias associated with Parkinson's disease, such as Lewy body disease (LBD), may be difficult as heightened cerebral amyloid plaque load can be found in both [11]. The usefulness of amyloid PET for differential diagnosis is supported by a recent meta-analysis confirming the associations between amyloid positivity as shown by amyloid imaging and clinical diagnosis [12]: Patients with the clinical diagnosis of dementia due to AD were 93% amyloid positive at 50 years of age and 79% at 90 years of age. Although this finding does not seem intuitive at first glance, there are several explanations as to why a smaller percentage of amyloid positivity prevalence can be found in older patients. The primary cause may be the occurrence of rivaling causes for dementia in older patients, resembling the clinical picture of AD. Although less likely, it may be possible that either AD in elderly patients is caused less by cerebral amyloid load or that the tracer binds less efficiently to amyloid of older patients. The mean prevalence estimate for patients with the clinical diagnosis of dementia due to AD was 88%. Patients with dementia due to Lewy bodies (DLB) were amyloid positive in 51% of cases, a figure compatible with the histopathological finding that around 50% of patients with DLB show amyloid depositions as discussed above. Patients with FTLD were amyloid positive only in 12% of cases, which is compatible with the histopathological finding that FTLD are caused by non-amyloid pathology only.

Despite these convincing results, clinicians are advised to carefully evaluate the results of an amyloid PET. They must judge whether the symptomatology is adequately explained by the brain disease revealed through amyloid PET or perhaps an additional cause.

2.2 Early Diagnosis

Amyloid PET may also be used as a means to diagnose AD at a very early stage as amyloid deposition precedes initial clinical symptoms by years if not decades [13], with estimates giving a time period of 20–30 years between the first discovery of an abnormal amyloid load and the beginning of dementia. In patients with mild cognitive impairment (MCI), a predisposing factor and frequent precursor of AD, a positive amyloid PET, for example, using the [11C] PiB compound, can be found in 50% [1]. In accordance with the assumption that amyloid precedes clinical symptoms, it was shown that amyloid positivity in clinically asymptomatic persons increases with age [12]. Usage of amyloid PET could therefore help discover the ailment and warn patients many decades prior to the outbreak of symptoms.

As there is currently no treatment available for AD at an asymptomatic stage, however, the advantages of early diagnosis are debatable. Nevertheless, it should be mentioned that early diagnosis would allow patients to plan better for their future circumstances regarding financial matters and caregiving and also inform themselves and relevant people about matters such as safety in everyday life, driving, and counseling options [14].

2.3 Excluding Alzheimer's Disease

As well as indicating a progression to AD, amyloid PET may also be used to rule out the disease as the cause of dementia: According to the guidance documents of the German Society of Nuclear Medicine and the American Society of Nuclear Medicine and Molecular Imaging [15, 16], it is highly unlikely to suffer from Alzheimer's disease without cerebral amyloid load in the form of plaques. Should the amyloid PET therefore be negative, one may assume with a high probability that AD is not the cause for the patient's dementia.

It may also help reassure patients who are concerned about developing the illness in the future. A survey by the Harvard School of Public Health showed that around 89% of patients stated that they would like to be informed whether or not AD was the cause for their impairments [14].

2.4 Limitations

Negative amyloid PET findings may be due to misdiagnosis; unsuccessful tracer production may also be possible in some cases [7]. Another potential limitation is the availability of tracer production/delivery and PET cameras. Furthermore, one should keep in mind that the examination involves radiation exposure (5.8–7.0 mSv).

2.5 Appropriate Use Criteria

Amyloid PET is already commonly clinically used and recommended in several guidelines (German Society of Nuclear Medicine (DGN) [16], European Association of Nuclear Medicine (EANM) [17], the American Society of Nuclear Medicine and Molecular Imaging (SNMMI) [15], German S3 Guideline Dementia [18]). In order to provide a framework for guiding clinicians who intend to apply amyloid PET, appropriate use criteria for this particular method were published in 2013 [15]. For patients suffering from a subjectively and objectively identified cognitive impairment, with AD as a likely, but as yet undecided by experts, diagnosis, and assuming that a positive or negative amyloid PET would provide assurance for the clinical diagnosis, the authors concluded that amyloid PET is a useful tool to draw on in the following situations:

1. The patient suffers from MCI in a persistent or progressively unexplained form.
2. Patients with the clinical criteria for possible AD but with unclear clinical presentation, e.g., an atypical clinical course or etiologically mixed.
3. Patient with early-onset dementia.

The following scenarios are deemed as inappropriate for the use of amyloid PET:

1. Patient fulfilling the clinical criteria for probable AD including typical age of onset.
2. To determine the severity of the dementing syndrome.
3. Asymptomatic persons with a positive family history of dementia or presence of apolipoprotein E (*APOE*) ε4 only.
4. The patient reports cognitive complaints which are, however, without substantiation on clinical examination (memory complainer).
5. In asymptomatic individuals.
6. For nonmedical use (e.g., legal insurance coverage or employment screening).

Unfortunately, the authors missed one important indication: identifying participants for anti-amyloid therapy. In persons or patients interested in participating in an anti-amyloid drug trial, amyloid positivity needs to be confirmed in advance. This may be achieved in vivo and noninvasively by amyloid PET.

Association of cerebral amyloid load and the apolipoprotein E 4 genotype and its implications, as well as an alternative method for assessing amyloid in vivo, shall be discussed in the notes section below.

3 Materials

Authorized radiopharmaceuticals include florbetapir (Amyvid®) [19], flutemetamol (Vizamyl®) [20], and florbetaben (Neuraceq®) [21]. All show a decreased signal-to-noise ratio compared to [^{11}C] PiB but are comparable with regard to accuracy: 97% concordance comparing PiB to florbetapir [22], 100% concordance comparing PiB to flutemetamol [23], and 100% concordance comparing PiB to florbetaben [24], respectively.

The comparatively long 110-min half-life of F-18-labeled PET compounds allows for the use of amyloid PET in regular clinical practice [15]. Production controls the quality of each product before distribution. Refer to individual products for storage information such as temperature and length. Information concerning their toxic attributes and consequent safety measures may also be found there. The compound [^{11}C]PiB is only available at special centers due to its short half-life of 20 min, which necessitates an on-site cyclotron.

4 Method

In the following, the method of amyloid PET shall be described using the [^{11}C]PiB compound as an example (Fig. 1). The procedure takes place as follows [9, 25]:

1. Examinations are performed on the Siemens ECAT HR PET scanner (CTI, Knoxville, TN) and follow a simplified image acquisition procedure [26] in order to avoid lengthy scanning procedures and to minimize the risk of movement artifacts in the cognitively impaired patients.
2. All patients are injected with 370 MBq [^{11}C]PiB at rest outside the scanner. Thirty minutes later, patients are placed in the scanner.
3. At 40 min postinjection, three 10-min frames are started and later summed into a single frame (40–70 min).
4. Acquisition is carried out in 3-D mode, and a transmission scan is carried out to allow for later attenuation correction.
5. Images are reconstructed, corrected of dead time, scatter, and attenuation, and standard uptake value ratio (SUVR) 40–70-min images are generated.
6. Images are spatially normalized using SPM2 [27]. The [^{11}C]PiB data are first coregistered to each individual's volumetric MRI and then automatically spatially normalized to the T1 MRI MNI template in SPM2 using warping parameters derived from the individual MRI normalization performed previously [28].

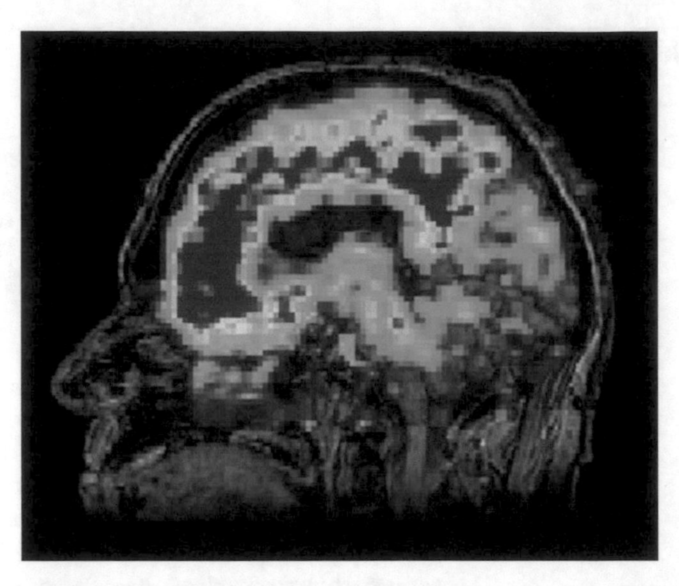

Fig. 1 Amyloid PET using 11C-PiB in a patient with dementia due to Alzheimer's disease overlaid on an MRI scan, sagittal view

7. For the assessment of an amyloid PET using F18 tracers, the FDA-approved evaluation algorithm should be used. Please compare the instructions accompanying the packaged tracers for further information on certified evaluation programs.
8. Concerning the evaluation of the PET scan, two different methods may be drawn upon: visual and fully automated analyses. The visual analysis may be aided by surface projection of tracer uptake pattern [29]. The automated analysis is based on thresholds of target (e.g., cerebral cortex) to reference tissue (e.g., cerebellar vermis) ratios.

The effectiveness of both methods to predict the progression of patients with MCI to manifest AD dementia within a 2-year time frame was compared by Grimmer et al. [30]. Both methods resulted in perfect negative predictive values (NPV) indicating that a negative amyloid PET rules out the development of manifest AD dementia within a manageable time frame. However, an MCI patient with a positive amyloid PET does progress to AD dementia in approximately 50% of cases, regardless of the method.

5 Associations with Other Biomarkers

As amyloid PET is not available everywhere, there is an alternative way of measuring cerebral amyloid load through analyzing the cerebrospinal fluid (CSF) obtained by spinal tap. Compared to markers in CSF, PET shows only one of the typical histopathological changes in AD [7]. However, one needs to keep in mind that β-amyloid measured in CSF is not interchangeable with its measurement in PET. The latter is a marker of lifetime amyloid deposition, while β-amyloid in CSF reflects current efflux rates. While both measurements may be used interchangeably with regard to diagnosis of AD [31–33], levels of amyloid in CSF and in PET are not associated throughout the brain but in brain areas adjacent to CSF only [34].

As mentioned above, there is an association between both the amount [35] and the progression of cerebral amyloid load [4] and the apolipoprotein E ε4 genotype (APOE-ε4) in AD. APOE-ε4 is a known risk factor for AD [36], the assumption being that APOE-ε4 transports amyloid through the brain-blood barrier [37] and is responsible for higher levels of cerebral hypometabolism in patients with AD that carry the gene as opposed to those who are ε4-negative [35]. A gene dose effect between the ApoE genotype and the uptake ratio of [^{11}C]PiB has been suggested [4], which lies in accordance with observations that there is a more rapid deterioration of clinical symptoms shown by ApoE ε4 carriers [36, 38]. The findings of such studies may be of help when designing future anti-amyloid treatment strategies [4].

References

1. Jansen WJ, Ossenkoppele R, Knol DL et al (2015) Prevalence of cerebral amyloid pathology in persons without dementia: a meta-analysis. JAMA 313:1924–1938
2. Braak H, .Braak E (1991) Neuropathological stageing of Alzheimer-related changes. Acta Neuropathol 82:239–259
3. Thal DR, Rüb U, Orantes M, Braak H (2002) Phases of Aß-deposition in the human brain and its relevance for the development of AD. Neurology 58:1791–1800
4. Grimmer T, Tholen S, Yousefi BH et al (2010) Progression of cerebral amyloid load is associated with the apolipoprotein E epsilon4 genotype in Alzheimer's disease. Biol Psychiatry 68:879–884
5. Klunk WE, Engler H, Nordberg A et al (2004) Imaging brain amyloid in Alzheimer's disease with Pittsburgh Compound-B. Ann Neurol 55:306–319
6. Ikonomovic MD, Klunk WE, Abrahamson EE et al (2008) Post-mortem correlates of in vivo PiB-PET amyloid imaging in a typical case of Alzheimer's disease. Brain 131:1630–1645
7. Grimmer T, Drzezga A, Kurz A (2010) Visualization of amyloid with positron emission tomography. Useful improvement in the diagnosis of dementia? Nervenarzt 81:602–606
8. Brun A, Englund B, Gustafson L et al (1994) Clinical and neuropathological criteria for frontotemporal dementia. The Lund and Manchester Groups. J Neurol Neurosurg Psychiatry 57:416–418
9. Drzezga A, Grimmer T, Henriksen G et al (2008) Imaging of amyloid plaques and cerebral glucose metabolism in semantic dementia and Alzheimer's disease. Neuroimage 39:619–633
10. Engler H, Santillo AF, Wang SX et al (2008) In vivo amyloid imaging with PET in frontotemporal dementia. Eur J Nucl Med Mol Imaging 35:100–106
11. McKeith IG, Dickson DW, Lowe J et al (2005) Diagnosis and management of dementia with Lewy bodies: third report of the DLB Consortium. Neurology 65:1863–1872
12. Ossenkoppele R, Jansen WJ, Rabinovici GD et al (2015) Prevalence of amyloid PET positivity in dementia syndromes: a meta-analysis. JAMA 313:1939–1949
13. Bateman RJ, Xiong C, Benzinger TL et al (2012) Clinical and biomarker changes in dominantly inherited Alzheimer's disease. N Engl J Med 367:795–804
14. Blendon RJ, Benson JM, Wikler EM et al (2012) The impact of experience with a family member with Alzheimer's disease on views about the disease across five countries. Int J Alzheimers Dis 2012:903645
15. Johnson KA, Minoshima S, Bohnen NI et al (2013) Appropriate use criteria for amyloid

PET: a report of the Amyloid Imaging Task Force, the Society of Nuclear Medicine and Molecular Imaging, and the Alzheimer's Association. J Nucl Med 54:476–490

16. Barthel, H., Meyer, P. T., Drzezga, A., Bartenstein, P., Boecker, H., Brust, P., Buchert, R., Coenen, H. H., La Fourgère, C., Gründer, G., Grünwald, F., Krause, B. J., Kuwert, T., Schreckenberger, M., Tatsch, K., Langen, K. J., and Sabri, O. (2015) Beta-amyloid-PET-Bildgebung des Gehirns – DGN-Handlungsempfehlung (S1-Leitlinie). www.awmf.org/uploads/tx_szleitlinien/031-052l_S1_Beta_Amyloid-PET-Bildgebung_Gehirn_2016-02.pdf
17. Minoshima S, Drzezga AE, Barthel H et al (2016) SNMMI procedure standard/EANM practice guideline for amyloid PET imaging of the brain 1.0. J Nucl Med 57:1316–1322
18. Deutsche Gesellschaft für Psychiatrie und Psychotherapie, Psychosomatik und Nervenheilkunde DGPPN and Deutsche Gesellschaft für Neurologie (DGN) (2016) S3-Leitlinie "Demenzen". https://www.dgppn.de/fileadmin/user_upload/_medien/download/pdf/kurzversion-leitlinien/REV_S3-leiltlinie-demenzen.pdf
19. https://www.accessdata.fda.gov/drugsatfda_docs/label/2012/202008s000lbl.pdf. 18 Apr 2017
20. https://www.accessdata.fda.gov/drugsatfda_docs/label/2016/203137s005lbl.pdf. 18 Apr 2017
21. https://www.accessdata.fda.gov/drugsatfda_docs/label/2014/204677s000lbl.pdf. 18 Apr 2017
22. Landau SM, Breault C, Joshi AD et al (2013) Amyloid-beta imaging with Pittsburgh compound B and florbetapir: comparing radiotracers and quantification methods. J Nucl Med 54:70–77
23. Vandenberghe R, Van Laere K, Ivanoiu A et al (2010) 18F-flutemetamol amyloid imaging in Alzheimer disease and mild cognitive impairment: a phase 2 trial. Ann Neurol 68:319–329
24. Villemagne VL, Mulligan RS, Pejoska S et al (2012) Comparison of 11C-PiB and 18F-florbetaben for Abeta imaging in ageing and Alzheimer's disease. Eur J Nucl Med Mol Imaging 39:983–989
25. Grimmer T, Henriksen G, Wester HJ et al (2009) Pittsburgh Compound B(PIB) uptake in positron emission tomography is associated with clinical severity of Alzheimer's disease. Neurobiol Aging 30(12):1902–1909
26. Lopresti BJ, Klunk WE, Mathis CA et al (2005) Simplified quantification of Pittsburgh Compound B amyloid imaging PET studies: a comparative analysis. J Nucl Med 46:1959–1972
27. Friston KJ, Holmes AP, Worsley KJ (1994) Statistical parametric mapping in functional imaging: a general linear approach. Hum Brain Mapp 2:189–210
28. Ziolko SK, Weissfeld LA, Klunk WE et al (2006) Evaluation of voxel-based methods for the statistical analysis of PIB PET amyloid imaging studies in Alzheimer's disease. Neuroimage 33:94–102
29. Minoshima S, Frey KA, Koeppe RA, Foster NL, Kuhl DE (1995) A diagnostic approach in Alzheimer's disease using three-dimensional stereotactic surface projections of fluorine-18-FDG PET. J Nucl Med 36:1238–1248
30. Grimmer T, Wutz C, Alexopoulos P et al (2016) Visual versus fully automated analyses of 18F-FDG and amyloid PET for prediction of dementia due to Alzheimer disease in mild cognitive impairment. J Nucl Med 57:204–207
31. Palmqvist S, Mattsson N, Hansson O (2016) Cerebrospinal fluid analysis detects cerebral amyloid-beta accumulation earlier than positron emission tomography. Brain 139:1226–1236
32. Fagan AM, Mintun MA, Shah AR et al (2009) Cerebrospinal fluid tau and ptau(181) increase with cortical amyloid deposition in cognitively normal individuals: implications for future clinical trials of Alzheimer's disease. EMBO Mol Med 1:371–380
33. Mattsson N, Insel PS, Landau S et al (2014) Diagnostic accuracy of CSF Ab42 and florbetapir PET for Alzheimer's disease. Ann Clin Transl Neurol 1:534–543
34. Grimmer T, Riemenschneider M, Forstl H et al (2009) Beta amyloid in Alzheimer's disease: increased deposition in brain is reflected in reduced concentration in cerebrospinal fluid. Biol Psychiatry 65:927–934
35. Drzezga A, Grimmer T, Henriksen G et al (2009) Effect of APOE genotype on amyloid plaque load and gray matter volume in Alzheimer disease. Neurology 72:1487–1494
36. Cosentino S, Scarmeas N, Helzner E et al (2008) APOE epsilon 4 allele predicts faster cognitive decline in mild Alzheimer disease. Neurology 70:1842–1849
37. Deane R, Sagare A, Hamm K et al (2008) apoE isoform-specific disruption of amyloid beta peptide clearance from mouse brain. J Clin Invest 118:4002–4013
38. Craft S, Teri L, Edland SD et al (1998) Accelerated decline in apolipoprotein E-epsilon4 homozygotes with Alzheimer's disease. Neurology 51:149–153

Check for updates

Chapter 10

Functional EEG Connectivity Alterations in Alzheimer's Disease

Florian Hatz and Peter Fuhr

Abstract

Quantitative EEG analyses, especially connectivity and graph analyses, are promising biomarkers in neurodegenerative disorders such as Alzheimer's disease. To develop reliable and valid biomarkers, many aspects in pre- and post-processing of EEG data are important.

Artifact detection and segment selection are done by visual inspection or automated routines, both with specific advantages and difficulties. Inverse solution is an elegant way of solving the issue of common source but may be somewhat imprecise in the absence of defined graphoelements such as epileptiform discharges. Connectivity analyses allow a description of interactions between brain regions, which may represent one of the earliest signs of cognitive decline.

Key words Alzheimer's disease, Dementia, Quantitative electroencephalography, Functional connectivity, Network analysis

1 Introduction

Quantitative EEG (qEEG) analyses, including functional connectivity analyses, are candidate biomarkers for various neurodegenerative disorders, especially to characterize physiological alterations related to cognitive deterioration [1–3]. Many different methods for quantification of EEG connectivities have been applied, most focusing on analyses in resting state. For resting state EEG recording, subjects sit or lay calm with eyes closed, not concentrating on anything yet without falling asleep. Analysis of these EEG recordings requires pre- and post-processing and is influenced by both technical and physiological artifacts.

Robert Perneczky (ed.), *Biomarkers for Preclinical Alzheimer's Disease*, Neuromethods, vol. 137,
https://doi.org/10.1007/978-1-4939-7674-4_10,

2 Preprocessing of EEG Data

Especially the selection of segments for analysis highly influences the results. The main technical artifacts of EEG arise from line noise (50 or 60 Hz) and from noisy or broken electrodes. Line noise can be reduced by notch filters, while noisy or broken electrodes must be detected and eliminated by replacing their signals using interpolation from nearby intact electrodes. High-density EEG recording is best for this procedure. The interpolation of signals is important, as the simple exclusion of noisy electrodes would change the number of evaluated electrodes per subject, thereby introducing difficulties for the statistical analysis [4]. Physiological artifacts arise mainly from eye movements, blinking, muscle activity, and heartbeats. They can be reduced or eliminated by selecting visually the least affected channels and segments. Physiological artifacts can also be isolated and eliminated by independent component analysis (ICA) or regression to electrooculographic (EOC) or electrocardiographic (ECG) signals [5]. These methods generally rely on a personal intervention involving a visual search for artifacts. That process is inherently subjective and hard to standardize across raters and studies. To eliminate the subjectivity, automated processing has been proposed. Nolan et al. [6] introduced the toolbox FASTER for the automated preprocessing of event-related potentials (ERP) (FASTER; Nolan et al. [6]) including an ICA for artifact reduction and an automated detection of bad channels, epochs, and ICA activations. The detection routines are based on median, kurtosis, correlation, and variance analyses of the channels. Other toolboxes such as Cartool [7] and Brainstorm [8] also include automated routines for bad-channel detection; no validation of their underlying routines is currently available. The widely known toolbox FieldTrip [9] uses frequency analysis for the automatic selection of bad channels. To our knowledge, the validity of this procedure is currently also not available, though it has been recently suggested that MEG processing should include frequency analysis of the recorded data for rating bad channels and segments [4]. Still another possibility is using the toolbox TAPEEG [10], which combines many of the previously mentioned methods and allows a fully automated preprocessing (sites.google.com/site/-tapeeg; see Fig. 1). Some important aspects of preprocessing include:

Fig. 1 (continued) from the resulting values, and signals with a *z*-value >3 are labeled as "bad." Results are summarized in a matrix (lower part of the figure). All signal periods labeled as "bad" by one of the methods are represented in *red*, the others in *green color*. For selection of bad channels or activations, the resulting matrix is evaluated horizontally (lower right matrix in figure, threshold: >70% bad signals). For selection of segments, the green squares (= good signals) in the resulting matrix are replaced by Peak2Min values (*brighter green colors* represent higher Peak2Min values), the *red squares* by zero values. Maximizing the summed values of the squares allows the automated selection of the optimal segments for further processing (lower left matrix in figure; *blue frame* represents a selected segment)

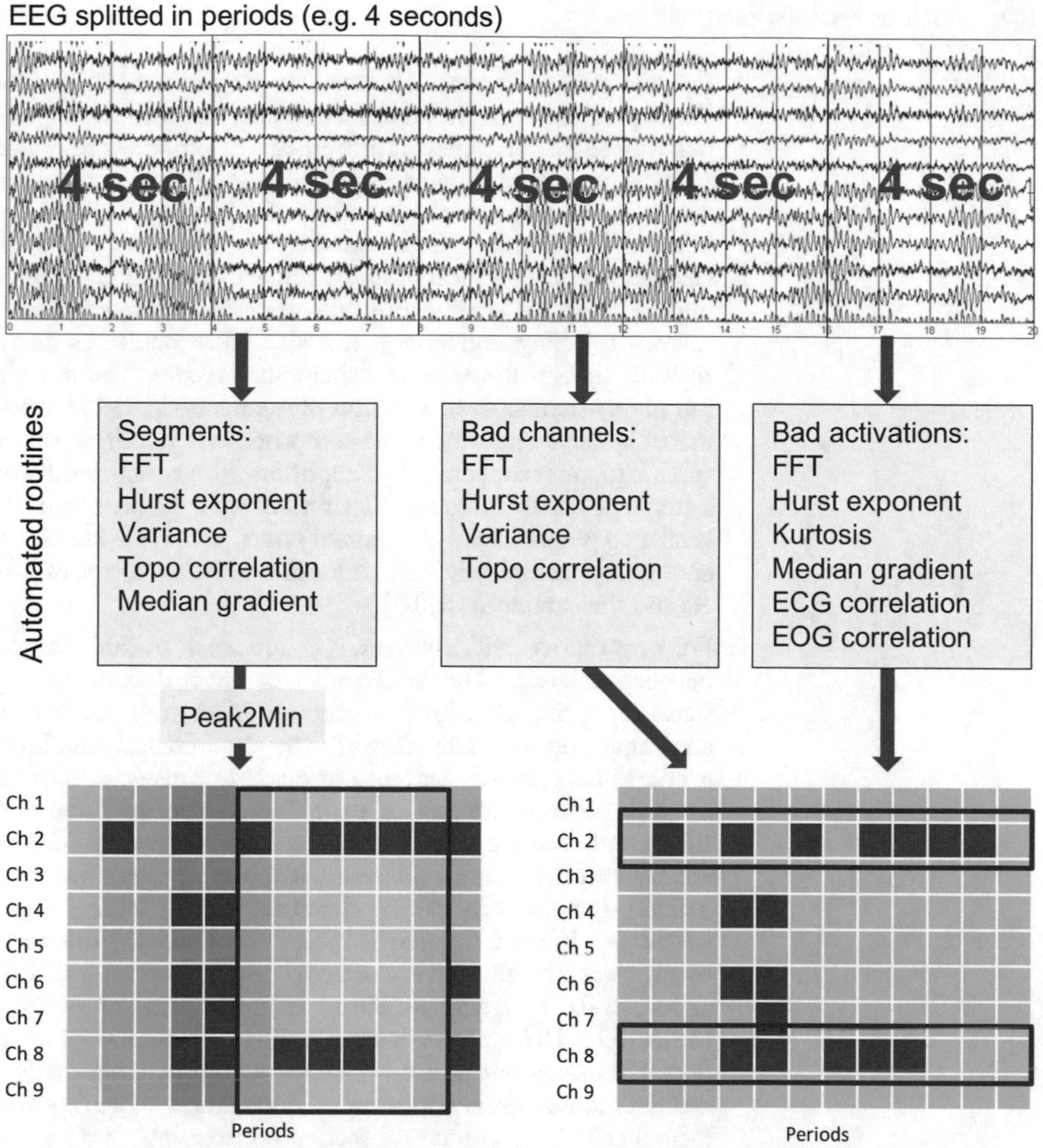

Fig. 1 Automated selection of segments, bad channels, and ICA activations in TAPEEG. TAPEEG includes routines to automatically select segments of EEG for processing and the detection of channels with high artifact load (bad channels). For artifact reduction, the automated selection of bad ICA activations is possible. For all three tasks, TAPEEG processes the EEG data similarly, using the workflow depicted in the figure. EEG data is first splitted into periods (e.g. 4 s). In every period, the signals (= channels) are analyzed using a set of predefined methods as listed in the central boxes of the figure. Every method evaluates the signals independently. FFT = Fast Fourier transformation using an 80% Hanning window. Band power 0.5–2 Hz > 70%, band power 15–30 Hz > 50%, and 50 Hz-power > 50% are labeled as bad. Hurst exponent = Hurst exponent is a measure of long-range dependence within a signal, with a range of 0–1. Human EEG has values around 0.7 [41]. Topo correlation = Correlation to surrounding channels (high correlation values are expected for high density EEG). Median gradient = Median value of all 1 timeframe differences in amplitude of every channel. Kurtosis = Kurtosis measures the peakedness of data and is high for ICA activations loading on one single electrode only. EOG correlation = Correlation to electrooculogram. ECG correlation = Correlation of peaks in signals to peaks in electrocardiogram. Peak2Min = Ratio of amplitude of peak frequency and minimal amplitude in lower frequency for every period in parieto-occipital electrodes (amplitude of IAF divided by amplitude of TF in). Peak2Min represents the amount of alpha activity and is positively correlated to alertness [10]. For Hurst exponent, variance, topo correlation, median gradient, kurtosis, ECG, and EOG correlation, *z*-values are calculated

1. Of a recording of several minutes of resting state EEG, usually only a part can be included in further analysis, while time periods with signs of sleepiness, movement, or other less frequent artifacts have to be eliminated. Visual rating has been the standard method but carries a problem of reliability, as raters are influenced by many unpredictable circumstances such as changes in concentration or interference with other rated EEG in close temporal relationship. In contrast, an automated routine for selecting segments is not able to recognize as many relevant EEG features as an experienced reader. The coarser but highly standardized selection of segments as used in automated systems might be a fair compromise if identification of specific graphoelements is less important. Regarding results of a frequency analysis, reliability of automated preprocessing is similar to preprocessing by trained raters, while it is less time-consuming, completely standardized, and independent of raters and their training [6, 10].
2. For quantitative EEG analyses, it is important to differentiate between task-related or evoked analyses and analyses in resting state. In resting state EEG recording, laboratory conditions are simple and well defined, while the physiological condition is not, as patients may be tense or relaxed or may concentrate on very different aspects of their life. In contrast, non-rest EEG conditions are more difficult to standardize across different laboratories, but the physiological condition is quite well defined. Brain activity can be characterized by looking at the power of different frequency bands. Traditionally, the total band power is subdivided in the delta (1–4 Hz), theta (4–8 Hz), alpha1 (8–10 Hz), alpha2 (10–13 Hz), and beta (13–30 Hz) bands [11]. Delta power is the dominant frequency of deep non-REM sleep but is also associated with functional cortical deafferentation during internal concentration when solving mental tasks [12]. Alpha and theta power respond in different and opposite ways. Relative theta power increases with increasing task demands, whereas relative alpha power decreases [13]. Attentional tasks and semantic memory demands lead to a selective suppression of alpha power in different "sub-bands." For resting state, increase of alpha2 power has been linked to an increased capacity to initiate new tasks [14]. Beta band power may also represent suppression of cortical activity, as shown for the motor cortex during motor activity [15].

Apart from separating the frequency according to the standardized frequency bands, frequency bands can be defined according to the individual alpha frequency (= peak frequency, IAF) and the transition frequency between theta and alpha power [14]:

Band power results defined by IAF represent the brain activity of a subject in a far more physiological way. However, results in groups

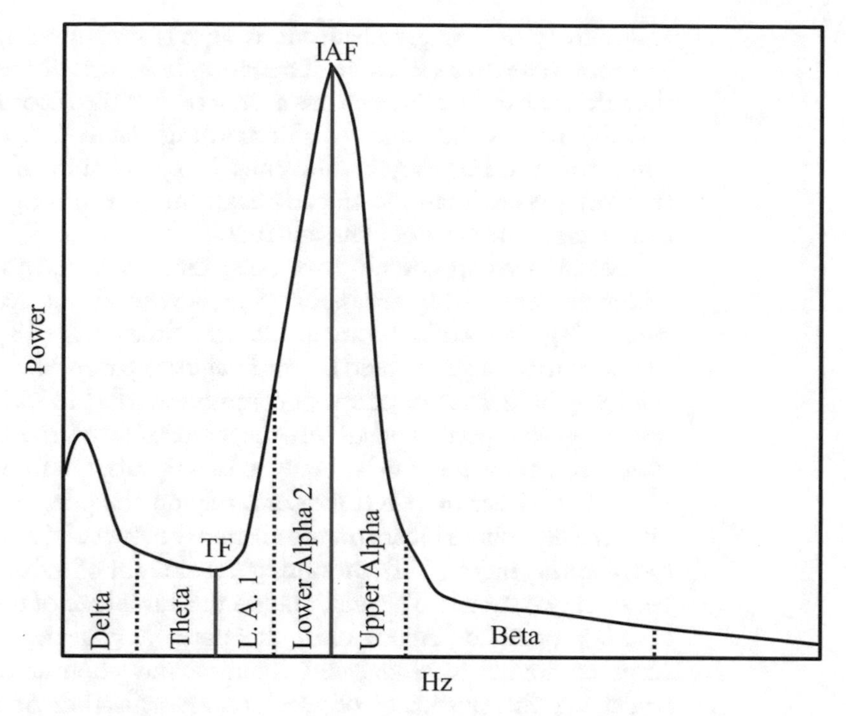

Fig. 2 Calculation of individual frequency bands according to Klimesch 1999. IAF = individual alpha frequency, TF = transition frequency between theta and alpha

of subjects may differ by applying individual or predefined bands, and an intraindividual shift in peak frequency may go undetected with the IAF method. Furthermore, the IAF method is not applicable when no peak is clearly identifiable.

3 Connectivity Analyses

In addition to the band powers of different brain regions, the interaction of brain activity between regions is also of interest. Connectivity can be characterized as "functional" or "effective". Functional connectivity denotes any kind of connection between two locations independently from direction or causal relation. For functional connectivity, the linkage can be direct, indirect, or due to a common trigger. Effective connectivity is defined as a connection where location A influences location B and refers to the influence that one neural system exerts over another [16]. Connectivities are best described separately for the different frequency bands, as in theory only cell populations that are synchronized in phase are able to communicate and interact [17]. This fact raises the problem of volume conduction and of the common source. Activity of deeper brain regions is recorded simultaneously by many electrodes, and activity at a single electrode is always measured against the common reference electrode. Therefore, brain activity recorded at the reference electrodes explains a large part of the recorded signal, and this part is the same for all recording electrodes. Measuring connectivity using simple coherence (= correlation of frequency spectra) overestimates connectivities, and the strong common source effect masks real connectivities.

Mainly two approaches have been used to solve this problem: Different connectivity measures were developed for reducing or eliminating effects of the common source. Nolte et al. [18] proposed a connectivity measure based on the imaginary part of the coherence, not being influenced by effects of common source [18]. However, as the imaginary part depends on the amplitude of the signal and magnitude of the phase delay between two signals, this measure is not ideal [19]. Other proposed measures rely on the phase information only, mainly using Hilbert transformation. Stam et al. [20] proposed the phase lag index (PLI), measuring the amount of stable phase lags between two signals. This method has the advantage of disregarding zero-lag phase difference and, therefore, annihilating effects of common source recording and volume conduction at the cost of neglecting the possibility of synchronous activation of two brain regions by a common pacemaker. Mutual information and the closely related synchronization likelihood [19] are measuring the degree of interaction between two or more time series, the former using entropy and the latter dynamical interdependencies. However, SL is also strongly influenced by the common source problem. Furthermore, all previously mentioned methods are not detecting the direction of information flow and thus are by definition measures of functional connectivity. Measures of effective connectivity such as Granger causality [21] and direct transfer function [22] are using multivariate auto-regressive models to estimate connectivity strength and direction between different brain regions.

4 Inverse Solutions

A second method to reduce the problems of common source and of volume conduction is the application of inverse solution methods. In theory, inverse solution methods are transposing the signals recorded at the surface back to their origins. A forward solution model is used to describe the electrical conductivity between different points in the head and the electrodes on the surface. Based on such models, different methods of inverse solution try to reconstruct the source of the signals recorded at the surface. Inverse solution methods carry some uncertainties. First, no perfect forward solution model exists. Forward solutions are calculated based on a MRI scan of the head, while conductivity values are defined for the different compartments.

The most important compartment is the skull, having the lowest conductivity. Unfortunately, a good estimation of the skull conductivity is difficult to obtain for single individuals. Therefore, all forward solutions can only be seen as a good approximation. Second, the methods to reconstruct sources based on signals recorded at the surface and a forward solution model have no unique solution. Mainly two methods are used today. One is the minimum norm estimation or its derivatives (as Loreta or sLoreta), trying to minimize the amount of power to explain all activity recorded at the surface while using a correction for deeper sources [23]. The other commonly used method is beamforming, using radar technology to locate the source of a signal recorded at many surface electrodes [24]. This method has a higher spatial resolution than minimum norm estimation but needs a very good estimation of the covariance matrix, which is often difficult to determine from the recorded EEG data. In summary, inverse solution transformation is interesting, allows analysis of source signals, and reduces the effect of common source and volume conduction. However, due to the methodological issues, results have always to be taken with caution, especially for analysis in resting state EEG. In contrast, in epileptology spikes simplify the source localization.

Newer methods of connectivity analysis are trying to take the non-stationarity of EEG recordings during resting state into account. Examples are the analysis of synchronization dynamics using the SL [25], the microstate segmented PLI [26], or the correlation of connectivities over time [27].

5 Graph Analysis

Based on results of connectivity analysis, graph analysis allows the characterization of networks by descriptive single measures. Common measures are gamma, lambda, and degree diversity. Gamma represents the amount of local connectivity, given by the number of connections between immediate neighbors of a location or "node"; lambda the efficiency of information flow, given by the shortest path between nodes; and degree diversity the scale-free distribution, given by a hierarchical organization of many nodes with few connections and few hubs with many. Higher scale-free distribution is linked to the amount of hubs in a network [28]. Stam et al. described [29] a model of physiological brain networks, showing the "normal brain" as an intermediate network state between a regular, a random, and a scale-free network with a small-world characteristic. According to Watts and Strogatz, the small-world network is a network with many local connections and a few long-distance connections, resulting in an enhanced signal propagation speed, computational power, and synchronizability [30].

6 Connectivity Analyses in Alzheimer's Disease

For Alzheimer's disease (AD) and its precursor, the mild neurocognitive disorder (DSM-5) due to AD, many studies analyzed alterations of connectivities and graph measures related to cognition. Alpha band connectivity correlates negatively [31–33], and theta and delta band connectivities correlate positively with the cognitive state [34, 35]. The networks derived from alpha band connectivities lose the small-world structure and evolve toward a more random network structure [31]. Primarily affected by this process are the hubs in the parieto-occipital and less the temporal regions [31].

7 Conclusions

Quantitative EEG analyses, especially connectivity analysis, are promising candidates for biomarkers. Test-retest reliability has been shown for PLI and msPLI [26, 36], and criterion validity has been documented by correlation between connectivity measures and clinical measures of cognition in patients with neurodegenerative diseases [34, 37–39]. The challenge will be to standardize analyzing methods and to document feasibility and validity of qEEG measures as possible diagnostic, prognostic, predictive, or response biomarkers [40].

References

1. Caviness JN, Hentz JG, Evidente VG, Driver-Dunckley E, Samanta J, Mahant P, Connor DJ, Sabbagh MN, Shill HA, Adler CH (2007) Both early and late cognitive dysfunction affects the electroencephalogram in Parkinson's disease. Parkinsonism Relat Disord 13:348–354
2. Fonseca LC, Tedrus GMAS, Letro GH, Bossoni AS (2009) Dementia, mild cognitive impairment and quantitative EEG in patients with Parkinson's disease. Clin EEG Neurosci 40:168–172
3. Roh JH, Park MH, Ko D, Park K-W, Lee D-H, Han C, Jo SA, Yang K-S, Jung K-Y (2011) Region and frequency specific changes of spectral power in Alzheimer's disease and mild cognitive impairment. Clin Neurophysiol 122:2169–2176. https://doi.org/10.1016/j.clinph.2011.03.023
4. Gross J, Baillet S, Barnes GR, Henson RN, Hillebrand A, Jensen O, Jerbi K, Litvak V, Maess B, Oostenveld R, Parkkonen L, Taylor JR, van Wassenhove V, Wibral M, Schoffelen J-M (2013) Good practice for conducting and reporting MEG research. Neuroimage 65:349–363. https://doi.org/10.1016/j.neuroimage.2012.10.001
5. Wallstrom GL, Kass RE, Miller A, Cohn JF, Fox NA (2004) Automatic correction of ocular artifacts in the EEG: a comparison of regression-based and component-based methods. Int J Psychophysiol 53:105–119. https://doi.org/10.1016/j.ijpsycho.2004.03.007
6. Nolan H, Whelan R, Reilly RB (2010) FASTER: fully automated statistical thresholding for EEG artifact rejection. J Neurosci Methods 192:152–162. https://doi.org/10.1016/j.jneumeth.2010.07.015
7. Brunet D, Murray MM, Michel CM (2011) Spatiotemporal analysis of multichannel EEG: CARTOOL. Comput Intell Neurosci 2011:813870. https://doi.org/10.1155/2011/813870
8. Tadel F, Baillet S, Mosher JC, Pantazis D, Leahy RM (2011) Brainstorm: a user-friendly application for MEG/EEG analysis. Comput Intell Neurosci 2011:879716. https://doi.org/10.1155/2011/879716
9. Oostenveld R, Fries P, Maris E, Schoffelen JM (2011) FieldTrip: open source software for advanced analysis of MEG, EEG, and invasive

electrophysiological data. Comput Intell Neurosci 2011:1

10. Hatz F, Hardmeier M, Bousleiman H, Rüegg S, Schindler C, Fuhr P (2014) Reliability of fully automated versus visually controlled pre- and post-processing of resting-state EEG. Clin Neurophysiol. https://doi.org/10.1016/j.clinph.2014.05.014
11. Deuschl G, Eisen A (1999) Recommendations for the practice of clinical neurophysiology: guidelines of the International Federation of Clinical Physiology. Clin Neurophysiol EEG Suppl 52, Chapter 1.5
12. Harmony T (2013) The functional significance of delta oscillations in cognitive processing. Front Integr Neurosci. https://doi.org/10.3389/fnint.2013.00083
13. Klimesch W (2012) Alpha-band oscillations, attention, and controlled access to stored information. Trends Cogn Sci 16:606–617. https://doi.org/10.1016/j.tics.2012.10.007
14. Klimesch W (1999) EEG alpha and theta oscillations reflect cognitive and memory performance: a review and analysis. Brain Res Rev 29:169–195. https://doi.org/10.1016/S0165-0173(98)00056-3
15. Miller KJ, Hermes D, Honey CJ, Hebb AO, Ramsey NF, Knight RT, Ojemann JG, Fetz EE (2012) Human motor cortical activity is selectively phase-entrained on underlying rhythms. PLoS Comput Biol. https://doi.org/10.1371/journal.pcbi.1002655
16. Friston KJ (2011) Functional and effective connectivity: a review. Brain Connect 1:13–36. https://doi.org/10.1089/brain.2011.0008
17. Fries P (2005) A mechanism for cognitive dynamics: neuronal communication through neuronal coherence. Trends Cogn Sci 9:474–480. https://doi.org/10.1016/j.tics.2005.08.011
18. Nolte G, Bai O, Wheaton L, Mari Z, Vorbach S, Hallett M (2004) Identifying true brain interaction from EEG data using the imaginary part of coherency. Clin Neurophysiol 115:2292–2307. https://doi.org/10.1016/j.clinph.2004.04.029
19. Stam CJ, van Dijk BW (2002) Synchronization likelihood: an unbiased measure of generalized synchronization in multivariate data sets. Phys Nonlinear Phenom 163:236–251. https://doi.org/10.1016/S0167-2789(01)00386-4
20. Stam CJ, Nolte G, Daffertshofer A (2007) Phase lag index: assessment of functional connectivity from multi channel EEG and MEG with diminished bias from common sources. Hum Brain Mapp 28:1178–1193. https://doi.org/10.1002/hbm.20346
21. Granger CWJ (1969) Investigating causal relations by econometric models and cross-spectral methods. Econometrica 37:424–438. https://doi.org/10.2307/1912791
22. Kamiński MJ, Blinowska KJ (1991) A new method of the description of the information flow in the brain structures. Biol Cybern 65:203–210
23. Pascual-Marqui RD (2002) Standardized low-resolution brain electromagnetic tomography (sLORETA): technical details. Methods Find Exp Clin Pharmacol 24(Suppl D):5–12
24. Hillebrand A, Barnes GR (2005) Beamformer analysis of MEG data. In: Magnetoencephalography. Academic Press, New York, pp 149–171
25. Betzel RF, Erickson MA, Abell M, O'Donnell BF, Hetrick WP, Sporns O (2012) Synchronization dynamics and evidence for a repertoire of network states in resting EEG. Front Comput Neurosci. https://doi.org/10.3389/fncom.2012.00074
26. Hatz F, Hardmeier M, Bousleiman H, Rüegg S, Schindler C, Fuhr P (2016) Reliability of functional connectivity of electroencephalography applying microstate-segmented versus classical calculation of phase lag index. Brain Connect 6:461–469. https://doi.org/10.1089/brain.2015.0368
27. Allen EA, Damaraju E, Eichele T, Wu L, Calhoun VD (2017) EEG signatures of dynamic functional network connectivity states. Brain Topogr. https://doi.org/10.1007/s10548-017-0546-2
28. de Haan W, Mott K, van Straaten ECW, Scheltens P, Stam CJ (2012) Activity dependent degeneration explains hub vulnerability in Alzheimer's disease. PLoS Comput Biol 8:e1002582. https://doi.org/10.1371/journal.pcbi.1002582
29. Stam CJ, van Straaten ECW (2012) The organization of physiological brain networks. Clin Neurophysiol 123:1067–1087. https://doi.org/10.1016/j.clinph.2012.01.011
30. Watts DJ, Strogatz SH (1998) Collective dynamics of "small-world" networks. Nature 393:440–442. https://doi.org/10.1038/30918
31. Stam CJ, De Haan W, Daffertshofer A, Jones BF, Manshanden I, van Cappellen van Walsum AM, Montez T, Verbunt JPA, de Munck JC, van Dijk BW, Berendse HW, Scheltens P (2009) Graph theoretical analysis of magnetoencephalographic functional connectivity in Alzheimer's disease. Brain 132:213
32. Stam CJ, van Cappellen van Walsum AM, Pijnenburg YAL, Berendse HW, de Munck JC, Scheltens P, van Dijk BW (2002) Generalized synchronization of MEG recordings in Alzheimer's disease: evidence for involvement of the gamma band. J Clin Neurophysiol 19:562–574

33. Vecchio F, Miraglia F, Piludu F, Granata G, Romanello R, Caulo M, Onofrj V, Bramanti P, Colosimo C, Rossini PM (2017) "Small World" architecture in brain connectivity and hippocampal volume in Alzheimer's disease: a study via graph theory from EEG data. Brain Imaging Behav 11:473–485. https://doi.org/10.1007/s11682-016-9528-3
34. Hatz F, Hardmeier M, Benz N, Ehrensperger M, Gschwandtner U, Rüegg S, Schindler C, Monsch AU, Fuhr P (2015) Microstate connectivity alterations in patients with early Alzheimer's disease. Alzheimers Res Ther 7:78. https://doi.org/10.1186/s13195-015-0163-9
35. Suárez-Revelo JX, Ochoa-Gómez JF, Duque-Grajales JE, Tobón-Quintero CA (2016) Biomarkers identification in Alzheimer's disease using effective connectivity analysis from electroencephalography recordings. Ing E Investig 36:50–57. 10.15446/ing.investig.v36n3.54037
36. Hardmeier M, Hatz F, Bousleiman H, Schindler C, Stam CJ, Fuhr P (2014) Reproducibility of functional connectivity and graph measures based on the phase lag index (PLI) and weighted phase lag index (wPLI) derived from high resolution EEG. PLoS One 9:e108648. https://doi.org/10.1371/journal.pone.0108648
37. Babiloni C, Triggiani AI, Lizio R, Cordone S, Tattoli G, Bevilacqua V, Soricelli A, Ferri R, Nobili F, Gesualdo L, Millán-Calenti JC, Buján A, Tortelli R, Cardinali V, Barulli MR, Giannini A, Spagnolo P, Armenise S, Buenza G, Scianatico G, Logroscino G, Frisoni GB, del Percio C (2016) Classification of single normal and Alzheimer's disease individuals from cortical sources of resting state EEG rhythms. Front Neurosci. https://doi.org/10.3389/fnins.2016.00047
38. Ranasinghe KG, Hinkley LB, Beagle AJ, Mizuiri D, Dowling AF, Honma SM, Finucane MM, Scherling C, Miller BL, Nagarajan SS, Vossel KA (2014) Regional functional connectivity predicts distinct cognitive impairments in Alzheimer's disease spectrum. Neuroimage Clin 5:385–395. https://doi.org/10.1016/j.nicl.2014.07.006
39. Triggiani AI, Bevilacqua V, Brunetti A, Lizio R, Tattoli G, Cassano F, Soricelli A, Ferri R, Nobili F, Gesualdo L, Barulli MR, Tortelli R, Cardinali V, Giannini A, Spagnolo P, Armenise S, Stocchi F, Buenza G, Scianatico G, Logroscino G, Lacidogna G, Orzi F, Buttinelli C, Giubilei F, Del Percio C, Frisoni GB, Babiloni C (2017) Classification of healthy subjects and Alzheimer's disease patients with dementia from cortical sources of resting state EEG rhythms: a study using artificial neural networks. Front Neurosci. https://doi.org/10.3389/fnins.2016.00604
40. Amur S, LaVange L, Zineh I, Buckman-Garner S, Woodcock J (2015) Biomarker qualification: toward a multiple stakeholder framework for biomarker development, regulatory acceptance, and utilization. Clin Pharmacol Ther 98:34–46. https://doi.org/10.1002/cpt.136
41. Bian N-Y, Wang B, Cao Y, Zhang L (2006) Automatic removal of artifacts from EEG data using ICA and exponential analysis. In: Proc. third int. conf. advnaces neural netw.—vol Part II. Springer-Verlag, Berlin, Heidelberg, pp 719–726

Part IV

Biomarker Discovery

Check for updates

Chapter 11

A Selected Reaction Monitoring Protocol for the Measurement of sTREM2 in Cerebrospinal Fluid

Amanda J. Heslegrave, Wendy E. Heywood, Kevin M. Mills, and Henrik Zetterberg

Abstract

Mass spectrometry plays an increasingly important role in the biomarker field with the advent of targeted proteomics. Tryptic peptides from a protein of interest can be used to create a targeted assay to interrogate cerebrospinal fluid (CSF) for biomarkers. Since heterozygous mutations in the *TREM2* gene have been associated with an increased risk of Alzheimer's disease, measuring this soluble protein in CSF has become a priority. This chapter demonstrates the development, optimization, and validation of a method to measure soluble TREM2 using a single reaction monitoring (SRM) targeted mass spectrometry assay.

Key words Proteomics, Biomarker, Alzheimer's disease, TREM2, Inflammation

1 Introduction

Since the mid-1990s, proteomics by mass spectrometry has been an extremely useful tool in the discovery of new biomarkers and for elucidating new disease mechanisms. Many papers have been published using untargeted proteomic (more recently referred to as hypothesis generating) MS workflows which can demonstrate and detect differences between disease and control cases in neurodegenerative disorders [1–3]. Until recently validation of possible biomarkers has mainly been possible by immunological means, such as western blot or enzyme-linked immunosorbent assay (ELISA). However, mass spectrometry is proving to be ever more useful in providing accurate and specific quantitative data on target proteins of interest in body fluids [4]. In particular this has proved of great interest to those researchers studying the composition of cerebrospinal fluid (CSF) to find biomarkers for Alzheimer's disease (AD) [3] and in the development of new tests for this disease.

Targeted proteomics (single reaction monitoring (SRM) detecting one analyte and multiple reaction monitoring (MRM), detecting more than one analyte) is a tandem mass spectrometry

Robert Perneczky (ed.), *Biomarkers for Preclinical Alzheimer's Disease*, Neuromethods, vol. 137,
https://doi.org/10.1007/978-1-4939-7674-4_11,

(MS/MS) technique usually performed on a triple quadrupole MS-based instrument. Tandem mass spectrometry has been used for more than 20 years [5, 6] but almost exclusively for the quantitation of small molecules, such as metabolites and drugs. However it is only recently that tandem mass spectrometry has been used more and more to quantify proteolytically generated peptides that act as surrogates of the corresponding intact proteins. The use of stable isotope-labelled peptide standards can make this method of quantitation not only extremely accurate but also more specific than conventional immunological-based assays.

This technique is generally used in conjunction with the hypothesis generating or proteomics-based technology such as label-free liquid chromatography-tandem mass spectrometry (LC-MS/MS) or label-based proteomic methods such as stable isotope labelling with amino acids in cell culture (SILAC) and tandem mass tagging (TMT). These types of methods provide protein identification and relative protein abundance using the spectral count, stable isotope labelling, or the total number of MS/MS spectra taken on peptides from a given protein in a given LC-MS/MS analysis. This is then linearly correlated with protein abundance but does not give absolute quantification [7–9]. For the development of targeted proteomics or peptide-based MS quantitation, the first step is proteolytic digestion of the proteins by a protease, typically trypsin, which allows the creation of fragments small enough to be analyzed within the mass range of triple quadrupole-based instruments. Trypsin cleaves at the carboxylic acid side of peptide bonds of lysine and arginine residues and is generally used due to its robustness, specificity, and the creation of peptides that are amenable to mass spectrometry. The peptides produced are now of optimal size and molecular weight and in the range of most quadrupole or ion trap based tandem mass spectrometers. Once the protein has been digested, it is possible to choose a "marker" peptide for that protein. This marker can then be weighed in the first mass analyzer (MS1) and its unique mass-to-charge (*m/z*) ratio recorded. The peptide is then fragmented in a collision cell by colliding it with argon gas to yield product ions which have their *m/z* ratio recorded in the second part of the mass spectrometer or MS2. This fragmentation is known as collision-induced dissociation (CID), and from this, a unique product ion is chosen as being the characteristic fingerprint of that peptide. Here we have two levels of identification for the peptide, firstly its intact mass—precursor ion—and secondly it is identifying peptide fragment mass, product ion. In method development more than one precursor and product ion will be chosen to optimize, and the most specific/sensitive will be chosen. Finally, a third and final level of identification is given by the elution time of the peptide from the UPLC chromatography column. Because in a complex biological sample digest, it is possible to get several peptides with the same precursor/product masses as your target, this third identifying step is very important. It is the use of fast chromatography-based systems such as UPLC with tandem mass spectrometry that provides extra specificity for the analysis of proteins that immunological-based assays cannot match. However, it should be noted that some immunological-based assays can be significantly more sensitive than MS-based analyses.

TREM2 is a receptor glycoprotein of 230 amino acids which belongs to the immunoglobulin superfamily. In the brain, TREM2 is expressed exclusively by myeloid cells which include microglia the brains resident immune cell [10]. It was known that homozygous mutations in this protein caused polycystic lipomembranous osteodysplasia with sclerosing leukoencephalopathy, also known as Nasu-Hakola disease [11, 12]. However, in 2013, it was discovered that heterozygous mutations in the TREM2 gene were noted as being significant risk factors for AD [13, 14].

This discovery led to intense interest in TREM2 and a requirement for sensitive assays to measure the soluble protein. In 2013 there were very few immunoassays available, and those that were available were not robust. Therefore, a different method of measuring sTREM2 was required, and this led to the development of an SRM technique that was published in early 2016 [15]. The benefit of this method is the huge scope for multiplexing as shown by Heywood et al. in [3]. Although this chapter is using sTREM2 as an example for the method, other interesting proteins can easily be multiplexed and augmented this way.

2 Materials

2.1 Mass Spectrometer

Targeted proteomics were performed on a Waters ultraperformance liquid chromatography system (Manchester, UK) coupled to a Waters XEVO TQ-S triple quadrupole mass spectrometer. The column used was a Waters CORTECS UPLC C18+ column, 90 Å 3 mm × 100 mm attached to a C18+ VanGuard pre-column.

2.2 Samples

CSF samples were collected from memory clinics in Sweden and the UK according to local ethics procedures. Samples were collected in the morning into polypropylene tubes (Sarstedt, Numbrecht, Germany) according to standard operating procedures; samples were then further aliquoted into polypropylene tubes, frozen, and stored at −80 °C within 2 h.

2.3 Reagents

All reagents are of mass spectrometry grade and prepared and stored at room temperature unless otherwise stated.

2.3.1 Digestion Reagents

Digestion buffer—100 mM Tris pH 7.8 containing 6 M urea, 2 M thiourea, and 2% amidosulfobetaine-14 (ASB-14). 2 M dithiothreitol (DTT) is made by dissolving 30 mg of DTT in 1 mL of 100 mM Tris pH 7.8. It is important that this solution is made up fresh on the day and kept on ice until use. Iodoacetamide (IAA) was prepared by dissolving 36 mg in 1 mL of 100 mM Tris pH 7.8; similarly this solution was made up fresh on the day and protected from light. Sequence grade trypsin (Promega, UK) is prepared by resuspending lyophilized 1 mg aliquots of the protease with 11 μL of 40 mM ammonium bicarbonate buffer. All reagents are from Sigma-Aldrich, Dorset, UK, unless otherwise stated.

2.3.2 Mass Spectrometry Reagents

MS mobile phases are A, LC-MS grade water with 0.1% formic acid (FA), and B, LC-MS grade ACN with 0.1% FA.

2.3.3 Peptides and Proteins

TREM2 peptide VLVEVLADPLDHR was synthesized by Generon (Maidenhead, UK). Yeast enolase standard was from Sigma-Aldrich (Dorset, UK).

3 Methods

Figure 1 shows a typical SRM development workflow, while Fig. 2 depicts an assay workflow.

3.1 SRM Set Up

To set up the SRM assay for sTREM2, we first identified a unique peptide from the protein sequence using Skyline, https://skyline.ms/project/home/software/Skyline/. Skyline is a freely available and open source Windows client application for building selected reaction monitoring (SRM)/multiple reaction monitoring (MRM) and parallel reaction monitoring (PRM—targeted MS/MS), although, it would be preferable to identify peptides from previous mass spec discovery experiments. Another useful resource is the Global Proteome Machine http://www.thegpm.org/. We checked that the peptide was completely unique to TREM2 and that none of the amino acids would be subject to posttranslational modification. This peptide was synthesized by Generon (Berkshire, UK) and then used to develop the assay.

For tuning of the peptide, standards are reconstituted in 50% ACN containing 0.1% FA before being infused into the mass spectrometer. The amount of peptide will need to be optimized but a suggested starting point is 100 μg/mL. A full scan can be performed initially to optimize tuning concentration, and the peptide is diluted accordingly. Once the concentration has been optimized, optimal collision and cone energies are obtained using the automatic "Intellistart" tuning software of the Xevo TQS mass spectrometry system. After the optimal fragmentation and MS parameters are obtained and entered into the SRM, the retention time of the peptide is assessed. In this case a 10 min chromatographic run time is used, which is shown in Table 1 (**Notes 1** and **2**).

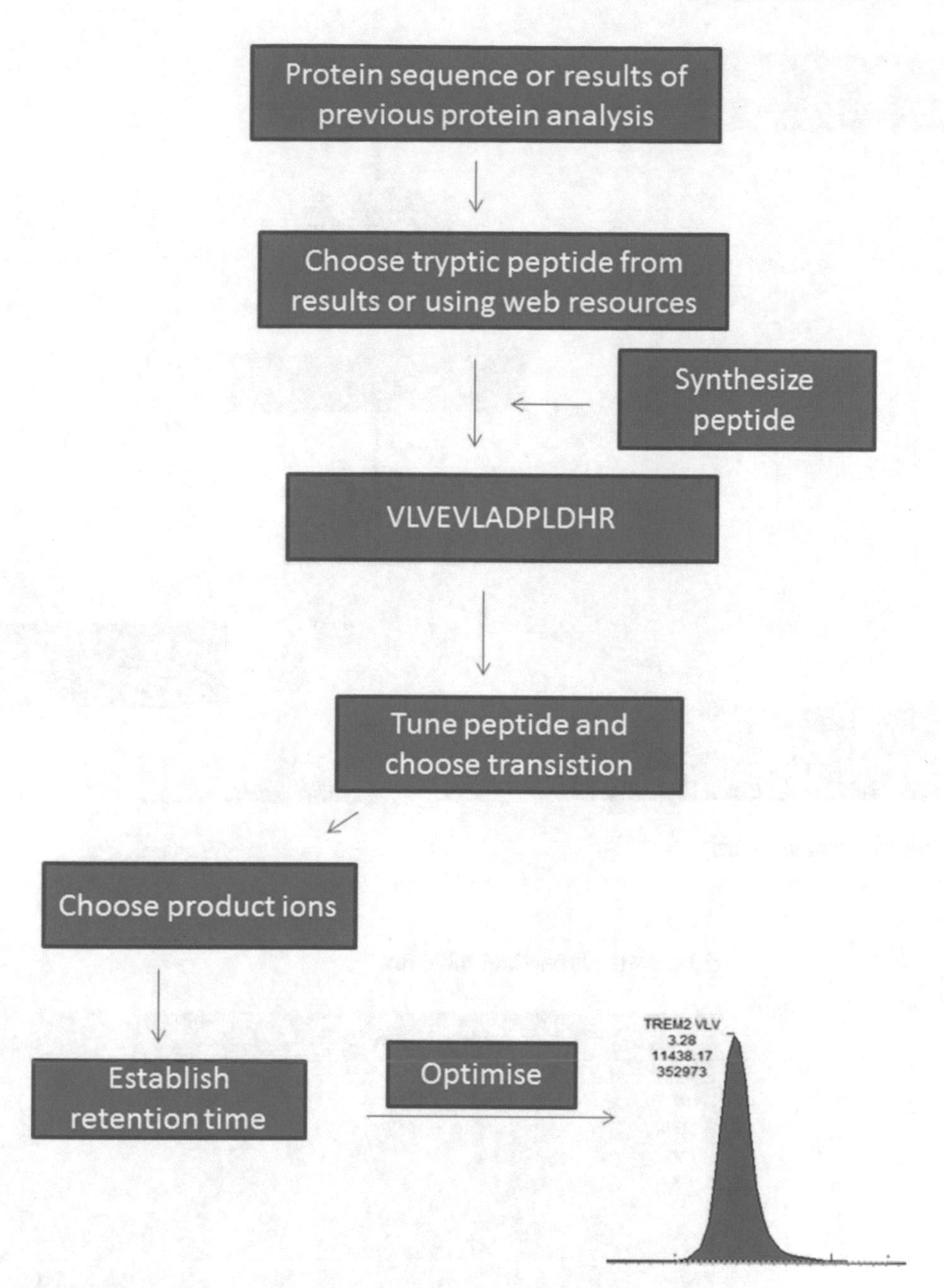

Fig. 1 Typical single reaction monitoring (SRM) development workflow

The advantage of using peptides generated from actual digests, under typical laboratory conditions, is that it allows the user to identify the peptides that are amenable to mass spectrometry, i.e., in silico determined tryptic fragments cannot predict chromatograph peak shape or fragmentation patterns. This often saves considerable time in method development (**Note 3**).

Once the assay has been developed using synthetic peptides, it is important to then ensure your peptide can be detected in the

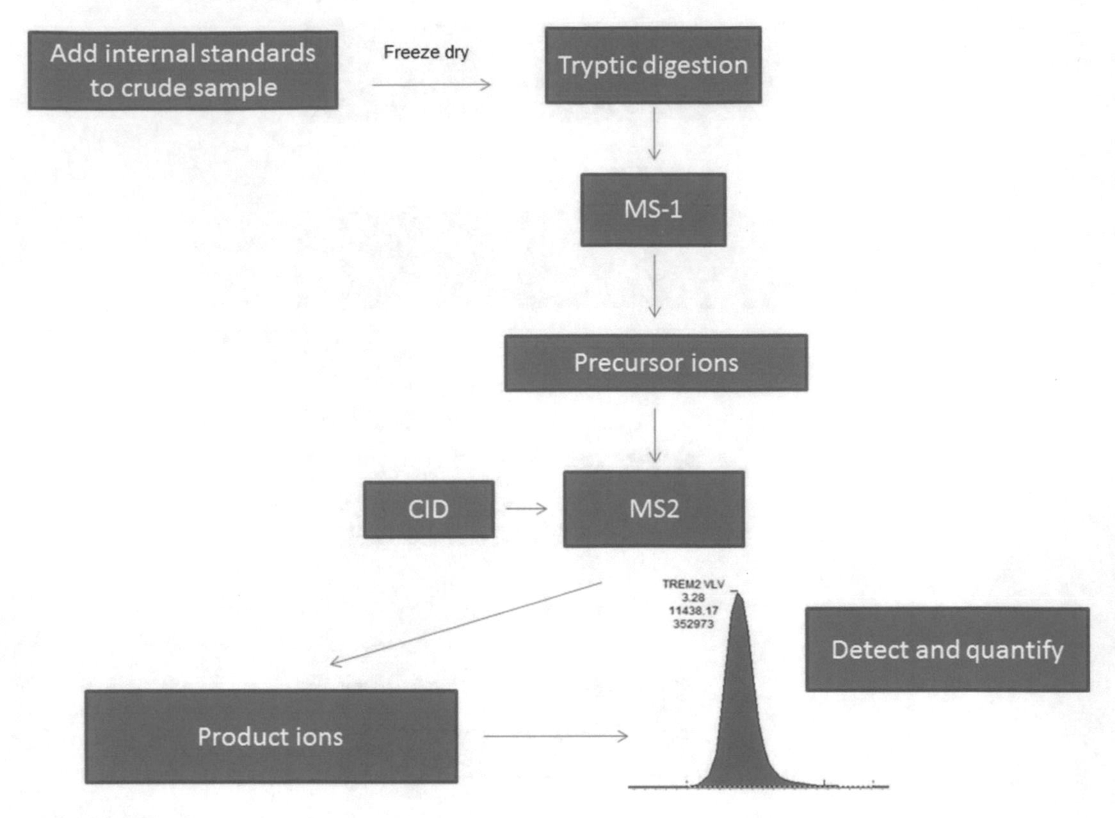

Fig. 2 Typical assay workflow

Table 1
10-minute chromatographic run

Time (min)	Flow rate (mL/min)	% A	% B	Gradient
Initial	0.8	97	3	0
0.2	0.8	97	3	6
7.00	0.8	60	40	6
7.01	0.8	0.1	99.9	6
8.00	0.8	0.1	99.9	6
8.01	0.8	97	3	6
10.00	0.8	97	3	1

appropriate sample matrix. In this case, the synthetic peptide is used to test this by spiking varying amounts in to CSF. It is also necessary to optimize the amount of your sample to dry down and the amount of digest to inject onto the column (**Notes 4–6**).

3.2 Sample Preparation

100 μL of each CSF sample is aliquoted into a fresh siliconized Eppendorf tube and freeze dried. The resulting proteins are then resuspended in 20 μL of digestion buffer and shaken at RT for 20 min. 1.5 μL of 2 M DTT is added to reduce disulfide bridges and the sample shaken for a further hour at RT. 3 μL of (2 M) IAA is added in order to carboamidomethylate any cysteines contained in the proteins, and the sample is then incubated at RT for a further 45 min in the dark. The sample is then diluted with 165.5 μL of LC-MS grade water and 10 μL of a solution containing 1 μg of trypsin. Samples are then vortexed and left in a water bath at 37 °C overnight. Before mass spectral analysis, centrifuge samples at 13,000 × *g* for 10 min to remove particulates before transfer into vials suitable for mass spectral analysis.

3.3 Mass Spectrometry

Using the chromatographic and mass spectral conditions developed and described above, the samples can be input into the sample manager and the analyses completed. During any analyses a standard curve of your peptide should also be included, and this should cover the range that the endogenous levels of peptides are expected to be present in your subject samples. Each standard curve is made up in pooled CSF in order that any matrix ion suppression is accounted for. It is also good practice to create quality controls as well using pooled CSF spiked with your peptide(s) in order that you can check the sensitivity of the assay at appropriate intervals (should fall within ±10% of established quantities). It is recommended to run QC samples at intervals of 10–15 subject samples. Internal standard values should also be monitored, and coefficients of variation (%CV) of these should also be used to monitor assay stability.

3.4 Data Analysis

Data analysis uses Waters MassLynx and TargetLynx V4.1 software to process the raw data. Methods can be set up so that the integrated peaks can be ratioed to the internal standard which can be yeast enolase in this case or a stable isotope-labelled peptide. Integration should be checked to confirm accuracy, and then response values can be interpolated from your standard curve.

4 Conclusion

In summary, the SRM method for the quantification of sTREM2 described in this chapter has been used to investigate levels of sTREM2 in a cohort of dementia subjects and controls and later validated on a separate cohort. More recently this protein has been found to be increased in early AD [16, 17] and highlights the importance of inflammation as an early event in this disease. This is a highly adaptable method, and multiplexing is a matter of adding suitably optimized MRM files as shown above.

5 Notes and Troubleshooting

1. For one transition the method time can be shortened, say 5 min, when multiplexing; it is suggested that MRM files should be grouped and split according to retention time, in order to maximize the dwell time of each ion.
2. Transitions should be selected on the best intensity and absence of interfering peaks.
3. Synthetic peptides that contain cysteine residues must be carboamidomethylated prior to infusion.
4. If matrix interference makes analyte detection a problem, then the MS resolution can be reduced; this results in an increase of sensitivity. Accurate mass will need to be determined for the synthetic peptide at the new resolution.
5. The least amount of sample to enable the analyte to be detected should be injected onto the column. Too much sample can lead to ion suppression effects. In a multiplex situation, it is worth grouping transactions according to the endogenous abundance of the analyte in the sample, in order that you can optimize the amount of sample to inject.

References

1. Conti A, Alessio M (2015) Comparative proteomics for the evaluation of protein expression and modifications in neurodegenerative diseases. Int Rev Neurobiol 121:117–152. https://doi.org/10.1016/bs.irn.2015.05.004
2. Sultana R, Boyd-Kimball D, Cai J et al (2007) Proteomics analysis of the Alzheimer's disease hippocampal proteome. J Alzheimer's Dis 11(2):153–164
3. Heywood WE, Galimberti D, Bliss E et al (2015) Identification of novel CSF biomarkers for neurodegeneration and their validation by a high-throughput multiplexed targeted proteomic assay. Mol Neurodegener 10:64. https://doi.org/10.1186/s13024-015-0059-y
4. Manwaring V, Heywood WE, Clayton R et al (2013) The identification of new biomarkers for identifying and monitoring kidney disease and their translation into a rapid mass spectrometry-based test: evidence of presymptomatic kidney disease in pediatric Fabry and type-I diabetic patients. J Proteome Res 12(5):2013–2021. https://doi.org/10.1021/pr301200e
5. Xu RN, Fan L, Rieser MJ et al (2007) Recent advances in high-throughput quantitative bioanalysis by LC-MS/MS. J Pharm Biomed Anal 44(2):342–355. https://doi.org/10.1016/j.jpba.2007.02.006
6. Kondrat RW, McClusky GA, Cooks RG (1978) Multiple reaction monitoring in mass sectrometry for diresct analysis of complex mixtures. Anal Chem 50:2017–2021
7. Bondarenko PV, Chelius D, Shaler TA (2002) Identification and relative quantitation of protein mixtures by enzymatic digestion followed by capillary reversed-phase liquid chromatography-tandem mass spectrometry. Anal Chem 74(18):4741–4749
8. Chelius D, Bondarenko PV (2002) Quantitative profiling of proteins in complex mixtures using liquid chromatography and mass spectrometry. J Proteome Res 1(4):317–323
9. Wang W, Zhou H, Lin H et al (2003) Quantification of proteins and metabolites by mass spectrometry without isotopic labeling

or spiked standards. Anal Chem 75(18): 4818–4826

10. Colonna M (2003) TREMs in the immune system and beyond. Nat Rev Immunol 3(6): 445–453. https://doi.org/10.1038/nri1106
11. Paloneva J, Manninen T, Christman G et al (2002) Mutations in two genes encoding different subunits of a receptor signaling complex result in an identical disease phenotype. Am J Hum Genet 71(3):656–662. https://doi.org/10.1086/342259
12. Kondo T, Takahashi K, Kohara N et al (2002) Heterogeneity of presenile dementia with bone cysts (Nasu-Hakola disease): three genetic forms. Neurology 59(7):1105–1107
13. Guerreiro R, Wojtas A, Bras J et al (2013) TREM2 variants in Alzheimer's disease. N Engl J Med 368(2):117–127. https://doi.org/10.1056/NEJMoa1211851
14. Jonsson T, Stefansson H, Steinberg S et al (2013) Variant of TREM2 associated with the risk of Alzheimer's disease. N Engl J Med 368(2):107–116. https://doi.org/10.1056/NEJMoa1211103
15. Heslegrave A, Heywood W, Paterson R et al (2016) Increased cerebrospinal fluid soluble TREM2 concentration in Alzheimer's disease. Mol Neurodegener 11:3. https://doi.org/10.1186/s13024-016-0071-x
16. Suarez-Calvet M, Araque Caballero MA, Kleinberger G et al (2016) Early changes in CSF sTREM2 in dominantly inherited Alzheimer's disease occur after amyloid deposition and neuronal injury. Sci Transl Med 8(369):369ra178. https://doi.org/10.1126/scitranslmed.aag1767
17. Suarez-Calvet M, Kleinberger G, Araque Caballero MA et al (2016) sTREM2 cerebrospinal fluid levels are a potential biomarker for microglia activity in early-stage Alzheimer's disease and associate with neuronal injury markers. EMBO Mol Med 8(5):466–476. 10.15252/emmm.201506123

Chapter 12

Soluble Amyloid Precursor Proteins in Blood: Methods and Challenges

Robert Perneczky and Panagiotis Alexopoulos

Abstract

Soluble amyloid precursor proteins (sAPP) are under investigation as novel biomarkers of Alzheimer's disease (AD) pathophysiology, but protein levels in cerebrospinal fluid do not seem to follow a consistent pattern, which limits their usefulness as diagnostic and prognostic tool. More recently, evidence has been presented for sAPP blood plasma concentration differences between patients with AD dementia and healthy control subjects. Blood can be easily accessed, which makes it an interesting target for biomarker discovery. This chapter presents a brief overview of sAPP in blood as a new AD biomarker candidate. Issues and challenges are discussed, both from a perspective of laboratory assessment and clinical application. A relevant technique for sAPP ascertainment is also provided.

Key words Alzheimer's disease, Dementia, Blood biomarker, Early diagnosis, Prognosis, Tau, Amyloid, Mild cognitive impairment

1 Introduction

Alzheimer's disease (AD) is characterized by a slowly progressive neurodegeneration, which as the disease progresses increasingly affects cognitive abilities and daily activities; dementia due to AD severely impairs the affected individuals' and their families' quality of life, and it ultimately leads to complete dependence upon care of others [1]. The first stage of AD, which can last several decades, is asymptomatic and preclinical. During the second stage, neurodegeneration becomes severe enough to affect cognitive performance, but normal daily activities remain largely intact; this pre-dementia clinical stage is usually referred to as mild cognitive impairment (MCI) [2]. Only in the last stage of AD, neurodegeneration becomes advanced enough to result in the typical amnestic-type dementia syndrome with significantly impaired everyday function [3]. Until recently, an AD diagnosis could only be established in clinical settings if evidence for dementia was available; however, more recently updated diagnostic criteria were presented,

Robert Perneczky (ed.), *Biomarkers for Preclinical Alzheimer's Disease*, Neuromethods, vol. 137,
https://doi.org/10.1007/978-1-4939-7674-4_12, © Springer Science+Business Media, LLC 2018

which include a stronger focus on a biologically defined diagnosis, with an emphasis on biomarkers, including imaging studies and cerebrospinal fluid (CSF) proteins. The National Institute on Aging-Alzheimer's Association (NIA-AA) guidelines, for example, conceptualize AD as a progressive disorder including all possible stages from asymptomatic to severely impaired [1, 4]. According to these guidelines, AD can be diagnosed in the MCI stage in case of a typical pattern of symptoms and if the presence of AD pathophysiology is evidenced by supporting biomarker findings. The according stage is termed MCI due to AD. This new approach implies that degenerative brain damage due to AD pathophysiological changes foregoes the onset of clinical disease by many years and that cerebral lesions may be present in individuals long before they develop dementia [5].

The hope for disease modification, rather than only symptomatic treatment, by one of the many compounds currently scrutinized in clinical trials as well as technological advances in diagnostic procedures and biomarker discovery, fuels the search for reliable markers of AD pathophysiology and clinical disease progression. Diagnostic and therapeutic progress depends on the ability to successfully diagnose the disease early in its course, before significant neural damage occurs, as well as to understand its natural evolution over time. The published literature in early clinical AD provides insight into the cognitive domains that are affected first (i.e., episodic memory and executive functioning) and associated biomarker changes. However, there remains an unmet need for prospectively collected real-world information regarding prognostic factors, disease course, functional decline, and disease burden in this early stage of the disease; furthermore, biological predictors and indicators of disease progression should be explored in unbiased, population-based cohorts. The majority of publications have so far focused on what may be termed "late" MCI, i.e., those most likely to transition to dementia within a year or two. The information is limited on what are the cognitive deficits and how these manifest and change at times more distal to the onset of AD dementia. Given these challenges, the need for current information on individuals at varying levels of risk for MCI due to AD would be useful to improve our understanding of the natural history and identify opportunities for intervention.

There is robust evidence that the so-called established CSF protein biomarkers amyloid-β $(A\beta)_{1-42}$, total-Tau (tTau), and phosphorylated-Tau $(pTau)_{181}$ [6] as well as functional and structural neuroimaging studies, including fluorodeoxyglucose positron emission tomography (PET) and magnetic resonance imaging of the mediotemporal lobe [7], have good diagnostic and prognostic accuracy. More recently developed CSF and imaging biomarker candidates show promise, such as PET tracers of Aβ [8] and tau [9]. Novel CSF protein markers, which mirror upstream events of Aβ generation, such as soluble amyloid precursor proteins (sAPP) [10], have also been studied, albeit with mixed results. Even if improved imaging or CSF markers of AD pathophysiology were successfully developed, their use in larger-scale studies would still be limited due to the associated costs, exposure to radiation, and invasiveness. Therefore, there is an urgent need to develop biomarkers which can be obtained with relative ease from peripheral body fluids such as blood to replace or assist the existing fluid and imaging markers.

2 Soluble Amyloid Precursor Protein in Blood as a New Biomarker Candidate

Blood sampling, in contrast to obtaining CSF, does not involve a laborious and invasive procedure, which makes the execution of repeated assessments and large-scale screening studies much easier. Blood is in constant exchange with all organs and tissue including the central nervous system; a wide range of physiological and physiopathological processes are therefore reflected in it; hence, blood is an optimal medium for biomarker discovery for a range of different health conditions [11]. Signaling proteins used by the brain to exert control over body functions can be detected in blood [12], and specific patterns of signaling protein changes may be associated with AD [13].

The sAPPα and sAPPβ, which are products of physiological and amyloidogenic cleavage of APP, respectively, are related to central upstream pathophysiological events in AD [14] (Fig. 1). Some earlier studies found increased sAPPβ CSF levels in AD vs. controls [15, 16] and stable vs. progressive MCI [10]. Other published reports do not support these results [17–20], including studies which nevertheless provide evidence for the potential usefulness of sAPPβ as a marker of target engagement in studies on inhibitors of the rate-limiting enzyme in the physiopathological cascade of Aβ generation, i.e., beta-site APP cleaving enzyme 1 (BACE1) [21].

More recently, results were published that indicate the possible usefulness of sAPPβ in blood plasma to support AD diagnosis. Significantly decreased plasma concentrations in patients with AD dementia were found as compared to age-matched healthy control subjects and patients with behavioral-variant frontotemporal dementia, i.e., a neurodegenerative condition with a different pathomechanism which does not typically involve APP-related changes; no concentration differences between the groups were described for sAPPα [22]. This study provides a first piece of evidence in support of sAPPβ in blood as a promising new biomarker candidate of AD, which may potentially improve the diagnostic accuracy of existing markers and also enable a less invasive diagnostic work-up. Further research is required to establish normal ranges and to replicate the results in independent cohorts including larger numbers of participants covering a wider spectrum of cognitive impairment. Our initial results also indicate that sAPPβ plasma concentrations as compared to healthy

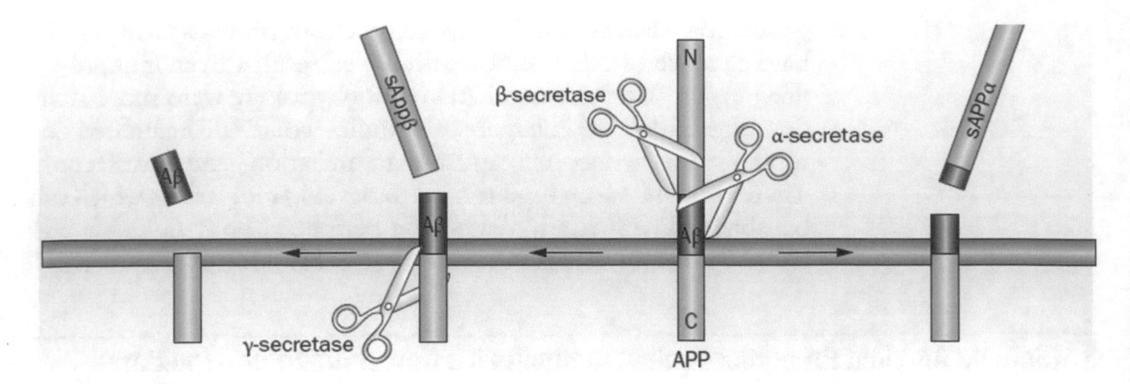

Fig. 1 Schematic diagram representing proteolytic cleavage of amyloid precursor protein (reproduced with permission from: R. Perneczky in [27]). The membrane-bound amyloid precursor protein (APP) is alternatively cleaved by either α-secretase, resulting in non-amyloidogenic fragments including soluble APP (sAPP)α or by β-secretase, resulting in sAPPβ and the harmful Aβ fragment via further cleavage by γ-secretase

controls are lower in MCI patients with an AD-typical PET metabolic pattern and comparable to plasma concentrations of AD dementia patients (manuscript in preparation).

3 Methods

3.1 General Pre-analytic Guidelines

See [23] for proposed consensus guidelines for the standardization of pre-analytic variables for blood-based biomarker studies in AD research.

1. Sample withdrawal should only be undertaken by qualified medical personnel. The identity of the patient should be checked, and the analysis order/study documentation should be filled out completely before withdrawal of the sample. Written informed consent by the patient or study subject must be available before any procedures are performed. Tubes should be clearly labelled beforehand and immediately closed after taking the sample to avoid contamination and evaporation.
2. Samples should optimally be taken after overnight fasting at a standard time in the morning, before any morning medications are taken. Closed blood withdrawal systems (e.g., Monovette, Vacuette) reduce the preanalytical influence on the results and the infection risk for medical personnel and are therefore preferable to syringes.
3. To avoid contamination, tubes without additive should be collected before tubes with additive. When withdrawing several tubes, the thinning tube should never be at the beginning (release of tissue factors on puncture).

4. Venous stasis should be around 20–30 mmHg under the systolic blood pressure so that arterial pulsation is maintained. During longer blood withdrawals (several tubes), the stasis should be relieved. Venous stasis should not last too long.
5. Blood should not be taken from horizontal veins or arterial entry sites. If this cannot be avoided, the entry site must be rinsed with 0.9% NaCl solution before withdrawal and a sufficient amount of blood (for adults 10–20 mL) thrown away. Directly before blood withdrawal, no medications should be injected through the catheter.
6. Blood withdrawal should not be performed with too fine cannulae since hemolysis can occur if cannulae are too fine. The blood should flow freely into the collection vessel. If the pressure is too low, hemolysis can occur. Samples should not be exposed to direct sunlight.

3.2 Collection of Blood Samples for Biomarker Testing

1. Whole blood samples are collected by venepuncture into commercially available anticoagulant-treated tubes (e.g., citrate or EDTA) and put into the freezer at 4 °C within 120 min after collection.
2. Plasma is isolated from blood cells (*see* **Note 1**) by centrifugation at 2500 × *g* for 15 min.
3. If the sample is not analyzed immediately, it should be apportioned into aliquots (e.g., 500 μL), and the aliquots should immediately be transferred to −20 °C and should within 2 weeks be transferred to −80 °C for long-term storage.

3.3 Considerations for Multicenter Research

1. Biomarker development and testing frequently involve collaboration between different centers and laboratories, which is associated with additional challenges and requires meticulous adherence to sample collection protocols. Multicenter studies require each individual laboratory to follow exactly the same protocol for plasma collection.
2. Aliquots of the samples are kept at −20 °C for not longer than 2 weeks or at −80 °C until use. Plasma samples must be shipped on dry ice to the central laboratory and are required to be immediately stored at −20 °C for not longer than 2 weeks or at −80 °C until further analysis (*see* **Note 2**).

3.4 Enzyme-Linked Immunosorbent Assay (ELISA)

Different methods have been established to quantify protein concentrations in body fluids. Here, ELISA will be discussed as an example, since it is a widely used method, which has been shown to be able to detect sAPP in human blood samples. ELISA is usually performed in duplicate and according to the commercial manufacturer's instructions (see [22] for an example of a sAPP assay). An ELISA includes coating, blocking, incubation, and detection according to the following general guidelines (specific assays may require adjustments to certain parameters):

1. Determine wells for reagent blank, test sample blank, test sample and diluted standard, and then coat the microwells with 100 μL appropriate diluted antigens (*see* **Note 3**).
2. Incubate the plate at room temperature (RT) for 2 h or at 4 °C after covering it with plate lid overnight.
3. Vigorously wash all unbound antigen off the plate to prevent false positive for at least three times with 300 μL wash buffer (*see* **Note 4**). Then block nonspecific binding by adding 200 μL of blocking buffer to each well. Incubate at RT for 1 h or at 4 °C overnight, and then wash plate as above.
4. Pipette 100 μL of diluted samples (*see* **Notes 2** and **5**) and standards to appropriate wells, incubate for 1 h at RT or at 4 °C overnight, and then repeat washing step.
5. Add 100 μL of second step antibody to wells and incubate for 1 h at RT; then repeat washing step.
6. For color development, 100 μL of chromogenic substrate is added to each well. Cover the plate and incubate for 15 min or until a suitable color has developed. The plate should preferably be protected against light during this incubation.
7. Determine the optical density of each well within 30 min through the bottom of the microwell plate using an automated or semiautomated photometer (ELISA reader) with appropriate wavelength). Determine the concentration of the samples from the standard curve using curve fitting software (*see* **Note 6**).

4 Conclusions

Proteomic analysis of blood samples is a powerful tool to identify novel, clinically more relevant biomarkers of AD. Several analytical platforms are available, with antibody-based techniques being particularly valuable if many samples have to be analyzed quantitatively. ELISA technology is at present the gold standard in quantifying protein markers. It combines the sensitivity of simple enzyme assays with the specificity of antibodies, being only capable of measuring the concentration of a single protein at a time, albeit in a highly specific and high-throughput manner. ELISA is able to detect proteins in complex matrices such as CSF and blood in the pg/mL range; research is simplified by the commercial availability of standardized off-the-shelf assays. Many researchers are competent in using ELISA, and they are a comparably low-cost solution. There is also no need for specialist equipment to run a standard

96-well assay. Having said that, there are also several shortcomings, including the relatively large sample volumes needed to run an ELISA), and restrictions in automatization due to the required wash steps, even though manual handling of samples may be prone to failure. A recent study across 14 clinical neurochemistry laboratories in Europe (Germany, Austria, and Switzerland) reported that the coefficient of variation CV (CV) of ELISAs was in the 20–30% range when using commercially available ELISA kits for AD diagnosis [24].

From reviewing the available literature on sAPP as a novel AD biomarker, two main conclusions arise. First, concentration changes of sAPPα/sAPPβ in CSF do not seem to follow a consistent pattern limiting their usefulness as diagnostic markers. Secondly, initial results relating to sAPPβ in blood are encouraging, but replication and further validation are required before firm conclusions can be drawn. Some evidence exists that a combination of the new and established biomarkers may improve the overall diagnostic accuracy [10], but more research in this field is needed. There also remains a need to standardize laboratory procedures to generate reliable results across laboratories in multicenter studies. Several quality control initiates have been launched worldwide to establish standardized protocols for biomarker assessment, including the Alzheimer's Association Global Biomarkers Consortium [25].

5 Notes

1. Compared with serum, plasma sampling needs no clotting time, and the isolation of cells and liquid phase is easily accomplished and, thus, is less time-consuming. Furthermore, plasma volume yield is about 10–20% higher compared to serum; the concentration of proteins in plasma is also greater than in serum, which contains clotting factors and related constituents [26].
2. Test samples should be measured soon after collection. For the stored frozen samples, it is important to avoid additional freeze-thaw cycles. When using frozen samples, it is recommended to thaw the samples at a low temperature and mix them completely by vortexing prior to use. Hemolyzed, icteric, or lipemic samples might invalidate certain tests.
3. Wells reserved for chromogen blanks should be left empty.
4. Always remove the wash buffer completely by tapping the pre-coated plate on paper towel. Do not wipe wells with paper towel.
5. Plasma samples are recommended to be diluted 1:50 to 1:100.

6. The dose-response may be nonlinear beyond the standard point and inaccurate; thus, do not extrapolate the standard curve beyond the highest point; diluted samples that are greater than the highest standard should be reanalyzed, and the results should be multiplied by the appropriate dilution factor.

References

1. Jack CR Jr, Albert MS, Knopman DS et al (2011) Introduction to the recommendations from the National Institute on Aging-Alzheimer's Association workgroups on diagnostic guidelines for Alzheimer's disease. Alzheimers Dement 7(3):257–262. https://doi.org/10.1016/j.jalz.2011.03.004. S1552-5260(11)00100-2 [pii]
2. Lopez OL (2013) Mild cognitive impairment. Continuum (Minneap Minn) 19(2 Dementia):411–424. https://doi.org/10.1212/01.CON.0000429175.29601.97
3. McKhann GM, Knopman DS, Chertkow H et al (2011) The diagnosis of dementia due to Alzheimer's disease: recommendations from the National Institute on Aging-Alzheimer's Association workgroups on diagnostic guidelines for Alzheimer's disease. Alzheimers Dement 7(3):263–269. https://doi.org/10.1016/j.jalz.2011.03.005. S1552-5260(11)00101-4 [pii]
4. McKhann GM (2011) Changing concepts of Alzheimer disease. JAMA 305(23):2458–2459. https://doi.org/10.1001/jama.2011.810.305/23/2458 [pii]
5. Giaccone G, Arzberger T, Alafuzoff I et al (2011) New lexicon and criteria for the diagnosis of Alzheimer's disease. Lancet Neurol 10(4):298–299.; author reply 300–301. https://doi.org/10.1016/S1474-4422(11)70055-2
6. Blennow K, Hampel H, Weiner M et al (2010) Cerebrospinal fluid and plasma biomarkers in Alzheimer disease. Nat Rev Neurol 6(3):131–144. https://doi.org/10.1038/nrneurol.2010.4
7. Frisoni GB, Fox NC, Jack CR Jr et al (2010) The clinical use of structural MRI in Alzheimer disease. Nat Rev Neurol 6(2):67–77. https://doi.org/10.1038/nrneurol.2009.215
8. Drzezga A, Grimmer T, Henriksen G et al (2008) Imaging of amyloid plaques and cerebral glucose metabolism in semantic dementia and Alzheimer's disease. NeuroImage 39(2):619–633. https://doi.org/10.1016/j.neuroimage.2007.09.020. S1053-8119(07)00806-3 [pii]
9. Scholl M, Lockhart SN, Schonhaut DR et al (2016) PET imaging of Tau deposition in the aging human brain. Neuron 89(5):971–982. https://doi.org/10.1016/j.neuron.2016.01.028
10. Perneczky R, Tsolakidou A, Arnold A et al (2011) CSF soluble amyloid precursor proteins in the diagnosis of incipient Alzheimer disease. Neurology 77(1):35–38. https://doi.org/10.1212/WNL.0b013e318221ad47
11. Chan K, Lucas D, Hise D et al (2004) Analysis of the human serum proteome. Clin Proteomics 1:101–226
12. Wyss-Coray T (2006) Inflammation in Alzheimer disease: driving force, bystander or beneficial response? Nat Med 12(9):1005–1015. https://doi.org/10.1038/nm1484.nm1484 [pii]
13. Ray S, Britschgi M, Herbert C et al (2007) Classification and prediction of clinical Alzheimer's diagnosis based on plasma signaling proteins. Nat Med 13(11):1359–1362. https://doi.org/10.1038/nm1653
14. Zhang H, Ma Q, Zhang YW et al (2012) Proteolytic processing of Alzheimer's beta-amyloid precursor protein. J Neurochem 120(Suppl 1):9–21. https://doi.org/10.1111/j.1471-4159.2011.07519.x
15. Wu G, Sankaranarayanan S, Wong J et al (2012) Characterization of plasma beta-secretase (BACE1) activity and soluble amyloid precursor proteins as potential biomarkers for Alzheimer's disease. J Neurosci Res 90(12):2247–2258. https://doi.org/10.1002/jnr.23122
16. Lewczuk P, Popp J, Lelental N et al (2012) Cerebrospinal fluid soluble amyloid-beta protein precursor as a potential novel biomarkers of Alzheimer's disease. J Alzheimers Dis 28(1):119–125. https://doi.org/10.3233/JAD-2011-110857
17. Rosen C, Andreasson U, Mattsson N et al (2012) Cerebrospinal fluid profiles of amyloid

beta-related biomarkers in Alzheimer's disease. Neuromolecular Med 14(1):65–73. https://doi.org/10.1007/s12017-012-8171-4

18. Brinkmalm G, Brinkmalm A, Bourgeois P et al (2013) Soluble amyloid precursor protein alpha and beta in CSF in Alzheimer's disease. Brain Res 1513:117–126. https://doi.org/10.1016/j.brainres.2013.03.019
19. Olsson A, Hoglund K, Sjogren M et al (2003) Measurement of alpha- and beta-secretase cleaved amyloid precursor protein in cerebrospinal fluid from Alzheimer patients. Exp Neurol 183(1):74–80
20. Zetterberg H, Andreasson U, Hansson O et al (2008) Elevated cerebrospinal fluid BACE1 activity in incipient Alzheimer disease. Arch Neurol 65(8):1102–1107. https://doi.org/10.1001/archneur.65.8.1102
21. Perneczky R, Alexopoulos P, Alzheimer's Disease neuroimaging Initiative (2014) Cerebrospinal fluid BACE1 activity and markers of amyloid precursor protein metabolism and axonal degeneration in Alzheimer's disease. Alzheimers Dement 10(5 Suppl):S425–S429.e1. https://doi.org/10.1016/j.jalz.2013.09.006
22. Perneczky R, Guo LH, Kagerbauer SM et al (2013) Soluble amyloid precursor protein beta as blood-based biomarker of Alzheimer's disease. Transl Psychiatry 3:e227. https://doi.org/10.1038/tp.2013.11
23. O'Bryant SE, Gupta V, Henriksen K et al (2015) Guidelines for the standardization of preanalytic variables for blood-based biomarker studies in Alzheimer's disease research. Alzheimers Dement 11(5):549–560. https://doi.org/10.1016/j.jalz.2014.08.099
24. Lewczuk P, Beck G, Ganslandt O et al (2006) International quality control survey of neurochemical dementia diagnostics. Neurosci Lett 409(1):1–4. https://doi.org/10.1016/j.neulet.2006.07.009
25. Carrillo MC, Blennow K, Soares H et al (2013) Global standardization measurement of cerebral spinal fluid for Alzheimer's disease: an update from the Alzheimer's Association Global Biomarkers Consortium. Alzheimers Dement 9(2):137–140. https://doi.org/10.1016/j.jalz.2012.11.003
26. Tammen H, Schulte I, Hess R et al (2005) Peptidomic analysis of human blood specimens: comparison between plasma specimens and serum by differential peptide display. Proteomics 5(13):3414–3422. https://doi.org/10.1002/pmic.200401219
27. Malpass K (2011) Alzheimer disease: a novel biomarker to detect early-stage Alzheimer disease. Nat Rev Neurol 7(8):420. https://doi.org/10.1038/nrneurol.2011.112

Check for updates

Chapter 13

Tau Imaging in Preclinical Alzheimer's Disease

Paul Edison

Abstract

Aggregated tau is a major neuropathological protein implicated in the pathophysiology of common neurodegenerative disorders, such as Alzheimer's disease (AD), Parkinson's disease without (PD) and with later dementia (PDD), Lewy body dementia (LBD), frontotemporal dementia (FTD), and corticobasal degeneration (CBD). Aggregated tau tangles, the result of hyperphosphorylation, is a pathological characteristic of a group of neurodegenerative conditions known as the tauopathies. In AD, it has been shown that the density of tau tangles closely correlates with neuronal dysfunction and cell death, which is not the case for β-amyloid. Until now, diagnostic and pathological demonstration of tau deposition has only been possible by invasive techniques such as brain biopsy or postmortem examination. The recent advances in the development of selective tau positron emission tomography (PET) tracers have allowed in vivo investigation of the presence and extent of tau pathology in patients suspected of having tauopathies and the role of tau in the early phases of neurodegenerative diseases. In this review, the role of aggregated tau will be discussed, as well as the challenges posed by, and the current status of, the development of selective tau tracers as biomarkers, and the new clinical information that has been uncovered, in addition to the opportunities for refining the diagnosis of tauopathies in the future.

Key words Tau imaging, Dementia, Neurodegenerative diseases, PET, Alzheimer's disease

1 Introduction

Alzheimer's disease (AD), Parkinson's disease without (PD) and with later dementia (PDD), Lewy body dementia (LBD), frontotemporal dementia (FTD), and corticobasal degeneration (CBD) are common neurodegenerative disorders. These neurodegenerative conditions are characterized by the disturbance of protein homeostasis, leading to the pathologic accumulation of protein, especially tau and β-amyloid (Aβ). Tau is a microtubule-associated protein necessary for the cytoarchitecture of the cells. However, aggregated tau, due to hyperphosphorylation, is a pathological characteristic of a group of neurodegenerative conditions known as the tauopathies [1]. The neuropathological substrates of AD are tau neurofibrillary tangles (NFTs) and Aβ plaques, while activated microglia, astrocytes, and neuropil threads also play a significant

Robert Perneczky (ed.), *Biomarkers for Preclinical Alzheimer's Disease*, Neuromethods, vol. 137,
https://doi.org/10.1007/978-1-4939-7674-4_13, © Springer Science+Business Media, LLC 2018

role in disease pathogenesis. Recent advances in selective tau tracer development for positron emission tomography (PET) imaging have, for the first time, allowed in vivo exploration of the presence and extent of tau pathology in patients suspected of having tauopathies.

2 Tau Protein in Tauopathies

Tau is a protein that exists mainly in axons and binds to microtubules to stabilize them. Microtubules are critical for maintaining cellular structural integrity [1–3]. Tau protein can occur as three-repeat (3R) or four-repeat (4R) forms. Equal numbers of 3R and 4R isoforms exist in the healthy adult human. In neurodegenerative diseases, the hyperphosphorylated form of tau causes its aggregation, resulting in neuronal dysfunction and death by reducing their stability. Aggregated tau in AD exists as paired helical filaments which further coalesce into neurofibrillary tangles (NFTs) [3–5]. These aggregates spread transynaptically throughout the brain, and it is shown that tau deposition correlates with the clinical progression of the disease. While Aβ and tau are the cardinal features of AD, it is clear that a single aggregated protein can be the underlying cause of several different clinical phenotypes or a single clinical phenotype can be caused by different aggregated proteins.

To date, research has focused primarily on the amyloid cascade hypothesis and this amyloid Aβ-centric approach to AD [2, 3]. Amyloid Aβ therapeutic approaches, particularly monoclonal antibodies against both soluble and fibrillary forms of amyloid Aβ beta, have failed in preventing clinical progression of AD, despite a reduction in the amyloid Aβ plaque burden [4]. However, there is some hope that these drugs modify disease progression if given early enough in the disease course, but long-term efficacy data in patients with prodromal AD are only starting to emerge [5, 6]. A recurring argument against the amyloid hypothesis is that despite the prominence of amyloid Aβ plaques, tau-associated pathologic changes are more closely correlated with disease severity [7]. Thus, tau may be an important protein in early detection and staging of AD. However, tau pathology is nonspecific and is a marker of neurodegeneration beyond AD in diseases such as FTD, traumatic brain injury, progressive supranuclear palsy (PSP), CBD, and chronic traumatic encephalopathy.

While amyloid Aβ deposition starts in the inferior frontal cingulate areas and then spreads to the association cortex, tau deposition takes place initially in the transentorhinal cortex (stages 1 and 2) in the stage of AD before symptoms are apparent, then spreads to the limbic, and symptoms then become evident [6]. As disease progresses, NFTs finally involve association cortical areas by Braak stages 5–6 areas at which time symptoms become severe. It is also shown that, at postmortem, occasional NFTs can be found in the brains of seemingly healthy 30-year-olds [7]. In vivo PET imaging studies have demonstrated that Aβ deposition can occur 10–20 years before the appearance of symptoms, and, as per the amyloid cascade neuroinflammation hypothesis, this amyloid Aβ deposition leads to a sequence of downstream events which result in tau hyperphosphorylation and aggregation leading to neuronal dysfunction. While the imaging studies support the theory that amyloid Aβ deposition takes place decades before symptoms appear, it is now recognized that Aβ deposition alone cannot account for the cognitive dysfunction present in AD. It has been shown that cognitively normal subjects with increased amyloid Aβ may have accelerated neurodegeneration and with progressive cognitive impairment, but the levels of amyloid Aβ do not correlate with cognitive function [8, 9]. However, studies have demonstrated that tau deposition correlates with cognitive function [10] and continues to progress throughout the course of disease, correlating with symptom severity [11]. It has also been demonstrated that injection of Aβ42 into transgenic tau mouse models leads to accelerated tau formation, and it spreads rapidly from the site of injection demonstrating that tau can propagate in neuronal networks [12, 13].

Different ultrastructural forms of tau can cause different disease phenotypes. Different diseases have different isoforms, and the brains of patients with AD contain equal amounts of 3R and 4R isoforms; Pick's disease is characterized by 3R isoform aggregation into Pick bodies, while CBD, PSP, and argyrophilic grain disorders contain aggregated 4R isoforms [1]. In addition to the above conditions, chronic traumatic encephalopathy (CTE), present in subjects who have suffered serial mild concussive brain injuries duc to, for example, American football, rugby, boxing, or frequent falls due to multiple etiology, is characterized histologically by tau deposition in areas of axonal injury [14]. Hence, it is essential to have a biomarker which could detect tau deposition in vivo. Studies have shown that the deposition of abnormal tau correlates closely with cell dysfunction and symptoms in contrast to amyloid Aβ deposition which occurs before symptoms appear. While amyloid Aβ may have an early and significant role in AD, the disappointing results from anti-amyloid therapies suggest that amyloid Aβ alone is not enough to cause neuronal damage, which again highlights the importance of imaging tau [3]. There are also suggestions that, similar to the presence of amyloid Aβ in the cognitively rate over time [15, 16], it is possible that tau deposition can occur early, and persistent tau deposition could lead to progressive disease. Tau imaging will allow us to evaluate whether tau can cause neuronal damage independently of amyloid Aβ production, and indeed whether tau production is associated with neuroinflammation.

3 Tau Imaging Agents

For an effective tau imaging agent, it must cross the blood-brain barrier, and should be taken up by neurofibrillary tangles, and should not bind to other aggregated proteins like amyloid. They should also be eliminated very quickly, and could be imaged using 11C and 18F radionuclides [2, 17], should have low nonspecific binding, and should be able to quantify by using simple quantitative methods. For clinical utility, these tracers should have the potential to differentiate easily between the patients and the controls even by visual inspection. Due to the predominantly intracellular nature of the tau, these agents should be able to engage readily with the targets and should have specific binding which increases with the density of the neurofibrillary tangles, ideally able to engage with the targets (tau tangles) within the cells which undergo predominant neurodegeneration. Additionally, these tracers should bind with different forms of posttranslational modifications. Whether differential binding to 3R or 4R is beneficial should still be evaluated.

Human tau is composed of six distinct splice variants that vary with respect to inclusion or exclusion of three alternatively spliced segments. In addition to phosphorylation and glycosylation, new modifications, such as the acetylation of amino groups, are being discovered. This could differentially influence the different binding affinity of closely related molecules. Such heterogeneity may contribute to the differential binding affinity for potential new tau tracers. Successful radiotracers must bind with high selectivity to tau and not be influenced by the amyloid or any other protein or substrates. It is also important that binding potential should represent the density of the tau (ideally receptor binding), and Bmax be proportional to the concentration of the tau protein.

Tau is an intracellular protein and is present in much lower quantities than amyloid, and the intracellular nature of the tau requires the tracer to cross the blood-brain barrier and the plasma membrane which will require tracers with low molecular weight. This is further compounded by the fact that tau is present in much lower quantities than amyloid, which will mean that a tau tracer should have high specific binding to tau and minimal binding to amyloid [18–22]. As with all PET tracers, it should be able to clear from the brain very rapidly and should not have active metabolites that enter the brain [19–22]. Figure 1 shows a number of tau PET imaging agents in current use.

3.1 [18F]T807 (AV-1451)

[18F]T807 (Flortaucipar) was originally tested in vivo and demonstrated high binding affinity and selectivity [23]. [18F]T807 demonstrated high specificity, with 25-fold selectivity for PHF-tau over Aβ, with rapid brain penetration and washout, and no active metabolites entering the brain [24]. While [18F]T807 was found

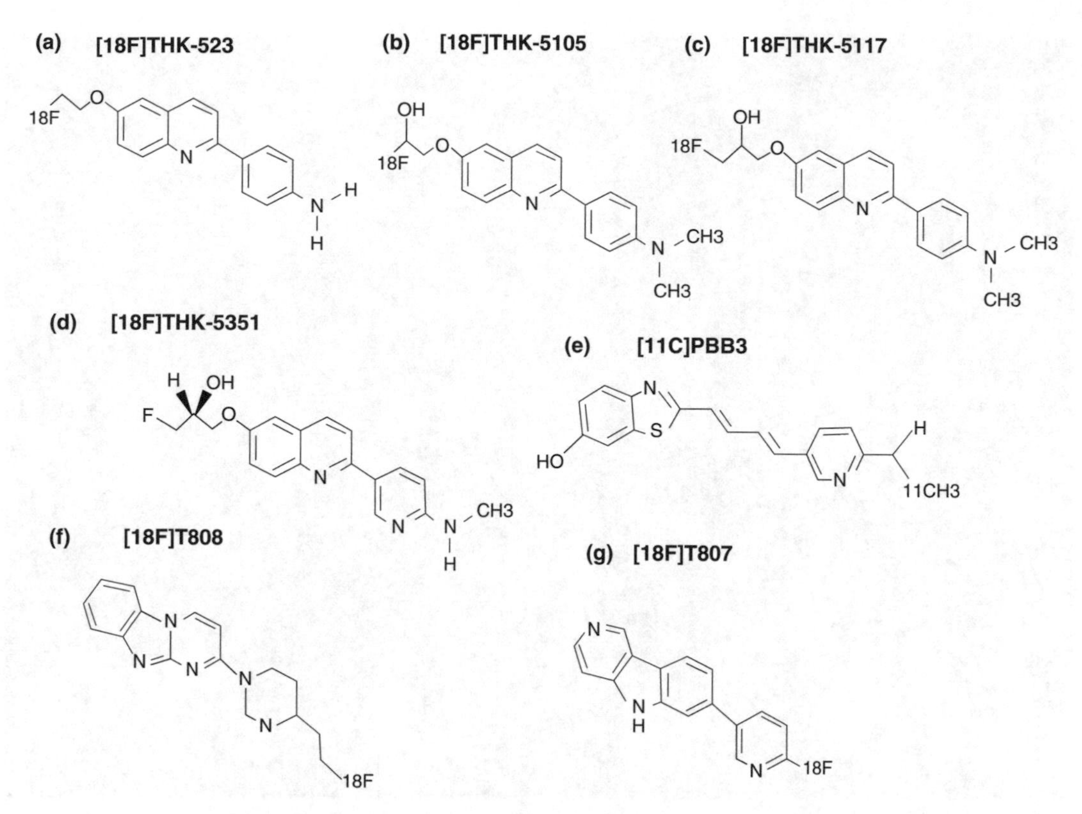

Fig. 1 Chemical structures of tau tracers in current use. (**a**) [18F]THK-523, (**b**) [18F]THK-5105, (**c**) [18F]THK-5117, (**d**) [18F]THK-5351, (**e**) [11C]PBB3, (**f**) [18F]T808, and (**g**) [18F]T807

promising, a closely related compound, [18F]T808, was thought to be unstable due to the defluorination. Additionally, [18F]T807 uptake closely correlated with histology. Initial in vivo studies in humans demonstrated good tracer kinetics, with a potential to discriminate patients who may have significant tau uptake compared to those who have not got any retention [25]. The preliminary work demonstrated significant deposition in AD compared to controls, and later studies further substantiated the original finding [26] (Fig. 2). It was further demonstrated that, in MCI subjects as the disease advances, there is progressive increase in tau deposition, and this was associated with a worse performance in cognitive function tests [27, 28]. It was also encouraging to see, in patients with posterior cortical atrophy, that there was a regional increase in the posterior cortex. However, consistent with the neuropathology, there was a regional predilection of the tau deposition. In PCA there was occipitotemporal and occipitoparietal uptake. Tau was also increased in FTD-tau and PSP, CBD, PD, dementia with Lewy bodies (DLB), and cerebral Aβ angiopathy (CAA), along with logopenic variant primary progressive aphasia, where the regional predilection differed depending on the specific site of involvement in different diseases

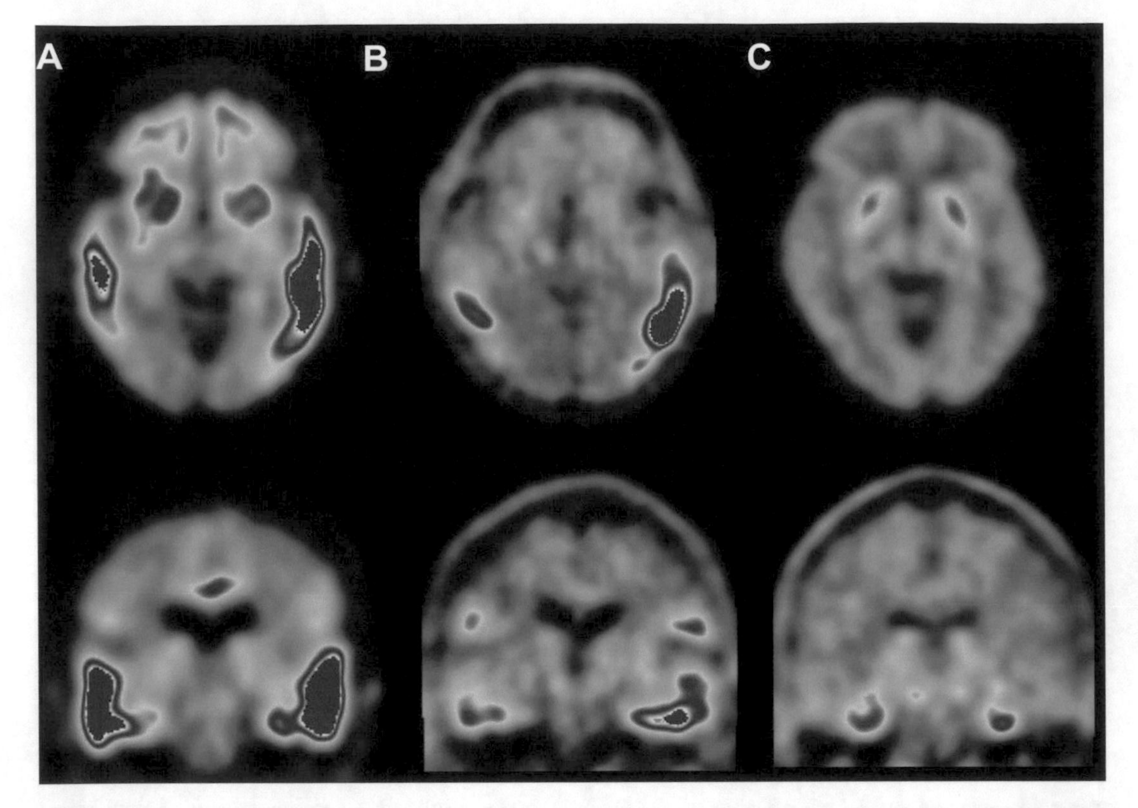

Fig. 2 [18F]AV-1451 in AD, MCI, and control subjects. Top panel demonstrates transverse section of tau images using [18F]AV1451 in Alzheimer's subjects (**a**), mild cognitive impairment subjects (**b**), and healthy controls (**c**). Bottom panel demonstrates the coronal sections for the corresponding subjects

[29, 30]. Other conditions where tau [31] is primarily driving the disease process like FTD, chronic traumatic encephalopathy (CTE), presents with different pattern of deposition of tau [32]. In cognitively normal elderly subjects, [18F]T807 binding has been shown to correlate with levels of CSF tau [33].

3.2 THK Compounds

Okamura and colleagues reported first [18F]THK compounds, of which [18F]THK523 was evaluated at the Tohuku University [34]. Pharmacokinetic studies showed excellent uptake and clearance in the brains of mice [35], and in vivo studies demonstrated significant uptake. While the uptake correlated well with cognitive function, there was significant uptake in white matter [36] making the tracer less favorable for clinical use. Two closely related compounds, [18F]THK5105 and [18F]THK5117, were then evaluated. In vivo studies showed that there was higher cortical retention in AD patients compared with healthy controls, and this correlated well with cognitive test results and brain volumes [37]. [18F]THK5117 uptake in AD and MCI subjects correlated well with cognitive performance, and there was some correlation between tau uptake and amyloid load. There was increased binding in the temporal region in all subjects with some correlation between the uptake and cerebral glucose metabolism [38].

In an attempt to improve the specificity of these compounds, [18F]THK5351 was recently developed, and preliminary studies in AD and healthy control subjects demonstrated very high affinity in classical regions with tau deposition. In further comparisons between [18F]THK5351 and [18F]THK5117, the former demonstrated faster kinetics, potentially allowing for better visualization [39]. However, recent studies have demonstrated that these compounds could be blocked by selegiline, indicating the nonspecificity of the tracer and the interaction with MAOB.

3.3 [11C]PBB3

2-((1E,3E)-4-(6-((11)C-methylamino)pyridin-3-yl)buta-1,3-dienyl)benzo[d]thiazol-6-ol ((11)C-PBB3) was another PET probe developed for tau imaging. It has been shown that [11C]PBB3 binds strongly to NFTs in AD brains. Fluorescence microscopy demonstrated labelling of non-ghost and ghost tangles along with detection of dystrophic neurites. Interestingly, radioligand binding to brain homogenates revealed multiple binding components with differential affinities for [11C]PBB3 with higher availability of binding sites on PSP tau deposits for [11C]PBB3 than [18F]AV-1451, and it is proposed that [11C]PBB3 can capture a wide range of tauopathies. Preliminary studies have demonstrated that there is significant binding in areas where there is presence of tau [40, 41]. These preliminary reports suggested that this could bind to other tauopathies, even with the absence of amyloid Aβ, and potentially could have a wide range of uses [40].

3.4 Other Compounds

Gobbi et al. have developed further high affinity binders, RO6931643, RO6924963, and RO6958948. These compounds bind with high affinity and specificity to the [3H]T808 binding site on tau aggregates [42]. Importantly, they lack affinity for Aβ plaques and showed low nonspecific binding to healthy brain tissue. Additionally, the compounds demonstrate macro- and microcolocalization of radioligand binding, rapid brain entry, and washout with well-tolerated metabolic patterns. These tracers are currently in clinical testing.

4 Conclusions

Tau imaging will allow us to visualize and quantify the tau aggregation in neurodegenerative diseases and is able to provide us with further information about the neuropathology of the disease process. This could revolutionize the field, as we are now able to understand more about the disease trajectory, and test different novel anti-tau therapeutic strategies. Tau imaging has the potential of becoming part of the diagnostic algorithm in the clinical practice.

References

1. Spillantini MG, Goedert M (2013) Tau pathology and neurodegeneration. Lancet Neurol 12(6):609–622. https://doi.org/10.1016/s1474-4422(13)70090-5
2. Lee HG, Perry G, Moreira PI et al (2005) Tau phosphorylation in Alzheimer's disease: pathogen or protector? Trends Mol Med 11(4):164–169. https://doi.org/10.1016/j.molmed.2005.02.008
3. Giacobini E, Gold G (2013) Alzheimer disease therapy—moving from amyloid-beta to tau. Nat Rev Neurol 9(12):677–686. https://doi.org/10.1038/nrneurol.2013.223
4. Grundke-Iqbal I, Iqbal K, Tung Y-C, Quinlan M, Wisniewsk H, Binder L (1986) Abnormal phosphorylation of the microtubule-associated protein tau in Alzheimer cytoskeletal pathology. Proc Natl Acad Sci U S A 83:4913–4917
5. Kosik K, Joachim C, Selkoe D (1986) Microtubule-associated protein tau is a major antigenic component of paired helical filaments in Alzheimer disease. Proc Natl Acad Sci U S A 83:4044–4048
6. Braak H, Braak E (1991) Neuropathological staging of Alzheimer related changes in Alzheimer's disease. Acta Neuropathol 82:239–259
7. Braak H, Braak E (1997) Frequency of stages of Alzheimer-related lesions in different age categories. Neurobiol Aging 18(4):351–357
8. Becker JA, Hedden T, Carmasin J et al (2011) Amyloid-beta associated cortical thinning in clinically normal elderly. Ann Neurol 69(6):1032–1042. https://doi.org/10.1002/ana.22333
9. Engler H, Forsby A, Almqvist O et al (2006) Two year follow up of amyloid deposition in patients with Alzheimer's disease. Brain 129:2856–2866
10. Gómez-Isla T, Hollister R, West H et al (1997) Neuronal loss correlates with but exceeds neurofibrillary tangles in Alzheimer's disease. Ann Neurol 41:17–24
11. Ingelsson M, Fukumoto H, Newell KL et al (2004) Early amyloid accumulation and progressive synaptic loss, gliosis and tangle formation in the Alzheimer's disease brain. Neurology 62:925–931
12. Gotz J, Chen F, van Dorpe J, Nitsch RM (2001) Formation of neurofibrillary tangles in P301l tau transgenic mice induced by Abeta 42 fibrils. Science 293(5534):1491–1495. https://doi.org/10.1126/science.1062097
13. Lewis J, Dickson DW, Lin WL et al (2001) Enhanced neurofibrillary degeneration in transgenic mice expressing mutant tau and APP. Science 293(5534):1487–1491. https://doi.org/10.1126/science.1058189
14. Gandy S, Ikonomovic M, Mitsis E, Elder G, Ahlers S, Barth J, Stone J, DeKosky ST (2014) Chronic traumatic encephalopathy: clinical-biomarkers correlations and current concepts in pathogenesis. Mol Degener 9:37
15. Villemagne VL, Pike KE, Chetelat G et al (2011) Longitudinal assessment of Abeta and cognition in aging and Alzheimer disease. Ann Neurol 69(1):181–192. https://doi.org/10.1002/ana.22248
16. Villemagne VL, Burnham S, Bourgeat P et al (2013) Amyloid B deposition, neurodegeneration, and cognitive decline in sporadic Alzheimer's disease: a prospective cohort study. Lancet Neurol 12:357–367. https://doi.org/10.1016/S1474-4422(13)70044-9
17. Laruelle M, Slifstein M, Huang Y (2003) Relationships between radiotracer properties and image quality in molecular imaging of the brain with positron emission tomography. Mol Imaging Biol 5(6):363–375
18. Villemagne VL, Furomoto S, Fodero-Tavoletto M et al (2012) The challenges of tau imaging. Future Neurol 7(4):409–421
19. Villemagne VL, Okamura N (2014) In vivo tau imaging: obstacles and progress. Alzheimers Dement 10(3 Suppl):S254–S264. https://doi.org/10.1016/j.jalz.2014.04.013
20. Okamura N, Harada R, Furumoto S et al (2014) Tau PET imaging in Alzheimer's disease. Curr Neurol Neurosci Rep 14(11):500. https://doi.org/10.1007/s11910-014-0500-6
21. Shah M, Catafau AM (2014) Molecular imaging insights into neurodegeneration: focus on tau PET radiotracers. J Nucl Med 55(6):871–874. https://doi.org/10.2967/jnumed.113.136069
22. Villemagne V, Fodero-Tavoletti M, Masters CL, Rowe C (2015) Tau imaging: early

progress and future directions. Lancet Neurol 14(1):114–124

23. Gao M, Wang M, Zheng QH (2014) Concise and high-yield synthesis of T808 and T808P for radiosynthesis of [(18)F]-T808, a PET tau tracer for Alzheimer's disease. Bioorg Med Chem Lett 24(1):254–257. https://doi.org/10.1016/j.bmcl.2013.11.025
24. Xia CF, Arteaga J, Chen G et al (2013) [(18)F]T807, a novel tau positron emission tomography imaging agent for Alzheimer's disease. Alzheimers Dement 9(6):666–676. https://doi.org/10.1016/j.jalz.2012.11.008
25. Chien DT, Bahri S, Szardenings AK et al (2013) Early clinical PET imaging results with the novel PHF-tau radioligand [F-18]-T807. J Alzheimers Dis 34(2):457–468. https://doi.org/10.3233/JAD-122059
26. Pontecorvo M, Devous Ml, Joshi A, Lu M, Siderowf A, Arora A, Mintun M (2015) Relationships between florbetapir PET amyloid and 18F AV-1451 (aka 18F-T807) PET tau binding in cognitively normal subjects and patients with cognitive impairments suspected of Alzheimer's disease. Human Amyloid Imaging Conference Book of Abstracts 2015 ID submission 98:131
27. Johnson K, Becker J, Sepulcre J et al. (2014) Tau PET: initial experience with F18 T807. Human Amyloid Imaging Conference Book of Abstracts 2014 ID submission 103:22
28. Mintun M, Devous M, Pontecorvo M, Joshi A, Siderowf A, Johnson K, Navitsky M, Lu M (2015) Potential for PET imaging tau tracer 18F-AV-1451 (also known as 18F-T807) to detect neurodegenerative progression in Alzheimer's disease. Human Amyloid Imaging Conference Book of Abstracts 2015 Submission ID 95:87
29. Ossenkoppele R, Schonhaut D, Baker S et al. (2015) Distinct [18F]AV1451 retention patterns in clinical variants of Alzheimer's disease. Human Amyloid Imaging Conference Book of Abstracts 2015 P55 Submission 113:62
30. Marquie M, Normandin M, Vanderburg C et al. (2015) Towards the validation of novel PET tracer T807 on postmortem human brain tissue samples. Abstract Online: Human Amyloid Imaging Book of Abstracts 2015 Submission ID 89:28
31. Mitsis EM, Riggio S, Kostakoglu L et al (2014) Tauopathy PET and amyloid PET in the diagnosis of chronic traumatic encephalopathies: studies of a retired NFL player and of a man with FTD and a severe head injury. Transl Psychiatry 4:e441. https://doi.org/10.1038/tp.2014.91
32. Dickerson B, McGinnis S, Gomperts S et al. (2015) [18F]T807 of frontotemporal lobar degeneration. Human Amyloid Imaging Conference Book of Abstracts 2015 P15 Submission ID 61:37
33. Chhatwal J, Schultz A, Marshall G et al. (2015) Entorhinal, parahippocampal, and inferior temporal F18-T807 SUVR correlates with CSF total tau and tau T181P in cognitively normal elderly. Human Amyloid Imaging Conference Book of Abstracts 2015 Submission ID 115:88
34. Okamura N, Suemoto T, Furumoto S et al (2005) Quinoline and benzimidazole derivatives: candiate probes for in vivo imaging of tau pathology in Alzheimer's disease. J Neurosci 25(47):10857–10862
35. Harada R, Okamura N, Furumoto S et al (2013) Comparison of the binding characteristics of [18F]THK523 and other amyloid imaging tracers to Alzheimer's disease pathology. Eur J Nucl Med Mol Imaging 40(1):125–132
36. Villemagne VL, Furumoto S, Fodero-Tavoletti MT et al (2014) In vivo evaluation of a novel tau imaging tracer for Alzheimer's disease. Eur J Nucl Med Mol Imaging 41(5):816–826
37. Okamura N, Furumoto S, Fodero-Tavoletti MT et al (2014) Non-invasive assessment of Alzheimer's disease neurofibrillary pathology using 18F-THK5105 PET. Brain 137(Pt 6):1762–1771. https://doi.org/10.1093/brain/awu064
38. Saint-Aubert L, Lemoine L, Marutle A et al. (2015) Relationship between post-mortem THK5117 binding and in-vivo PET biomarkers uptake in Alzheimer's disease. Human Amyloid Imaging Conference Book of Abstracts 2015 ID submission 94:94
39. Harada R, Okamura N, Furumoto S et al. (2015) First-in-human PET study of a novel tau tracer [18F]THK5351. Human Amyloid Imaging Conference Book of Abstracts 2015 Submission ID 24
40. Maruyama M, Shimada H, Suhara T et al (2013) Imaging of tau pathology in a tauopathy mouse model and in Alzheimer patients compared to normal controls. Neuron 79(6):1094–1108. https://doi.org/10.1016/j.neuron.2013.07.037
41. Shimada H, Higuchi M, Ikoma Y et al (2013) In vivo visualization of tau pathology. Alzheimers Dement 9(4):P101–P102. https://doi.org/10.1016/j.jalz.2013.05.172
42. Gobbi LC, Knust H, Körner M et al (2017) Identification of three novel radiotracers for imaging aggregated tau in Alzheimer's disease with positron emission tomography. J Med Chem 60(17):7350–7370

Check for updates

Chapter 14

Retinal Imaging in Early Alzheimer's Disease

Tom MacGillivray, Sarah McGrory, Tom Pearson, and James Cameron

Abstract

Changes in the brain that lead to Alzheimer's disease are thought to start decades before cognitive symptoms emerge. If biomarkers for these early stages could be identified, it would contribute to a more accurate estimation of an individual's risk of developing disease and enable the monitoring of high-risk (presymptomatic) persons as well as providing the means for assessing the efficacy of new interventions. The retina links to the visual processing and cognitive centers of the brain, but it is also an extension of the brain sharing embryological origins as well as a blood supply and nerve tissue. It therefore has huge potential as a site for biomarker investigation through easy, noninvasive imaging and computational image analysis to reveal valuable information about microvascular health, deposition, and neurodegenerative damage. Capturing reliable longitudinal data pertaining to the onset of Alzheimer's disease is a key target, but a high degree of standardization is necessary if the potential of the retina is to be fully realized. Our goal is to provide the reader with guidelines on how to execute robust retinal imaging and analysis for neuroretinal biomarker discovery and to highlight advantages and limitations of the techniques.

Key words Alzheimer's disease, Dementia, Retinal imaging, Non-invasive, Biomarker, Fundus, Blood vessel, Neurodegeneration

1 Introduction

The retina is the light-sensing inner surface of the human eye and is comprised of layers of tissue consisting of neurons and supporting cells, interconnected by synapses (see Fig. 1). The retinal ganglion cell (RGC) layer, the innermost cellular layer, projects its axons across the inner retina, called the retinal nerve fiber layer (RNFL), through the optic nerve to the visual processing and cognitive centers of the brain. Developmentally and anatomically, the retina is an extension of the brain sharing embryonic origins as well as features, besides nerve tissue, such as small blood vessels and a blood-tissue barrier [1]. It also has the advantage of being easier to image than the brain and with far superior resolution—resolvable detail in images of the eye being at a micron level compared to millimeters with conventional magnetic resonance (MR) imaging. The retina is

Robert Perneczky (ed.), *Biomarkers for Preclinical Alzheimer's Disease*, Neuromethods, vol. 137,
https://doi.org/10.1007/978-1-4939-7674-4_14,

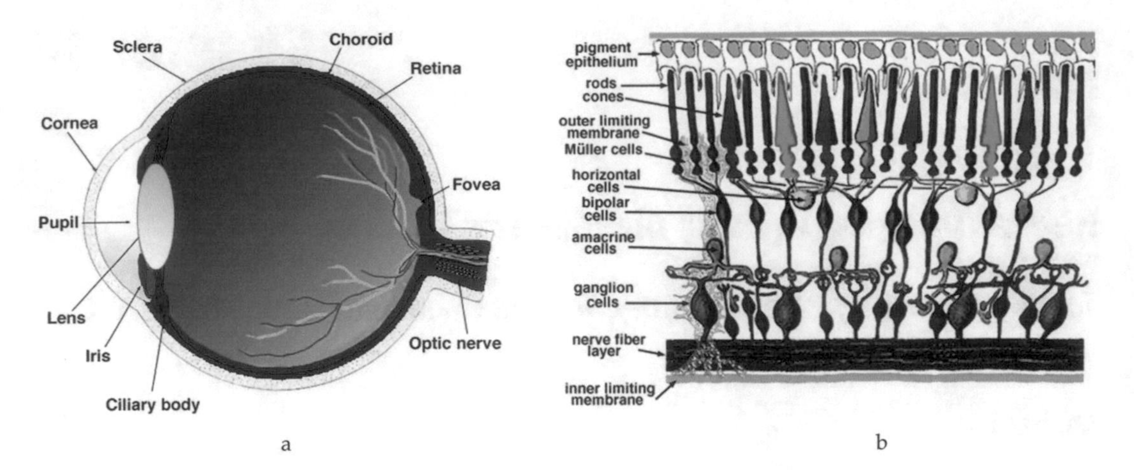

Fig. 1 Anatomy of the human eye [26]. (**a**) Diagram showing the main structures. (**b**) Illustration of the organization of the retinal layers and cells

therefore an appealing research site with the potential to reveal important aspects of brain health (see Fig. 1).

Cerebral small vessel disease is part of Alzheimer's disease (AD) etiology [2, 3]. Damage to the lining of the small vessels in the brain sees failure of this barrier and leakage of material into tissue, which causes further injury. However, these vessels are difficult to visualize directly in vivo. Due to the homology between the retinal and cerebral microvasculature, the condition of the retinal vasculature is a proxy for cerebral microvascular health [4]. Measuring geometric features of the vascular pattern visible in retinal imaging (such as vessel widths, tortuosity, and network branching complexity) infers the health of the retinal microvascular system and therefore the state of the cerebral microvasculature. Thus, the retina offers opportunity for changes or abnormalities in its microvasculature to act as surrogate markers of microvascular pathology in AD. Indeed, such changes have previously been reported in patients with AD (the most consistent finding being a sparser retinal vascular network), but the results across studies have been variable with inconsistencies most likely due to the application of assorted methodologies [5]. Standardizing methods and measurements will likely advance biomarker discovery in this area.

Deposition of amyloid β (Aβ) protein in the brain is a key pathological feature of AD. Aβ also accumulates in the inner layers of the retina [6]. Novel methods are emerging for detection in the back of the eye such as fluorescence imaging of labeled Aβ plaques [7]. However, these remain in their infancy with continuing developments in the technology and initial studies largely confined to nonhuman subjects. Aβ is also found in drusen, a marker of age-related macular degeneration (AMD) predominantly occurring in the macula at the center of the retina [8]. Drusen are abnormal accumulations of extracellular material that build up in the retinal pigmented epithelium (RPE) layer and appear as white or yellow spots when imaging the fundus or surface of the retina. The progression of the formation of drusen in the peripheral retina was more prevalent in patients with AD in comparison to age-matched controls [9]. Preliminary findings, albeit on a small cohort ($n = 29$), suggest a nascent imaging-derived biomarker for similar plaque formation in the brain that merits further investigation.

Optical coherence tomography (OCT) yields cross-sectional images of the internal retinal structure, and the high resolution of newer systems feature automated measurement of the different cellular layers. Quantification of RNFL thickness (and diminution over time) is a marker of axonal loss, and thinning occurs in AD and its preceding stage of mild cognitive impairment [10]. Neuronal loss can be accessed indirectly from macular thickness measurements (as 30–35% of the retina thickness in the macular is comprised by the RGCs and their fibers) and, with recent advancements in the technology, thickness mapping of the ganglion cell complex (GCC) layer which comprises RNFL, RGC, and the inner plexiform layer [11, 12]. Hence, OCT presents additional opportunity for neuroretinal biomarker discovery.

OCT angiography (OCT-A) is a recent advancement in retinal imaging that enables visualization of the microvasculature within the retina using a fast capture of OCT over a small volume and then deciphering the component of the returning signal related to blood flow [13]. This modality images very small vessels that are otherwise only visible with more invasive fluorescein angiography imaging. With small vessel disease linked to AD, OCT-A may reveal valuable information pertaining to the earliest signs of changes in the retina and brain.

We now present an overview of established retinal imaging modalities (see Sect. 2) and their analysis methodologies (see Sect. 3) that are currently viable for biomarker discovery in AD.

2 Materials

2.1 Setup Prior to Imaging

Table 1 gives a summary of the principal retinal imaging modalities we use for biomarker discovery in AD. Retinal imaging is a patient-friendly tool that is noninvasive, quick, and easy to perform. Most people are familiar with the typical imaging procedure, as many of them will have attended the high street optometrist for routine eye health checks. For each instrument (see Note 1), the person being imaged is asked to sit with his or her chin on a rest and with their forehead placed firmly against a stabilizing bar (see Note 2). Before proceeding, the instrument operator checks that the person is comfortable, adjusting the height of the instrument table if necessary. The operator also adjusts the headrest frame to ensure that the individual's eye is in line with the height adjustment mark (usually a horizontal white line). The person is

Table 1

A summary of the principal retinal imaging modalities

Modality	Light source (nm)	Field of view (°)	Resolution (μm to pixel)	Description
Optical biometry[a]	780	–	–	Interferometry measures the eye's axial length, distance from front to back
Fundus camera[b]	white flash	45	7 ([a])	Color photograph showing blood vessels, optic nerve head, and macula
SLO[c]	820	30	5 ([a])	Infrared laser scans retina to produce an image similar to a fundus camera
UWF-SLO[d]	532 and 633	135	10–15 ([a])	Two lasers scan a larger field of view showing areas of peripheral retina
OCT[c]	870	10–30	3.9–7 ([a]), 14([t])	Infrared light penetrates retina and interferometry resolves tissue layers
OCT[e]	840	10–30	5[a], 15 ([t])	
OCT-A[e]	840	3–10	5 ([a])	Differences in backscattered OCT between sequential scans at same cross-section used to construct a map of blood flow

Included is resolution, with axial ([a]) and transverse ([t]) values where appropriate
[a]IOL Master (Carl Zeiss Meditec, Germany)
[b]Canon CR-DGi (Canon USA Inc., Lake Success, NY)
[c]Heidelberg SPECTRALIS (Heidelberg Engineering, Dossenheim, Germany)
[d]Optos Daytona (Optos Plc, Dunfermline, UK)
[e]RTVue XR 100 Avanti (Optovue, Inc., Fremont, CA)

then asked to look at a fixation light (see Note 3) while the operator focuses and aligns the system before instigating the image acquisition, which involves the instrument illuminating the retina with its light source and then capturing the reflected light.

We image both eyes (see Note 4) with each of the devices and without administering drops for dilation (see Note 5). Conducting several imaging protocols one after the other can see an individual become fatigued resulting in a decline of image quality. While pupil dilation might help mitigate this effect, the ability to conduct imaging without it amounts to a real advantage as administering eye drops adds considerable complexity to the protocols. For instance, individuals must have their intraocular pressure and anterior chamber (the fluid-filled space inside the eye between the iris and the cornea) assessed as a precaution for acute angle-closure glaucoma, which is more common in older people. There is also a restriction on driving and a sensitivity to bright light, which extends for several hours after administering drops.

Where possible, we perform all the imaging of an individual for a particular study visit on the same day. In addition to the modalities described below, we use optical biometry to measure axial length, i.e., the distance from the front of the eye to the back. Axial length may influence the measurement of some retinal parameters, such as vessel widths, and thus needs adjusted for in data analysis. However, the significance of this measurement error is uncertain, and along with other variables such as instrument optics/magnification, it may be that future focus will be on retinal metrics more independent of optical and refractive variables like global vascular branching complexity.

We highlight retinal imaging's susceptibility to the presence of media opacities, such as cataract, which are more common in the elderly population. The presence of opacities in the eye can obscure the retinal detail, although experienced operators may be able to adjust the instrument positioning and operation to achieve useful images.

2.2 Fundus Imaging

Fundus imaging generates a 2D image of the retina directly behind the pupil. A fundus camera achieves this with a specialized low-power microscope, an attached digital camera, and a flash of white light for illumination, which creates a color photograph of the retina. We use a Canon CR-DGi (Canon USA Inc., Lake Success, NY) to image an area showing the optic disk (OD) or optic nerve head, macula, and blood vessels (see Fig. 2a). We standardize by centering the image acquisition midway between the centers of the OD and macula to give a digital fundus photograph suitable for quantitative analysis of key vascular and structural features. Repeating the same acquisition when capturing further images within or between individuals reduces uncertainties when comparing measurements.

A scanning laser ophthalmoscope (SLO) such as in the Heidelberg SPECTRALIS (Heidelberg Engineering, Dossenheim, Germany) uses an infrared (IR) laser beam and confocal microscope arrangement to scan across the surface of the retina and form a fundus picture. This particular SLO images at a slightly higher resolution than the fundus camera but captures a smaller area (see Table 1; Fig. 2b). The main purpose of the SLO imaging in the SPECTRALIS is to guide the operator to the desired location for OCT scanning and to enable accurate visit-to-visit registration for repeat imaging sessions. Yet, the clarity and resolution of the SLO images lend themselves to additional computational analysis of the retinal vasculature, especially the larger vessels where small changes or abnormalities might be more feasible to detect than with the fundus camera. We center imaging on the optic nerve head.

To image more of the retina surface in a single capture, an ultra-widefield SLO (UWF-SLO) employs scanning mirrors to direct laser light to regions of the back of the eye that are inaccessible with a fundus camera or standard SLO. We use an Optos Daytona (Optos

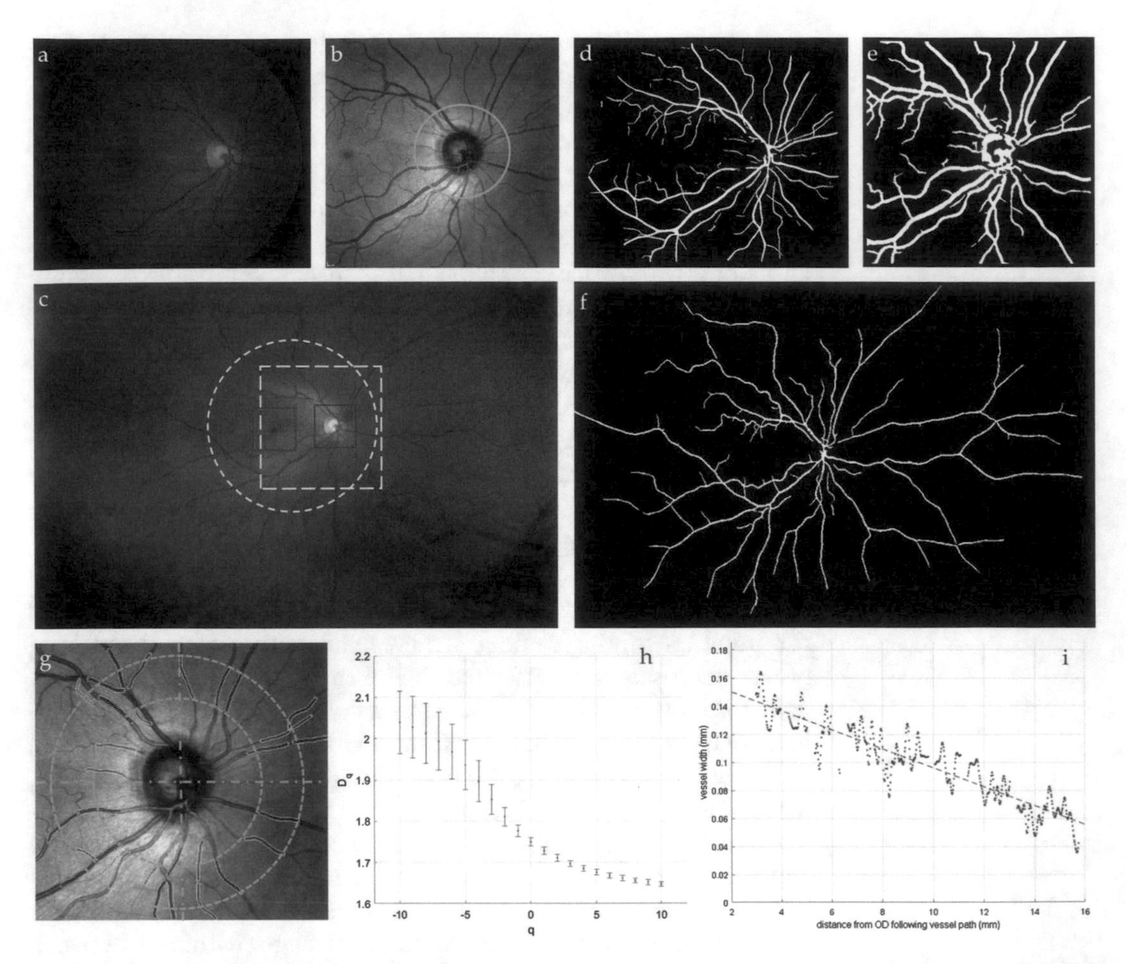

Fig. 2 Analyzing the fundus. (**a**) Color fundus camera image of a right eye showing the region containing the OD, macula, and blood vessels. (**b**) SLO imaging of the same eye centered on the OD and acquired with an IR light source, which generates a grayscale image. Green circle denotes standard positioning of peripapillary OCT acquisition around the optic nerve head. (**c**) UWF-SLO view centered on the macula, imaging a larger region of the retina than the fundus camera (dashed yellow circle) and SLO (dashed yellow square). The standard positioning of OCT-A acquisitions centered on the OD and macula is also denoted (solid magenta squares). (**d–f**) Retinal blood vessels automatically detected by computerized image processing techniques in each of the three fundus imaging modalities. (**g**) SLO image overlaid with the standard measurement zone B (dashed cyan circles) used in the analysis of blood vessel widths. The six largest arterioles (red) and venules (blue) and their individual widths (yellow) are used to calculate CRAE, CRVE, and AVR. (**h**) Generalized dimension spectrum (D_q) calculated for the vasculature detected in the fundus camera image and used to quantify the complexity/sparsity of the vascular branching pattern. For each value of q, the calculation of D_q from the detected vessels is repeated a number of times to give a mean value and associated error. (**i**) Width measurements (black dots) performed along the path of a vessel extracted from the UWF-SLO image with straight line fit (dashed black line) to quantify tapering as the vessel heads out toward the periphery

Plc, Dunfermline, UK) to facilitate the study of features and changes in the periphery that we cannot see with the other modalities, though at the sacrifice of a lower resolution for an increase in image field (see Table 1; Fig. 2c). Image acquisition is centered on the macula.

2.3 Optical Coherence Tomography and Angiography

OCT enables cross-sectional imaging of the internal retinal structures (with a limited tissue depth <2 mm). An optical beam directed at the retina and interferometry to resolve the back-scattered light signals and determine the depth of reflecting structures is an axial scan or A-scan. Scanning the beam across the retina to acquire numerous A-scans builds up the 2D image. The Heidelberg SPECTRALIS uses spectral domain OCT (SD-OCT) in which the depth information is computed by analyzing the interference signal based on the wavelength of returning light. We capture a circular peripapillary OCT scan in order to measure changes in structure around the OD, particularly in RNFL thickness (see Fig. 3a–d). In addition, we capture a volume scan centered on the macula to measure retinal and RGCL layer thickness. The RTVue XR Avanti (Optovue, Inc., Fremont, CA), also a SD-OCT, maps RNFL and GCC layer thickness with an OD-centered OCT acquisition (see Fig. 3e, f).

The RTVue XR Avanti has an angiographic mode that reveals the very small vessels at different depths in the retina. We perform

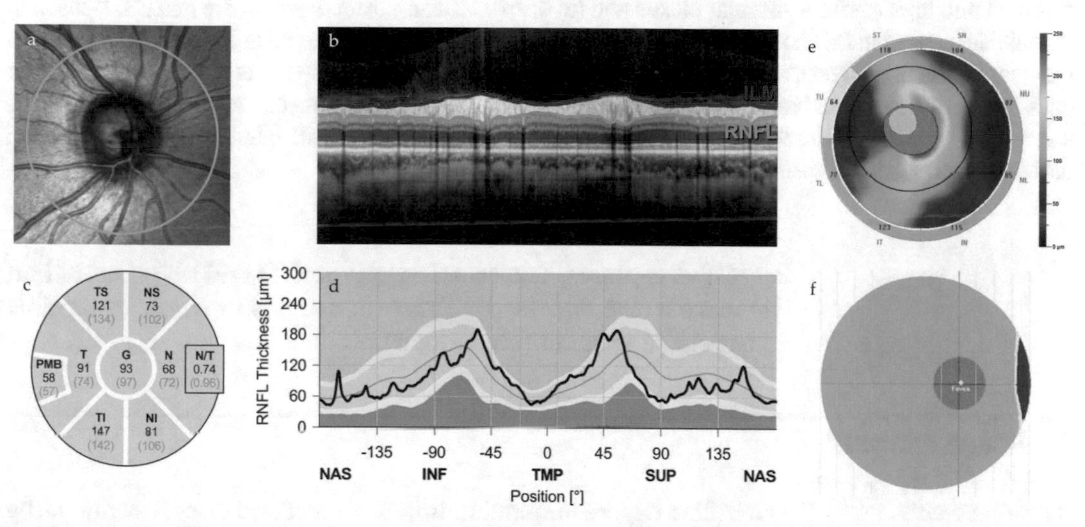

Fig. 3 Analyzing the retinal layers. (**a**) SLO image focused on OD, showing the location of the peripapillary OCT scan acquisition (green circle). (**b**) OCT image showing the retinal layers and delineation of the ILM (red line) and RNFL border (cyan line), which gives RNFL thickness. (**c**) Evaluating RNFL thickness using the standard TSNIT map which subdivides the measurements and color codes them compared to normative values. (**d**) RNFL thickness profile (black line) with color coding emphasizing statistical significance in comparison to normal limits—green indicates within normal limits; blue, borderline above; yellow, borderline below; and red, below. The person imaged here shows mostly green, whereas we might expect more yellow and red for a person with AD progression. (**e**) Map of RNFL thickness around the OD and (**f**) GCC thickness, also color coded to emphasize the presence of any abnormally thin areas. Again, in AD progression, thinner or more red areas might be anticipated

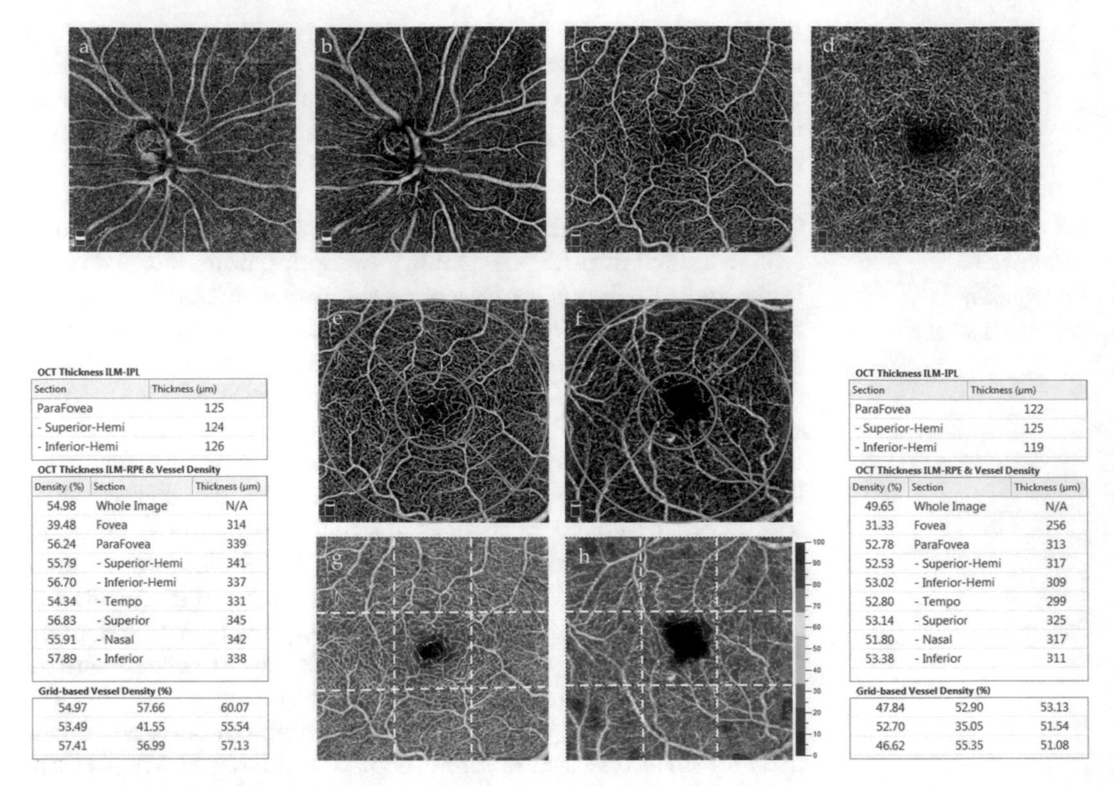

OCT Thickness ILM-IPL

Section	Thickness (μm)
ParaFovea	125
- Superior-Hemi	124
- Inferior-Hemi	126

OCT Thickness ILM-RPE & Vessel Density

Density (%)	Section	Thickness (μm)
54.98	Whole Image	N/A
39.48	Fovea	314
56.24	ParaFovea	339
55.79	- Superior-Hemi	341
56.70	- Inferior-Hemi	337
54.34	- Tempo	331
56.83	- Superior	345
55.91	- Nasal	342
57.89	- Inferior	338

Grid-based Vessel Density (%)

54.97	57.66	60.07
53.49	41.55	55.54
57.41	56.99	57.13

OCT Thickness ILM-IPL

Section	Thickness (μm)
ParaFovea	122
- Superior-Hemi	125
- Inferior-Hemi	119

OCT Thickness ILM-RPE & Vessel Density

Density (%)	Section	Thickness (μm)
49.65	Whole Image	N/A
31.33	Fovea	256
52.78	ParaFovea	313
52.53	- Superior-Hemi	317
53.02	- Inferior-Hemi	309
52.80	- Tempo	299
53.14	- Superior	325
51.80	- Nasal	317
53.38	- Inferior	311

Grid-based Vessel Density (%)

47.84	52.90	53.13
52.70	35.05	51.54
46.62	55.35	51.08

Fig. 4 Analyzing the retina using OCT-A. (**a**) Nerve head segment and (**b**) radial peripapillary capillary (RPC) segment and (**c**) superficial vascular plexus and (**d**) deep vascular plexus region in the macula. Measuring retinal thickness from OCT and vessels density from OCT-A using a TSNIT map (blue zones) in the superficial vascular plexus of (**e**) a normal eye and (**f**) an eye of individual who has suffered a stroke (and so thought to have deteriorated vascular brain health). (**g**–**h**) Color coding of vessel density measurements performed using a 3 × 3 grid (white dashed lines) to subdivide the superficial vascular plexus into squares. The individual believed to have poor vascular brain health exhibits a lower vessel density

2 OCT-A captures, one centered on the OD and one centered on the macula (see Fig. 4a–d). This procedure is the most susceptible to patient movement (see Note 6).

3 Image Analysis

3.1 Overview

We utilize expert human grading to detect and classify retinopathy (i.e., type, number, size, and location), including drusen, in our retinal imaging. Computerized software tools that exploit image processing and analysis techniques enable further quantitative and reproducible measurements of structural features such as the vasculature visible in fundus images and the different retinal layers in OCT [14, 15]. Such computer-assisted methods measure variations or abnormalities in the retina that might be imperceptible to or missed by a human grader, in an objective and repeatable fashion. In this section, we focus on the key quantitative analysis pertaining to the modalities previously described in Sect. 2.

3.2 Fundus Imaging (See Fig. 2)

Our group has developed software (VAMPIRE; Vascular Assessment and Measurement Platform for Images of the REtina, Universities of Edinburgh and Dundee, UK) to extract parametric information from fundus imaging [16–18]. Retinal vessel segmentation is the precursor to the measurement of features such as vessel width, tortuosity, and complexity of the vascular network [19]. We perform this for fundus camera, SLO, and UWF-SLO images (see Fig. 2d–f). Additional semiautomatic classification distinguishes venules from arterioles, removes non-vessel artifacts, and detects the center and boundary of the OD along with center of the macula. The operator can correct any of these labels through a user interface that allows them to interact with an image being analyzed.

VAMPIRE software places the standard set of circular measurement zones on a fundus camera or SLO image—Zone B is a ring 0.5–1 OD diameters away from the center, and Zone C is a ring extending from OD boundary to 2 OD diameters away. Measurements of the widths of the six widest arterioles and venules crossing Zone B (see Fig. 2g) are used to calculate the central retinal arteriolar equivalent (CRAE) and central retinal venular equivalent (CRVE) and give the arteriole to venule ratio (AVR; CRAE/CRVE) [20]. Similarly, the tortuosity (i.e., how much a vessel twists and turns) of the six widest arterioles and venules crossing Zone C is measured [21] and then further summarized by calculating median arteriolar and venular values along with the standard deviations and ranges.

The complexity (or sparsity) of the vascular branching pattern visible with fundus imaging can be quantified using fractal analysis to yield a fractal dimension that combines contributions of individual vessel parameters into a single global value [14]. Variations in the fractal dimension are an indicator of deviation away from normative branching arrangements and so a potential marker of disease [22]. We use a particular approach called multifractal analysis, which generates several fractal dimensions at different "scales" (see Fig. 2h), and we record dimensions D_0, D_1, and D_2 as these have previously been reported as likely sensitive markers of small vascular changes [23]. Application of the analysis is to vessels appearing in Zone C and separated into arteriolar and venular measurements.

With UWF-SLO, it is not only possible to assess retinopathy and the vasculature in the peripheral retina but to explore novel parameters for biomarker discovery in addition to conventional Zone B–C measurements (see Note 7). For instance, we can analyze the main vessels over a longer path distance than is achievable with other fundus modalities and look at how the width changes as the vessel extends out toward the periphery (see Fig. 2i). An abnormal decline in the vessel's width (quantified by fitting a model such as a straight line) may indicate an unhealthy vessel.

Computational analysis of features appearing in fundus imaging such as the retinal vasculature has the ability to return large numbers of measurements. However, further endeavor is required to develop these into effective summary parameters that characterize the state or health of the retina and then to investigate which are relevant to brain changes and AD.

3.3 Optical Coherence Tomography and Angiography (See Fig. 3)

The major task in analyzing OCT is accurate segmentation of retinal layers as thickness changes are an important indicator of disease status. Modern OCT instruments such as the Heidelberg SPECTRALIS and RTVue XR Avanti are equipped with software to automatically delineate the borders of the internal limiting membrane (ILM) and the RNFL, as the distance between these two boundaries gives the measurement of RNFL thickness (see Fig. 3b). Thickness of the RNFL is evaluated using the standard TSNIT (temporal, superior, nasal, inferior, temporal) mapping approach which subdivides or regionalizes the measurement and color codes statistical significance compared to a database of normative values (see Fig. 3c). RNFL thickness also presented as a profile plot with color coding again underlining statistical significance in comparison to normal limits (see Fig. 3d) or as a thickness map (see Fig. 3e). Further delineation of boundaries in OCT enables mapping of other layers such as the GCC (see Fig. 3f).

While these quantitative OCT mapping options present exciting opportunities for biomarker discovery, the pathological processes behind retinal thinning in AD are incompletely understood, and further investigations are required to determine the use of retinal OCT as a potential surrogate marker of neurodegeneration (Fig. 4).

OCT-A works by comparing consecutive scans at the same location to detect flow using motion contrast and enable demarcation of blood vessels. As a means of quantifying OCT-A, the instrument calculates vessel density parameters, i.e., the percentage area occupied by the larger appearing vessels and the microvasculature in particular regions. This can be the entire scan area as well as standard regions such as TSNIT sectors or a grid-based division of the image (see Fig. 4e–g). For each scanned region, the software calculates the vessel densities and thicknesses in various layers of the retina and optic nerve head.

OCT-A offers potentially exciting avenues to explore in the context of neuroretinal biomarker discovery, particularly around small vessels and AD. However, caution is urged in the interpretation and analysis of OCT-A imaging. The modality detects flow above a threshold level, and so not all vessels may appear in the images such as blocked or leaky ones. In addition, it is more susceptible to image artifact caused by patient motion or reflections from internal retinal structures that may not always be obvious to the untrained eye or computer algorithm.

4 Conclusions

Monitoring for the development and progression of changes in the retina via imaging has the potential to yield biomarkers for the onset of brain changes and deterioration linked to AD. The retinal imaging protocols described here help support the conduct of experimental medicine studies to understand and investigate the in vivo pathophysiology of small disease, deposition and neurodegeneration in AD, and to identify indicative retinal changes.

We highlight that the ease-of-use of modern retinal imaging devices does not remove the importance of detailed and systematic review of the captured images, for quality, registration, segmentation errors, and other important processing steps that are essential before data extraction. Trained and experienced operators are vital to both the capture of suitable images and the subsequent measurements.

Future challenges for researchers in this field include distinguishing normal retinal changes due to aging from those linked to failing brain health and investigating whether changes in the retina relate to other recognized brain changes such as abnormalities on MR imaging, declining cognitive scores, and measures from biological samples such as blood and spinal fluid. Thus, an essential target is reliable longitudinal data, not just in the eye but other non-retinal markers, in appropriately powered prospective studies.

We envisage that, with concerted investigative effort, over the next few years, retinal imaging and analysis will contribute biomarkers for AD and that this could include presymptomatic screening.

5 Notes

1. Set up and positioning for the Optos device is different to the other machines. The person sits in front of it, holds the handgrips on both sides of the instrument, and places their head against the faceplate to look through a small eyehole. They move themselves into the correct position for imaging by adjusting themselves against the faceplate until a target light they can see turns from blue or red to green.
2. Having the person's forehead placed firmly against the stabilizing bar of the headrest frame helps keep their head very still and minimizes movement that can result in images that are unsuitable for analysis.
3. A fixation light is a visual cue inside the device for a person to look at. This directs imaging to particular parts of the retina. It can be a different colored light for different instruments but is most commonly blue, green, or orange.

4. Increasing evidence suggest some elements of the visual system are asymmetric [24]. We therefore image both eyes in order to determine the impact that this might have on associations of retinal parameters with other measurements related to brain changes and AD.
5. Pupil dilation with eye drops is not necessary with the imaging protocols described here. A darkened room is usually sufficient for a person's pupils to enlarge and to take a picture with the fundus camera. The other instruments generally work well in low levels of background lighting. With the exceptions of the fundus camera and UWF-SLO, the illuminating light source is not strong and causes only small amounts of pupil constriction. Between acquisitions, we ask a person to replenish their tear film by either closing or blinking their eyes. Asking them to fixate with the other eye (i.e., the one not under examination) on a distant target (and not the internal fixation light) can also help to enlarge the pupil.
6. The procedure involves two 3–4 s bursts of data capture per imaging location on the retina during which the person must stay extremely still and stare at the fixation light. Any slight movement introduces noticeable "streaking" artifacts into the resulting OCT-A image.
7. In the analysis of UWF-SLO to measure features, given the larger surface of imaged retina, it is crucial to account for the curvature of the eye; otherwise measurements will be invalid [25].

Acknowledgments

Support from the Engineering and Physical Sciences Research Council (EPSRC) (grant number EP/M005976/1), the Medical Research Council (MRC) (grant number MR/L015994/1), the Alzheimer's Research UK Scotland Network Centre, the University of Edinburgh Innovation Initiative Grants scheme, the Edinburgh and Lothians Health Foundation, Optos, and SINAPSE (Scottish Imaging Network: A Platform for Scientific Excellence) is gratefully acknowledged. We also thank the Computer Vision and Image Processing Group at the University of Dundee, NHS Lothian R&D, the Edinburgh Clinical Research Facility, Edinburgh Imaging, and the Anne Rowling Regenerative Neurology Clinic.

References

1. London A, Benhar I, Schwartz M (2013) The retina as a window to the brain-from eye research to CNS disorders. Nat Rev Neurol 9(1):44–53. https://doi.org/10.1038/nrneurol.2012.227
2. Knopman DS (2006) Dementia and cerebrovascular disease. Mayo Clin Proc 81(2):223–230. https://doi.org/10.4065/81.2.223
3. Reitz C, Brayne C, Mayeux R (2011) Epidemiology of Alzheimer disease. Nat Rev Neurol 7(3):137–152. https://doi.org/10.1038/nrneurol.2011.2
4. Patton N, Aslam T, Macgillivray T et al (2005) Retinal vascular image analysis as a potential screening tool for cerebrovascular disease: a rationale based on homology between cerebral and retinal microvasculatures. J Anat 206(4):319–348. https://doi.org/10.1111/j.1469-7580.2005.00395.x
5. McGrory S, Cameron JR, Pellegrini E et al (2017) The application of retinal fundus camera imaging in dementia: a systematic review. Alzheimers Dement (Amst) 6:91–107. https://doi.org/10.1016/j.dadm.2016.11.001
6. Ratnayaka JA, Serpell LC, Lotery AJ (2015) Dementia of the eye: the role of amyloid beta in retinal degeneration. Eye (Lond) 29(8):1013–1026. https://doi.org/10.1038/eye.2015.100
7. Koronyo-Hamaoui M, Koronyo Y, Ljubimov AV et al (2011) Identification of amyloid plaques in retinas from Alzheimer's patients and noninvasive in vivo optical imaging of retinal plaques in a mouse model. Neuroimage 54(Suppl 1):S204–S217. https://doi.org/10.1016/j.neuroimage.2010.06.020
8. Anderson DH, Talaga KC, Rivest AJ et al (2004) Characterization of beta amyloid assemblies in drusen: the deposits associated with aging and age-related macular degeneration. Exp Eye Res 78(2):243–256
9. Aslam A, Peto T, Barzegar-Befroei N et al (2014) Assessing peripheral retinal drusen progression in Alzheimer's dementia: a pilot study using ultra-wide field imaging. Invest Ophthalmol Vis Sci 55(13):659–659
10. Thomson KL, Yeo JM, Waddell B et al (2015) A systematic review and meta-analysis of retinal nerve fiber layer change in dementia, using optical coherence tomography. Alzheimers Dement (Amst) 1(2):136–143. https://doi.org/10.1016/j.dadm.2015.03.001
11. Cunha LP, Almeida AL, Costa-Cunha LV et al (2016) The role of optical coherence tomography in Alzheimer's disease. Int J Retina Vitreous 2:24. https://doi.org/10.1186/s40942-016-0049-4
12. Lisboa R, Paranhos A Jr, Weinreb RN et al (2013) Comparison of different spectral domain OCT scanning protocols for diagnosing preperimetric glaucoma. Invest Ophthalmol Vis Sci 54(5):3417–3425. https://doi.org/10.1167/iovs.13-11676
13. de Carlo TE, Romano A, Waheed NK et al (2015) A review of optical coherence tomography angiography (OCTA). Int J Retina Vitreous 1:5. https://doi.org/10.1186/s40942-015-0005-8
14. Patton N, Aslam TM, MacGillivray T et al (2006) Retinal image analysis: concepts, applications and potential. Prog Retin Eye Res 25(1):99–127. https://doi.org/10.1016/j.preteyeres.2005.07.001
15. Abramoff MD, Garvin MK, Sonka M (2010) Retinal imaging and image analysis. IEEE Trans Med Imaging 3:169–208. https://doi.org/10.1109/RBME.2010.2084567
16. MacGillivray T (2012) VAMPIRE: vessel assessment and measurement platform for images of the retina. In: Tan EYKNJH, Acharya UR, Suri JS (eds) Human eye imaging and modeling. CRC Press, Boca Raton, FL
17. Cameron JR, Ballerini L, Langan C et al (2016) Modulation of retinal image vasculature analysis to extend utility and provide secondary value from optical coherence tomography imaging. J Med Imaging (Bellingham) 3(2):020501. https://doi.org/10.1117/1.JMI.3.2.020501
18. Pellegrini E, Robertson G, Trucco E et al (2014) Blood vessel segmentation and width estimation in ultra-wide field scanning laser ophthalmoscopy. Biomed Opt Express 5(12):4329–4337. https://doi.org/10.1364/BOE.5.004329
19. Trucco E, Giachetti A, Ballerini L et al. (2015) Morphometric measurements of the retinal vasculature in fundus images with vampire. In: Biomedical image understanding. John Wiley & Sons, Inc., pp 91–111. doi:https://doi.org/10.1002/9781118715321.ch3
20. Knudtson MD, Lee KE, Hubbard LD et al (2003) Revised formulas for summarizing retinal vessel diameters. Curr Eye Res 27(3):143–149
21. Lisowska A, Annunziata R, Loh GK et al (2014) An experimental assessment of five indices of retinal vessel tortuosity with the RET-TORT public dataset. Conf Proc IEEE Eng Med Biol Soc 2014:5414–5417. https://doi.org/10.1109/EMBC.2014.6944850

22. Doubal FN, MacGillivray TJ, Patton N et al (2010) Fractal analysis of retinal vessels suggests that a distinct vasculopathy causes lacunar stroke. Neurology 74(14):1102–1107. https://doi.org/10.1212/WNL.0b013e3181d7d8b4
23. Stosic T, Stosic BD (2006) Multifractal analysis of human retinal vessels. IEEE Trans Med Imaging 25(8):1101–1107
24. Cameron JR, Megaw RD, Tatham AJ et al (2017) Lateral thinking – interocular symmetry and asymmetry in neurovascular patterning, in health and disease. Prog Retin Eye Res. https://doi.org/10.1016/j.preteyeres.2017.04.003
25. Croft DE, van Hemert J, Wykoff CC et al (2014) Precise montaging and metric quantification of retinal surface area from ultra-widefield fundus photography and fluorescein angiography. Ophthal Surg Lasers Imaging Retina 45(4):312–317. https://doi.org/10.3928/23258160-20140709-07
26. Kolb H (1995) Simple anatomy of the retina. In: Kolb H, Fernandez E, Nelson R (eds) Webvision: the organization of the retina and visual system. University of Utah Health, Salt Lake City, UT

Check for updates

Chapter 15

In Vivo Volumetry of the Cholinergic Basal Forebrain

Michel J. Grothe, Ingo Kilimann, Lea Grinberg, Helmut Heinsen, and Stefan Teipel

Abstract

Degeneration of cortically projecting acetylcholine-containing neuronal populations within the basal forebrain cholinergic system (BFCS) is a central pathogenetic aspect of Alzheimer's disease (AD) and forms the rationale for the use of cholinomimetics as antidementive treatment. The role of the cholinergic deficit in AD pathophysiology has mainly been studied in experimental animal models and in neuropathologic examinations of human autopsy data of advanced disease stages. Interactions between cholinergic deficits and accumulation of cortical amyloid pathology point to a relevant role of BFCS degeneration for the preclinical stage of AD. The advent of novel computational morphometry techniques for the automated analysis of high-resolution MRI data allows studying AD-related atrophy in the living human brain with ever-increasing temporal and regional detail. Combining these morphometry techniques with recently developed stereotactic mappings of the BFCS provides a method for automated MRI-based volumetry that can sensitively assess degenerative changes of the BFCS in vivo. Here, we outline the general methodological approach for MRI-based BFCS volumetry and describe the specifics of different image processing choices and analysis strategies. We further discuss possibilities and limitations of this method for studying BFCS degeneration in the course of AD, with a special emphasis on using MRI-based BFCS volumetry as an imaging biomarker for defining the preclinical disease stage.

Key words Cholinergic basal forebrain, Nucleus basalis Meynert, Substantia innominata, MRI, Morphometry, Amyloid pathology

1 Introduction

Decades of neuropathologic and pharmacologic research implicate a cholinergic neurotransmission deficit in the pathophysiology of Alzheimer's disease (AD) that is caused by severe neurofibrillary degeneration of cortically projecting cholinergic neurons of the basal forebrain cholinergic system (BFCS), most notably in the nucleus basalis Meynert (NBM) [1–4]. Consistent observations of a correlation between the extent of cholinergic loss at autopsy and antemortem dementia severity led to the "cholinergic hypothesis" of AD [5] and form the rationale for the use of cholinomimetic drugs as antidementive treatment in AD [6]. Moreover, cholinergic system degeneration was found to be coupled to the accumulation

Robert Perneczky (ed.), *Biomarkers for Preclinical Alzheimer's Disease*, Neuromethods, vol. 137,
https://doi.org/10.1007/978-1-4939-7674-4_15,

of cortical amyloid pathology [7–10], particularly in the preclinical stage of AD as indicated by cortical amyloid pathology in the absence of cognitive impairments at time of death [11–13]. This contrasts with the well-described degenerative changes in medial temporal and neocortical brain regions, which typically follow accumulation of amyloid pathology with a temporal delay and demarcate the transition to clinical stages of the disease [14, 15].

Together these data point to a great potential of using MRI-based in vivo measurements of BFCS degeneration as AD biomarker with relevance to the preclinical, at risk of AD dementia, stage. However, the complex anatomy of the BFCS and the indistinguishability of cholinergic cell clusters on high-resolution structural MRI scans place important constraints on the in vivo structural analysis of this brain region [16, 17]. A first approach to in vivo measurement of cholinergic degeneration in AD focused on manual measurement of a circumscribed part of the BFCS (usually referred to as "substantia innominata" in this context) that can be consistently traced by the use of the anterior commissure as external anatomic landmark [17]. This approach is limited in that it only assesses a small and relatively unspecific subsection of the BFCS, which consists of several groups of distinct cholinergic cell clusters that extend approximately 20 mm in anterior-posterior direction [18–21] (Fig. 1). These subdivisions of the BFCS are characterized by a differential cortical innervation preference [22, 23], show differential associations with specific cognitive and behavioral functions [24, 25], and vary in their vulnerability to AD-related neurodegeneration [3]. Thus, although this imaging marker can readily detect the severe BFCS degeneration present in AD dementia [17, 26–28], its sensitivity for subtle degenerative processes at prodromal disease stages may be limited, and studies using this marker in early disease stages have reported conflicting results [26, 29, 30].

As an extension to the manual volumetry approach, we have developed automated techniques for BFCS morphometry that are based on stereotactic mappings of the BFCS into MRI standard space in combination with computational methods for image processing and regional volume extraction [18, 21, 31–34]. While still being indirect in vivo measures of cholinergic degeneration based on volumetric changes on MRI, these techniques allow a much more comprehensive assessment of total and subregional BFCS changes in the course of AD. Over the last years, these techniques have been used in several independent cohort studies by us and other international research groups, providing novel insights into the differential course of BFCS degeneration during aging and AD pathogenesis and how these changes relate to the accumulation of cortical amyloid pathology and the emergence of cognitive impairments [35–44]. Thus, in these studies it could be shown that although the BFCS is highly vulnerable to the normal aging process, AD-associated degeneration of this region greatly exceeds the age-related changes and can already be detected at prodromal

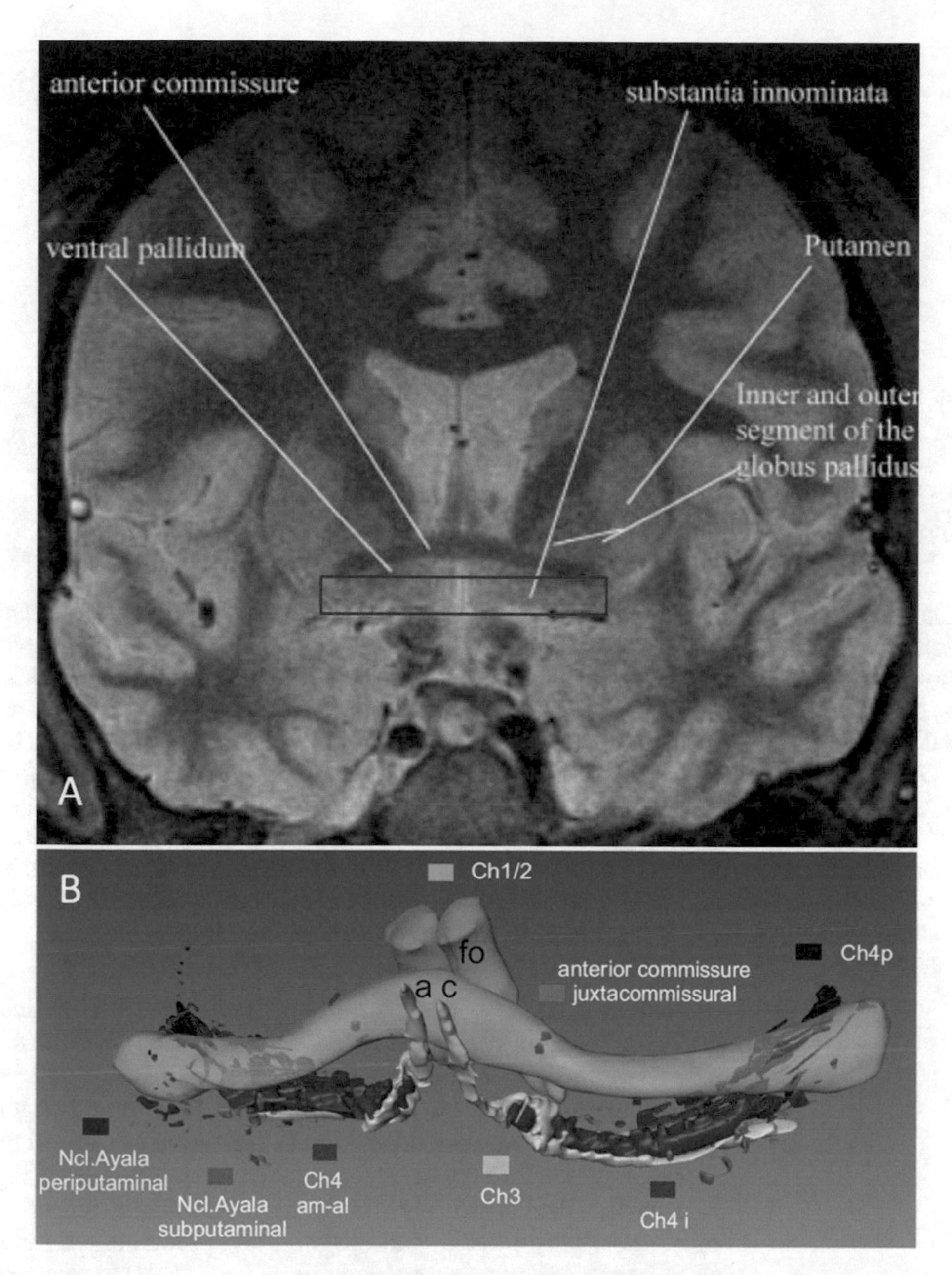

Fig. 1 Overview of substantia innominata and cholinergic basal forebrain. (**a**) A coronal proton-weighted MRI scan at the level of the anterior commissure illustrating the anatomic location of the substantia innominata and neighboring structures (Adapted from [21] with permission to reprint granted by Oxford University Press). (**b**) Computer-assisted 3D reconstruction of the basal forebrain cholinergic nuclei based on histology of a postmortem brain. Cholinergic nuclei stretch in anterior-posterior direction from the medial septum (Ch1) over the vertical and horizontal limb of the diagonal band of Broca (Ch2 and Ch3) to the nucleus basalis (Ch4). *ac* anterior commissure, *fo* fornix

stages of the disease. Coinciding with histopathologic evidence for a subregion-specific BFCS vulnerability to AD-related neurodegeneration [3], in vivo BFCS atrophy appears to manifest first in the posterior subdivision of the NBM and spreads anteriorly to affect

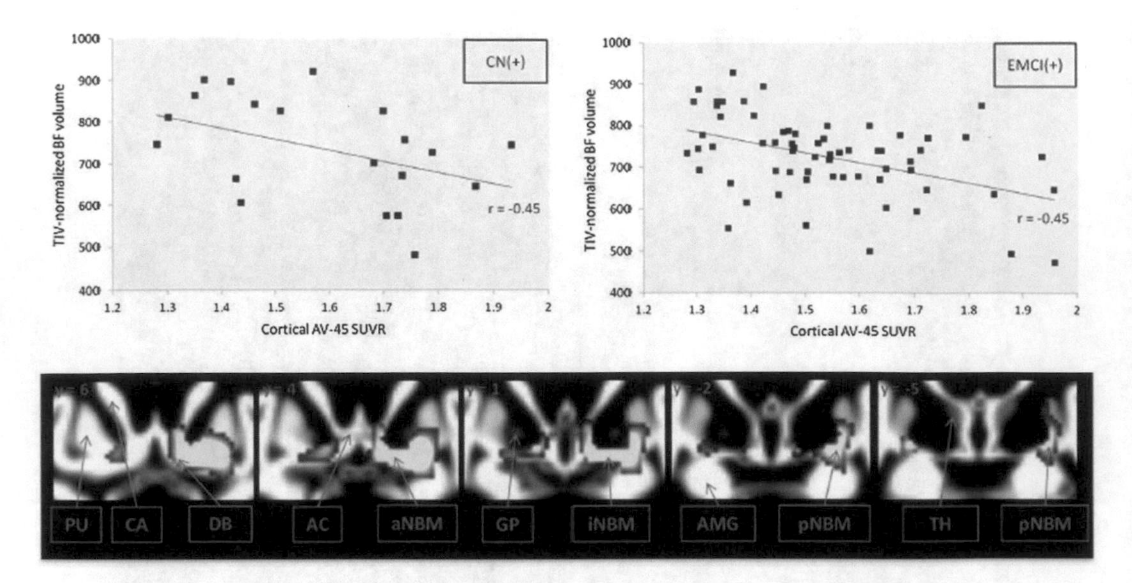

Fig. 2 In vivo association between basal forebrain volume and cortical amyloid load. Scatter plots with linear regression lines illustrate the association between cortical amyloid load measured in vivo by AV45-PET standard uptake value ratio (SUVR) (x-axis) and cholinergic basal forebrain volume extracted using MRI-based volumetry (y-axis) in cognitively normal (CN+; left) and early mild cognitive impairment (EMCI+; right) individuals with detectable amyloid pathology. Figures in the bottom row show voxel-wise effects of a linear regression of cortical AV45-PET SUVR on regional gray matter volume in the combined CN(+) and EMCI(+) subgroups. Effects are superposed on coronal sections of the reference template and magnified to better depict the cholinergic basal forebrain. Blue numbers indicate MNI space coordinates. Note the relative regional specificity of the amyloid effect for the cholinergic basal forebrain nuclei. *AC* anterior commissure; *AMG* amygdala; *DB* diagonal band of Broca; *CA* caudate; *GP* globus pallidus; (*a*, *i*, *p*) *NBM*, anterior, intermediate, posterior nucleus basalis Meynert; *PU* putamen; *TH* thalamus (Figure adapted from [36] with permission to reprint granted by Elsevier)

the whole NBM and the rostral cholinergic nuclei as individuals progress to clinically manifest AD. Importantly, by combining BFCS volumetry with PET- or CSF-derived biomarkers of amyloid pathology, the distinct association between cholinergic degeneration and cortical amyloid deposition in presymptomatic and predementia stages of AD could be reproduced in vivo [36–38] (Fig. 2). Accordingly, BFCS atrophy was found to be a significant predictor of an underlying amyloid pathology among predemented individuals, whereas hippocampus atrophy was not, suggesting a distinct potential of BFCS volumetry as an imaging biomarker for the characterization of preclinical AD [35].

2 Materials

2.1 Stereotactic Maps of the BFCS

Generation of MRI-based stereotactic atlases of the BFCS has been hindered by the lack of imaging contrast for the distinct cholinergic cell clusters on common high-resolution structural MRI scans. However, their localization can be mapped indirectly to stereotactic

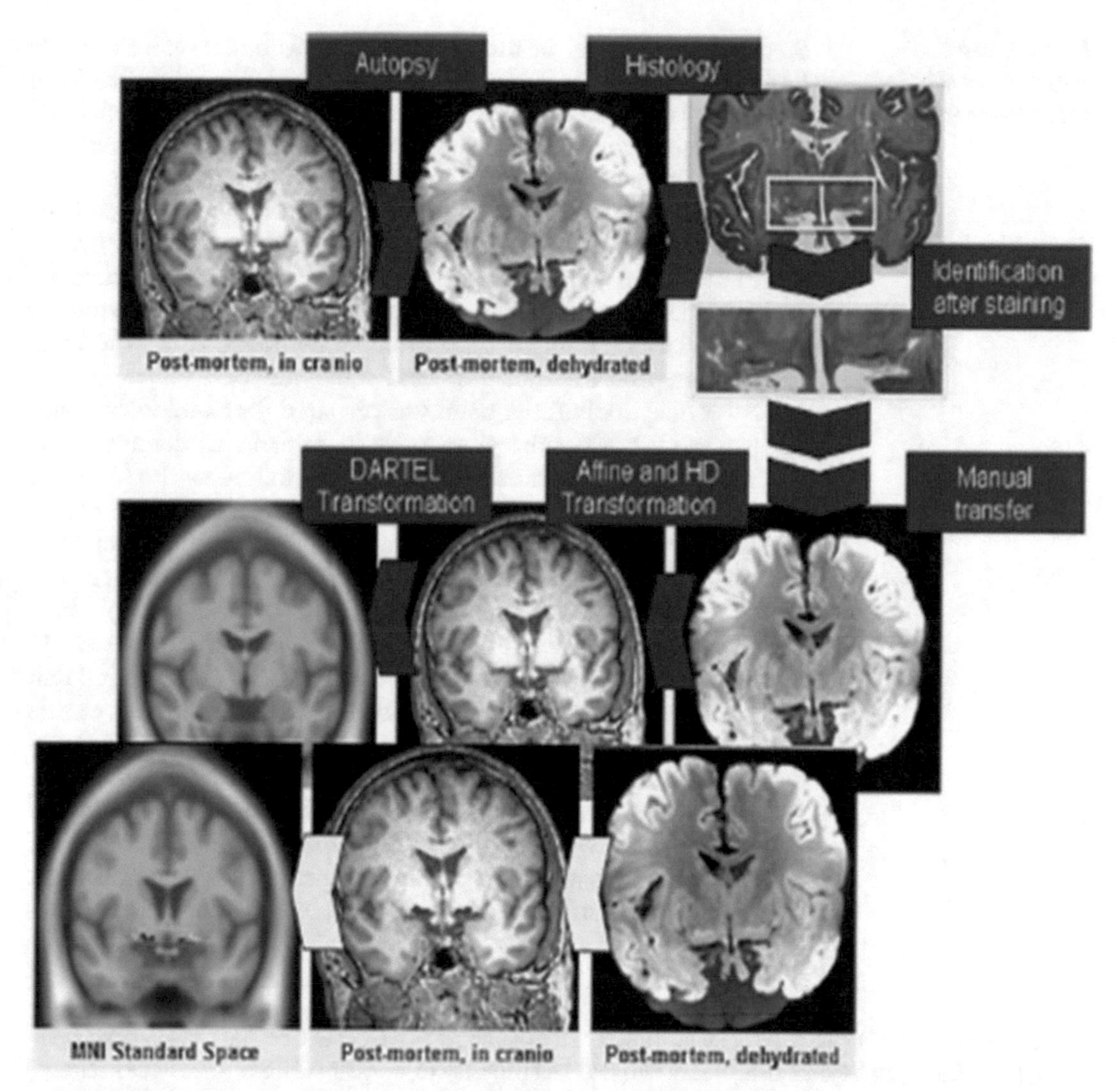

Fig. 3 Stereotactic mapping based on histology and postmortem MRI. The upper part of the figure shows the postmortem processing with MRI scans of the brain in cranio (i.e., in situ) and after formalin fixation. Afterward the subregions were identified by histological staining and manually transferred into the MRI of the dehydrated brain in the alcohol space (color-coded mask at the lowest row of the figure). MRI scans were transferred first into the MRI space of the in cranio scan and then into MNI standard space. The combined transformation matrices were used to transfer the BFCS mask from alcohol space to MNI space via MRI in cranio space (Reprinted from [34], with permission from IOS Press)

standard space using histologic delineation in autopsy data and subsequent spatial normalization to stereotactic standard space using a corresponding postmortem MRI scan (Fig. 3) [18, 21, 34]. In the following we will describe this process based on our latest development of a stereotactic BFCS atlas [34].

2.1.1 Acquisition of Postmortem MRI and Histologic Preparation

Stereotactic mapping of the BFCS nuclei was based on an autopsy brain specimen of a non-demented 56-year-old man who died from a myocardial infarction. Results from a standardized and validated questionnaire interview with an informant of the brain donor were reviewed by a behavioral neurologist and two geriatricians and considered to be without any evidence for cognitive decline or psychiatric illnesses prior to death [45]. A cerebral MRI scan was performed in situ 15 h after death. Afterward the brain was removed from the skull and placed into formalin deposit for 3 months and subsequently dehydrated by upscaling ethanol storage. Before and after the brain dehydration procedure, further MRI scans were performed.

After dehydration, the brain was prepared for histological staining [46, 47]. BFCS subregions were identified and delineated on digital pictures of the stained slices (400 μm) following the Mesulam nomenclature [22, 48]. The basal forebrain refers to a brain region whose central parts are formed by the anterior perforate substance in humans. This region is included in the "unnamed medullary substance" (substantia innominata) described by Reil [49], which extends parallel to the optic nerve on the basis of the forebrain. In their comprehensive studies, Heimer et al. [50, 51] summarized that a high number of fiber systems and nuclei, including the ventral striatum with ncl. accumbens, ventral pallidum, and extended amygdala, are crowded in the substantia innominata. The NBM is distinguished by predominantly big hyperchromic neurons. Individual neurons can already be identified at low magnification as small blue dots after staining with gallocyanin, a classical Nissl stain (Fig. 4). The cholinergic nature of many of these big hyperchromic neurons

Fig. 4 (continued) level of the anterior compartment of the ncl. accumbens [52] and central parts of the paraterminal gyrus. Bar 1 mm, valid for all figures of the plate. Left hemisphere, 58-year-old male, myocardial infarct. (**b**) Plane of section 2 mm caudal to (**a**) passing ventral pallidum extending into caudal compartment of ncl. accumbens. (**c**) Plane of section 4 mm caudal to (**a**) comprising the caudal parts of the anterior perforate substance and the bed nucleus of the stria terminalis. (**d**) Plane of section 11 mm caudal to (**a**) comprising the optic tract with supraoptic nucleus. (**e**) Plane of sections 16 mm caudal to (**a**) comprising crura cerebri and central nucleus of the amygdaloid complex. (**f**) Horizontal gallocyanin-stained section through the left hemisphere of a 60-year-old female, myocardial infarct. Arrowheads = branches of the anterolateral central arteries (Aa. centrales anterolaterales, perforant arteries); white arrowhead in figure (**c**) = perforating branch of central artery subdividing Ch4am from Ch4al; asterisks = artifacts caused by either vacuum embedding of tissue, leakage of the blood-brain barrier, or condensed galactolipids during alcohol dehydration of brain tissue. *a c* anterior commissure, *acc* ncl. accumbens, *acj* juxtacommissural cells of the anterior commissure [53, 54], *a l* ansa lenticularis, *a p* ansa peduncularis, *Ayala pp* periputaminal part of Ayala's nucleus, *Ayala sp* subputimale part of Ayala's nucleus, *B st* bed nucleus of the stria terminalis, *caud* caudate nucleus, *Ce* central nucleus of the amygdala, *Ch1* medial septum, *Ch2* vertical limb of the diagonal band of Broca, *Ch3* horizontal limb of the diagonal band of Broca, *Ch4al* anterolateral part of Ch4, *Ch4am* anteromedial part of Ch4, *Ch4p* posterior part of Ch4, *cr* crus cerebri, *c t* claustrum temporale or endopiriform nucleus, *d B* diagonal band of Broca, *g p* globus pallidus, *gpe* external pallidum, *gpi* internal pallidum, *g pt* paraterminal (subcallosal) gyrus, *I C* insula magna of Calleja, *i ol* olfactory islands, *put* putamen, *s o* supraoptic nucleus, *s t* stria terminalis, *s th* subthalamic nucleus, *tr o* optic tract, *v p* ventral pallidum

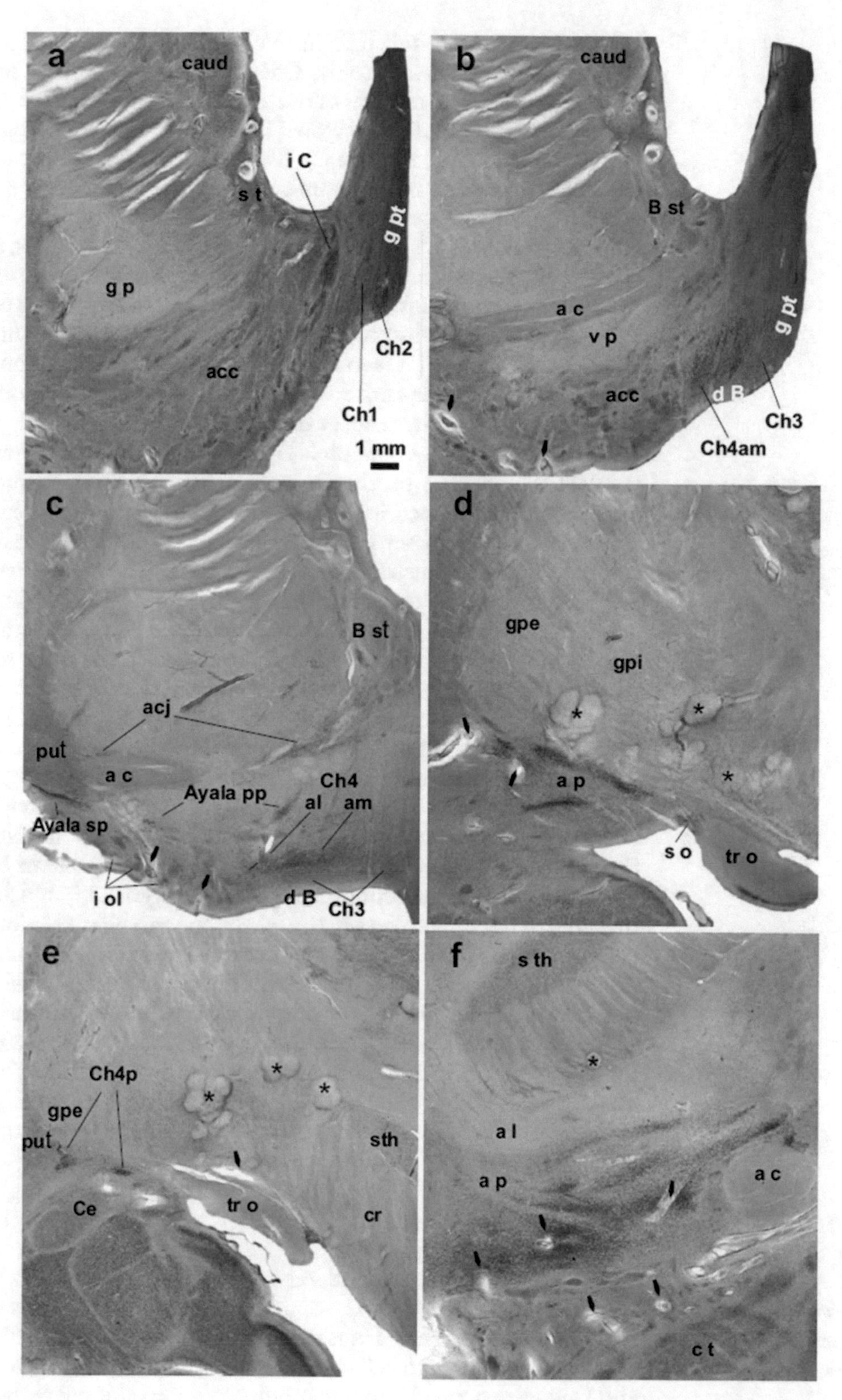

Fig. 4 Histology of the cholinergic basal forebrain nuclei. (**a**) Coronal gallocyanin (Nissl)-stained section at the

caused Mesulam to dub particular clusters and strands of NBM neurons Ch4am, Ch4al, Ch4iv, Ch4id, and Ch4p. Similar hyperchromic/cholinergic neurons of the medial septum, the vertical, and the horizontal limb of the diagonal band of Broca were designated Ch1, Ch2, and Ch3, respectively [22, 48]. Ayala's nucleus subputaminalis and smaller hyperchromic neurons closely attached to certain regions of the anterior commissure (juxtacommissural) were not included in Mesulam's terminology [53–56]. Taken together, these cholinergic neurons and clusters form the BFCS that arches from the septum through Broca's diagonal band and deep regions of the anterior perforate substance (Fig. 1b). Topographically, anterior commissure and BFCS show a more or less close relationship. Therefore, the former can serve as a landmark for localization and assessment of the rostrocaudal extent of the latter.

An interesting observation is that the different components of the BFCS appear to be closely associated with fibers. Although fibers remain unstained in Nissl sections, clear spaces in between and overall arrangement of BFCS neurons reflect course, caliber, and density of intranuclear fibers and fascicles. A few thin fibers are associated with Ch1 neurons, whereas three to four fiber fascicles appear to pierce Ch2 (Fig. 4a). A mixture of a few thick and numerous thinner fibers guide neuronal arrangement in the most rostral parts of Ch4am (Fig. 4b), whereas numerous small fibers and less numerous small neurons are characteristic of Ch3. The cell-rich compact Ch4al/am sector is pierced by thin fibers that apparently enter or leave the cluster from its dorsomedial extreme (Fig. 4c). The conspicuous fiber tract of the ansa peduncularis separates Ch4i into a dorsal and ventral part (Fig. 4d). Horizontal sections through the ansa peduncularis illustrate the conspicuous components of the ansa peduncularis that connect thalamic, amygdaloid, and hypothalamic regions and prove the close topographic relationship between ansa lenticularis and Ch4i cell clusters (Fig. 4f). In addition to a close spatial relationship with fibers and tracts, NBM clusters are arranged around perforating branches of the arteriae striatales anterolaterales. They form conspicuous sleeves around the latter (Fig. 4f). One more or less invariably penetrating artery was described by Mesulam to separate Ch4al from Ch4am (Fig. 4c, white arrowhead). In our experience, Ayala's subputaminal nucleus was always closely attached to a medium caliber perforating artery that coursed in an oblique medio-lateral direction through the basal forebrain (Fig. 4c, left arrowhead).

2.1.2 Mapping from Histologic Slices to Stereotactic Standard Space Via Postmortem MRI

The BFCS subregions identified on the histological slices were manually transferred from the digital pictures into the corresponding MR slices of the dehydrated brain. Due to the small size of some of the delineated cell clusters, clusters corresponding to Ch1 and Ch2 were merged into one subdivision (Ch1-2), and within the NBM we only considered a distinction between an anterior-intermediate (Ch4a-i) and a posterior subdivision (Ch4p). Transformation of the delineations from the space of the dehydrated brain into the space of the in situ brain scan was based on an initial 12-parameter affine transformation followed by a high-dimensional nonlinear registration between the two brain scans. The improved gray/white matter contrast of the in situ brain scan enabled the use of the highly accurate high-dimensional DARTEL (Diffeomorphic Anatomical Registration Through Exponentiated Lie Algebra) registration method [57, 58] for the final transformation from in situ space into Montreal Neurological Institute (MNI) standard space.

The "in situ mask" described above is one of three currently available stereotactic maps of the BFCS that have been developed over the last years (Fig. 5). These maps primarily differ in the number of analyzed postmortem brains, the delineation criteria and detail used for histologic definition of the BFCS and its subdivisions, as well as the methods and computational algorithms used for stereotactic mapping into MNI standard space. A first map published in 2005 [21] was also based on a single (but separate) postmortem brain specimen and used a delineation system based on Mesulam's nomenclature, although the final map did not distinguish between distinct subdivisions within the BFCS. Due to methodical limitations, the histologic delineations were only performed within one brain hemisphere and were mirrored to the contralateral hemisphere after stereotactic normalization to MNI space based on a 12-parameter affine transformation. Zaborszky and colleagues later refined the stereotactic BFCS mapping by generating probabilistic maps of different BFCS subdivisions based on the analysis of ten postmortem brains [18]. Stereotactic mapping into MNI space was performed using a low-dimensional nonlinear registration to the single-subject MNI space template, which was applied directly to the postmortem MRI of the fixed brain specimen. In contrast to the single-subject mappings, the probabilistic maps encode valuable information about the interindividual variability of BFCS anatomy. However, due to inherent limitations of the stereotactic mapping procedure, these probabilities are also influenced by spatial registration inaccuracies in addition to interindividual variability in regional anatomy. The BFCS atlas as originally published includes probabilistic maps of different BFCS subdivisions termed Ch1-2, Ch3, Ch4, and Ch4p, although the delineation criteria may have slightly differed from those originally proposed by Mesulam, particularly in the definition of Ch3 and its distinction from Ch4am (see Sect. 2.1.1, Fig. 4). However, the maps that have finally been made publicly available through the SPM Anatomy toolbox [59] only include merged versions of Ch1-2/Ch3 and Ch4/Ch4p, and these also show some anatomic differences to the original mappings [18] (Fig. 5 C + D).

2.2 Structural MRI Data for Automated BFCS Morphometry

Requirements for the structural MRI data used for automated BFCS volumetry match the general recommendations for automated morphometry approaches [60]. The structural MRI data

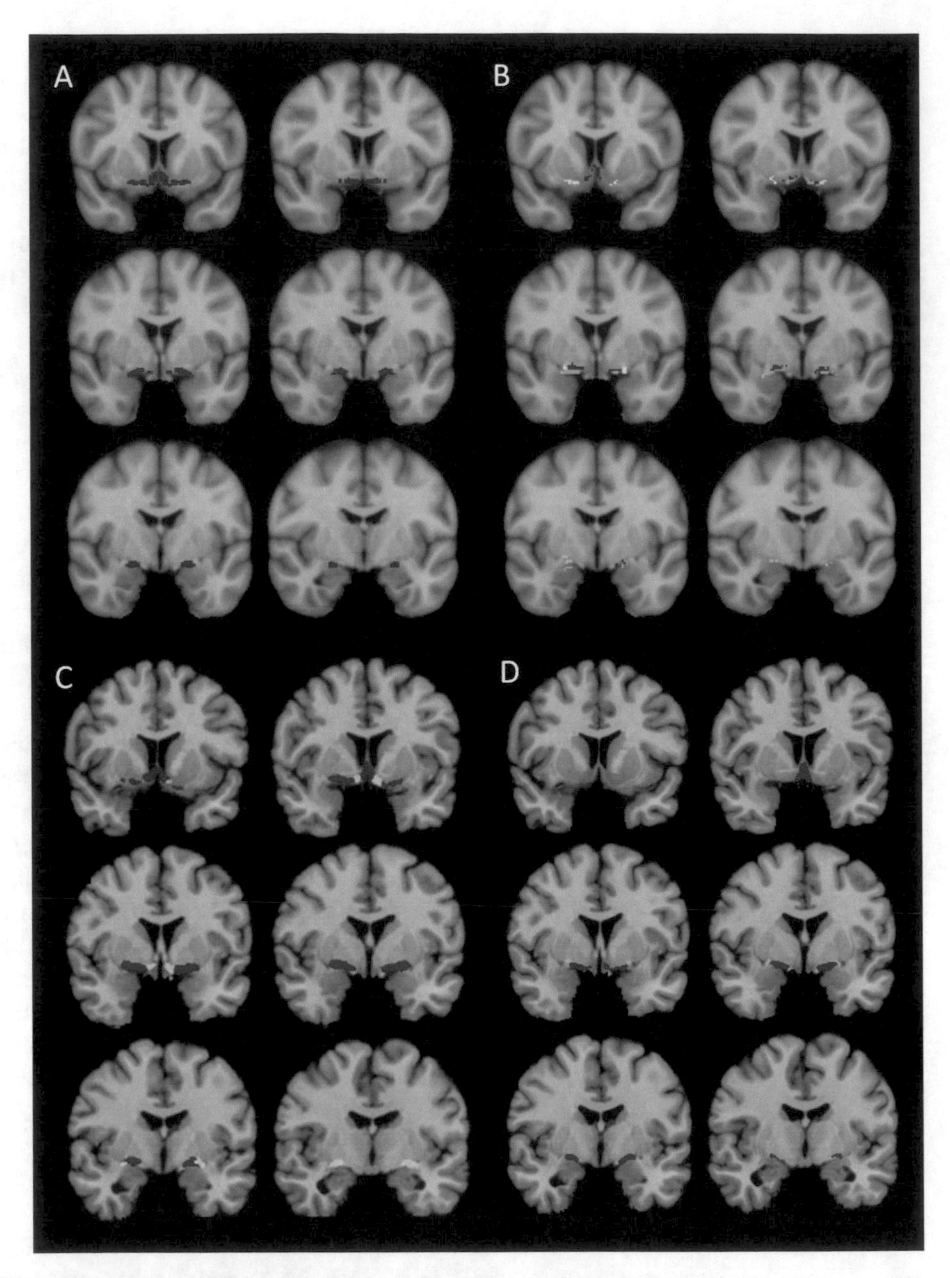

Fig. 5 Stereotactic maps of the cholinergic basal forebrain. Illustrations of the stereotactic maps of the cholinergic basal forebrain in MNI space as published by [21] (**a**), [34] (**b**), and [18] (**c**). Panel (**d**) shows the stereotactic maps from [18] that have finally been made publicly available through the SPM Anatomy toolbox [59]. Panel (**c**) is taken from the original publication in [18] (with permission to reprint granted by Elsevier), and panels (**a**), (**b**), and (**d**) have been produced by projecting the stereotactic maps onto coronal sections with identical MNI space coordinates as in (**c**) (running from $y = 8$ in the upper left corner to $y = -7$ in the lower right corner). The probabilistic maps in (**c**) and (**d**) are represented as discrete maximum probability maps. Note that these maps are designed to match the single-subject MNI space template, whereas maps in (**a**) and (**b**) correspond to standard MNI space (MNI152 average template)

should be based on a high-resolution three-dimensional T1-weighted imaging sequence with an approximate image resolution of 1 × 1 × 1 mm isotropic voxel size, such as provided by magnetization prepared (MP-RAGE) or spoiled gradient echo (IR-SPGR) acquisitions. While acquisitions at 3 T magnetic field strength are generally preferable due to their higher signal-to-noise ratio and contrast-to-noise ratio between brain tissues, the method also sensitively detects AD-related BFCS atrophy using MRI scans acquired at 1.5 T [32, 33, 39].

3 Methods

3.1 Automated Volumetry Approaches

The availability of detailed stereotactic maps of the BFCS renders this brain region accessible to "atlas-based" automated volumetry approaches. The overarching principle of these volumetry approaches is to spatially transform ("normalize") the individual native space MRI scan into the standardized stereotactic space where the BFCS atlas is defined in (typically MNI space) and use this transform to calculate individual BFCS volumes. After matching the native space scan to a stereotactic standard space template, the regional information of an individual's anatomy is encoded in the transformation parameters that describe the deformations necessary for the nonlinear matching (represented by vector fields, also called "deformation" fields). The Jacobian determinant map of these vector fields specifically reflects the regional volumetric changes, i.e., the amount of dilation or shrinkage that occurred at each voxel during matching to the stereotactic standard space template.

In the most widely used "voxel-based morphometry" approach to automated volumetry, the native space MRI scan is first segmented into broad classes of tissue types (gray matter, white matte, cerebrospinal fluid), and then the gray matter tissue of interest is spatially transformed into stereotactic standard space, and voxel values of the spatially transformed gray matter maps are finally scaled ("modulated") by the Jacobian determinant to reflect the amount of regional gray matter tissue present before warping. The sum of modulated gray matter voxel values within a given region of interest (ROI) defined in template space (here the BFCS atlas) represents the gray matter volume of this region in the individual MRI scan (Fig. 6) [60].

In an alternative approach, sometimes referred to as "Jacobian integration" or "tensor-based morphometry" [60, 61], the voxel values are directly summed up on the Jacobian determinant maps instead of the modulated gray matter maps. This approach produces individual native space volumes of a given ROI independent of tissue type and is equivalent to the volume calculation of a ROI mask that has been transformed from template space to the individual's native space based on the inverse spatial transformation parameters [62].

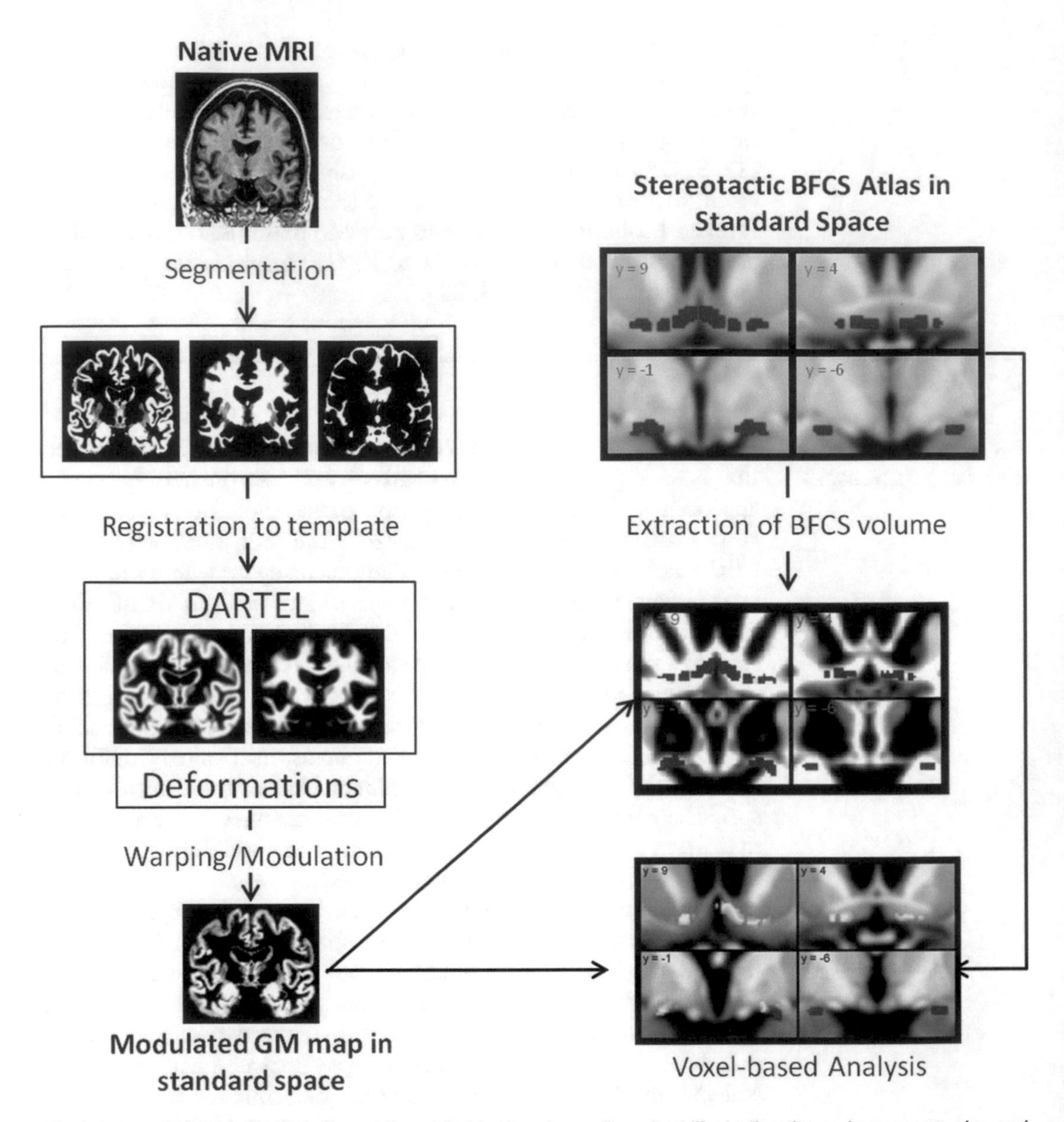

Fig. 6 Automated analysis of cholinergic basal forebrain volume. Flowchart illustrating the main preprocessing and analysis steps for automated volumetry of the cholinergic basal forebrain. For details, see descriptions in the text

3.2 Preprocessing of Structural MRI Scan

According to the automated volumetry pipeline outlined above, preprocessing of the structural MRI scans principally consists of tissue-type segmentation and spatial normalization to the stereotactic template space where the BFCS mask is defined in. Given the relatively small spatial extent of the BFCS and its subdivisions, the whole approach critically depends on the accuracy of the spatial normalization procedure, aiming for a perfect fit of the native space scan to the anatomy of the normalization template. Thus, best results can be achieved using new-generation high-dimensional nonlinear spatial registration algorithms, such as DARTEL [57] or SyN [63], which have been demonstrated to outperform more traditional (and still more commonly used) affine or low-dimensional nonlinear registration algorithms [58].

In addition to the choice of registration algorithm, accuracy of the spatial normalization procedure may also depend on the target template used for spatial normalization. While the most commonly used templates for spatial normalization are MNI standard space templates, these may not be the optimal targets for spatial normalization of scans derived from individuals of advanced age and with neurodegenerative disease. Given that the neuroanatomy of these individuals may markedly deviate from the characteristics of MNI space templates reflecting a healthy adult population, larger deformations are necessary in the registration process, thereby increasing the propensity for registration errors. Moreover, differences in the amount of anatomic deviation from the MNI space templates between patients and healthy age-matched controls may result in a systematic group bias in registration accuracy [64, 65]. Thus, the accuracy and homogeneity of the spatial normalization may be improved for these populations by using group- or population-specific normalization templates [33, 39, 65]. However, this also comes with the drawback of having to spatially transform the stereotactic BFCS masks that are originally designed to match MNI standard space, and the optimal trade-off between possible improvements in inter-subject registration accuracy and potential anatomic misregistration errors during spatial transformation of the BFCS ROI is not clear a priori and likely depends on the studied population.

The routine used for automated tissue-type segmentation may also have a marked influence on the automated assessment of BFCS volume. Automated tissue-type segmentation routines of the most widely used software packages are usually optimized for detecting gray/white matter contrasts in cortical brain regions, whereas the cholinergic nuclei of the BFCS are scattered throughout an anatomically complex subcortical brain region consisting of several other neuronal populations as well as traversing white matter fiber bundles. Moreover, some of the theoretical assumptions about tissue-type distribution and corresponding intensity contrasts markedly differ between available segmentation routines, resulting in notable differences in segmentation outcomes [64, 66]. Thus, depending on the routine used for tissue-type segmentation, more or less parts of the BFCS will actually be segmented as gray matter. This potentially limits the applicability of a standard "voxel-based morphometry" approach to BFCS volumetry and renders "Jacobian integration/tensor-based morphometry" a suitable, tissue-independent, alternative in this context [31, 32, 37]. However, at least the tissue segmentation routine implemented in SPM's VBM8 toolbox [67] typically segments all or most parts of the BFCS map as gray matter, and in previous work, we found that these BFCS gray matter volume measurements appeared to be more sensitive for AD-related changes than regional volumes extracted through Jacobian integration [33].

In summary, for most settings we now generally recommend a processing pipeline for automated BFCS volumetry that is based on a "voxel-based morphometry" approach as outlined above, in combination with VBM8-type tissue segmentation [67] and high-dimensional spatial normalization using SPM's DARTEL algorithm [57].

3.3 Analysis of Cholinergic Basal Forebrain Volumes

Following the automated volumetry approach described above, individual volumes of the BFCS or its subdivisions can be derived by summing up Jacobian determinant voxel values or modulated gray matter voxel values within respective ROI masks derived from stereotactic BFCS maps (see Sect. 2.1). In order for these measures to reflect the extent of atrophic change rather than interindividual differences in head size, the individual volumes further need to be scaled to the total intracranial volume before subjecting them to statistical analysis [62, 68]. Alternatively, interindividual differences in head size may be accounted for directly within the statistical model that is used for assessing the diagnostic or prognostic potential of BFCS volume as imaging biomarker for AD [68].

Detailed neuropathologic studies have pointed to a differential vulnerability of BFCS subregions to AD-related neurodegeneration, being markedly more pronounced in the NBM and particularly its posterior subdivision compared to anterior medial nuclei [3]. While assessment of subregional BFCS volumes principally allows capturing such subregional differences using the outlined in vivo volumetry approach, inference will be limited to the idiosyncrasies of the delineation system and detail used for defining BFCS subdivisions in the available stereotactic maps [18, 21, 34, 69]. A complementary and spatially unbiased approach for determining subregional specificity for AD-related neurodegeneration within the BFCS is to assess volumetric differences on the voxel level. Thus, instead of extracting and comparing individual volumes of predefined BFCS subdivisions, diagnostic group differences may be assessed for every voxel within a wider BFCS ROI using mass univariate statistical testing (Fig. 6). Although this approach does not directly produce individual values that could be used as an imaging biomarker, the resulting voxel-wise map of AD-specific BFCS atrophy may be used in independent cohorts for extracting subregional BFCS volume with a potentially higher sensitivity for capturing AD-related changes [37].

4 Conclusions

The outlined method for automated MRI-based BFCS volumetry provides a means to study BFCS degeneration in the course of aging and AD in vivo. Studies using this method could provide important in vivo evidence for long-standing neuropathologic observations of a correlation between the degree of BFCS degeneration and the severity of antemortem cognitive impairments in AD as well as in other neurologic dementing diseases, including Lewy body disorders and primary progressive aphasias [20, 32, 40, 70–76]. In contrast to autopsy studies, which are usually confined to relatively advanced disease stages and limited sample sizes, MRI-based BFCS volumetry may be particularly useful for studying early disease stages and their distinction from the normal aging process [33, 34, 38, 39, 77]. Most interestingly, using combined in vivo BFCS volumetry and PET- or CSF-based biomarker assessments of amyloid pathology in large-scale cohort studies could consistently reproduce the initial neuropathological evidence for a distinct association between BFCS degeneration and cortical amyloid pathology in the preclinical phase of AD [11–13, 35–38] (Fig. 2). By contrast, established MRI-based measures of medial temporal lobe atrophy did not show similar associations with amyloid pathology at this early disease stage, indicating that BFCS volumetry may be a particularly useful MRI-based biomarker for characterizing preclinical AD. Of course, in the existing cross-sectional autopsy and neuroimaging studies, the direction of the amyloid-BFCS interaction remains unknown. Preclinical studies in cultured cells as well as in experimental animal models could provide evidence both for a pronounced neurotoxic effect of amyloid pathology on cholinergic basal forebrain neurons [78–81], as well as for an acceleration of cortical amyloid accumulation due to cholinergic denervation [82–84], possibly caused by dysfunctional clearance mechanisms [85]. Assessment of the relationships between amyloid pathology and BFCS degeneration through in vivo neuroimaging and biomarker data provides the unique possibility to study the temporal relation between these pathologic hallmarks of AD longitudinally and may thus provide entirely novel insights into the role of these interactions during natural disease progression.

An important limitation of MRI-based in vivo volumetry of the BFCS is that the volumetric measurement is only an indirect marker of cholinergic system integrity. Although the use of histology-based stereotactic atlases of the BFCS increases the correspondence of the analyzed brain region with the localization of cholinergic cell clusters in the basal forebrain, one cannot exclude that the volumetric measurement may also reflect change in other non-cholinergic neuronal populations. For example, cholinergic neurons in the basal forebrain are intermingled, and closely interact, with cortically projecting GABAergic neurons, which are also increasingly being recognized for their critical role in basal forebrain-mediated modulation of cortical activity and cognition [86]. In addition, the arrangement of the different BFCS components is closely associated with traversing fiber populations as well as with perforating arteries (see Sect 2.1.1). Thus, age- or disease-related decreases in fiber number and diameter as well as increases in vessel diameter could have an impact on in vivo BFCS volumetry. Hence, while the consistently observed associations with amyloid pathology and emerging cognitive deficits underline the pathologic and clinical relevance of MRI-based BFCS volumetry in AD, the implications of these volumetric alterations for cholinergic neurotransmission need to be studied in more detail. Pharmacologic challenge paradigms [87, 88] and multimodal imaging studies employing PET-based imaging tracers of cholinergic neurotransmission [89–91] may be suitable analytic approaches to further investigate the neurotransmitter specificity of volumetric BFCS changes on structural MRI.

With respect to the potential of MRI-based BFCS volumetry as an imaging biomarker for preclinical AD, one has to keep in mind that the outcome of the volumetry approach is sensitive to several methodological choices and there is currently no consensus on the optimal analytic pipeline for this rather experimental technique. Variability may arise through differences in the spatial accuracy and regional detail of the stereotactic atlases used to map the anatomic location of the BFCS, as well as differences in the automated algorithms used for tissue-type segmentation and spatial normalization of the MRI images. In addition to the diagnostic or prognostic potential under experimental conditions, a widespread clinical use of a biomarker will ultimately also depend on the harmonization and standardization of the respective measurement protocol, such as is currently being done on an international basis for hippocampus volumetry as the best established MRI-based imaging biomarker for AD so far [92, 93].

References

1. Whitehouse PJ, Price DL, Struble RG et al (1982) Alzheimer's disease and senile dementia: loss of neurons in the basal forebrain. Science 215(4537):1237–1239
2. McGeer PL, McGeer EG, Suzuki J et al (1984) Aging, Alzheimer's disease, and the cholinergic system of the basal forebrain. Neurology 34(6):741–745
3. Vogels OJ, Broere CA, ter Laak HJ et al (1990) Cell loss and shrinkage in the nucleus basalis Meynert complex in Alzheimer's disease. Neurobiol Aging 11(1):3–13
4. Perry EK, Blessed G, Tomlinson BE et al (1981) Neurochemical activities in human temporal lobe related to aging and Alzheimer-type changes. Neurobiol Aging 2(4):251–256
5. Bartus RT, Dean RL 3rd, Beer B et al (1982) The cholinergic hypothesis of geriatric memory dysfunction. Science 217(4558):408–414
6. Birks J (2006) Cholinesterase inhibitors for Alzheimer's disease. Cochrane Database Syst Rev 1:CD005593. https://doi.org/10.1002/14651858.cd005593
7. Perry EK, Tomlinson BE, Blessed G et al (1978) Correlation of cholinergic abnormalities with senile plaques and mental test scores in senile dementia. Br Med J 2(6150):1457–1459
8. Ikonomovic MD, Klunk WE, Abrahamson EE et al (2011) Precuneus amyloid burden is associated with reduced cholinergic activity in Alzheimer disease. Neurology 77(1):39–47. https://doi.org/10.1212/WNL.0b013e3182231419
9. Beach TG, McGeer EG (1992) Senile plaques, amyloid beta-protein, and acetylcholinesterase fibres: laminar distributions in Alzheimer's disease striate cortex. Acta Neuropathol 83(3):292–299
10. Arendt T, Bigl V, Tennstedt A et al (1985) Neuronal loss in different parts of the nucleus basalis is related to neuritic plaque formation in cortical target areas in Alzheimer's disease. Neuroscience 14(1):1–14
11. Beach TG, Honer WG, Hughes LH (1997) Cholinergic fibre loss associated with diffuse plaques in the non-demented elderly: the preclinical stage of Alzheimer's disease? Acta Neuropathol 93(2):146–153
12. Potter PE, Rauschkolb PK, Pandya Y et al (2011) Pre- and post-synaptic cortical cholinergic deficits are proportional to amyloid plaque presence and density at preclinical stages of Alzheimer's

disease. Acta Neuropathol 122(1):49–60. https://doi.org/10.1007/s00401-011-0831-1
13. Beach TG, Kuo YM, Spiegel K et al (2000) The cholinergic deficit coincides with Abeta deposition at the earliest histopathologic stages of Alzheimer disease. J Neuropathol Exp Neurol 59(4):308–313
14. Young AL, Oxtoby NP, Daga P et al (2014) A data-driven model of biomarker changes in sporadic Alzheimer's disease. Brain 137(Pt 9):2564–2577. https://doi.org/10.1093/brain/awu176
15. Jack CR Jr, Knopman DS, Jagust WJ et al (2013) Tracking pathophysiological processes in Alzheimer's disease: an updated hypothetical model of dynamic biomarkers. Lancet Neurol 12(2):207–216. https://doi.org/10.1016/s1474-4422(12)70291-0
16. Huckman MS (1995) Where's the chicken? AJNR Am J Neuroradiol 16(10):2008–2009
17. Hanyu H, Asano T, Sakurai H et al (2002) MR analysis of the substantia innominata in normal aging, Alzheimer disease, and other types of dementia. AJNR Am J Neuroradiol 23(1):27–32
18. Zaborszky L, Hoemke L, Mohlberg H et al (2008) Stereotaxic probabilistic maps of the magnocellular cell groups in human basal forebrain. NeuroImage 42(3):1127–1141. https://doi.org/10.1016/j.neuroimage.2008.05.055
19. Halliday GM, Cullen K, Cairns MJ (1993) Quantitation and three-dimensional reconstruction of Ch4 nucleus in the human basal forebrain. Synapse 15(1):1–16. https://doi.org/10.1002/syn.890150102
20. Teipel SJ, Flatz W, Ackl N et al (2014) Brain atrophy in primary progressive aphasia involves the cholinergic basal forebrain and Ayala's nucleus. Psychiatry Res 221(3):187–194. https://doi.org/10.1016/j.pscychresns.2013.10.003
21. Teipel SJ, Flatz WH, Heinsen H et al (2005) Measurement of basal forebrain atrophy in Alzheimer's disease using MRI. Brain 128(Pt 11):2626–2644. https://doi.org/10.1093/brain/awh589
22. Mesulam MM, Mufson EJ, Levey AI et al (1983) Cholinergic innervation of cortex by the basal forebrain: cytochemistry and cortical connections of the septal area, diagonal band nuclei, nucleus basalis (substantia innominata), and hypothalamus in the rhesus monkey. J Comp Neurol 214(2):170–197. https://doi.org/10.1002/cne.902140206
23. Zaborszky L, Csordas A, Mosca K et al (2013) Neurons in the basal forebrain project to the cortex in a complex topographic organization that reflects Corticocortical connectivity patterns: an experimental study based on retrograde tracing and 3D reconstruction. Cereb Cortex. https://doi.org/10.1093/cercor/bht210
24. Muir JL (1997) Acetylcholine, aging, and Alzheimer's disease. Pharmacol Biochem Behav 56(4):687–696
25. Pappas BA, Bayley PJ, Bui BK et al (2000) Choline acetyltransferase activity and cognitive domain scores of Alzheimer's patients. Neurobiol Aging 21(1):11–17
26. George S, Mufson EJ, Leurgans S et al (2011) MRI-based volumetric measurement of the substantia innominata in amnestic MCI and mild AD. Neurobiol Aging 32(10):1756–1764. https://doi.org/10.1016/j.neurobiolaging.2009.11.006
27. Hanyu H, Tanaka Y, Shimizu S et al (2005) Differences in MR features of the substantia innominata between dementia with Lewy bodies and Alzheimer's disease. J Neurol 252(4):482–484. https://doi.org/10.1007/s00415-005-0611-8
28. Gao FQ, Pettersen JA, Bocti C et al (2013) Is encroachment of the carotid termination into the substantia innominata associated with its atrophy and cognition in Alzheimer's disease? Neurobiol Aging 34(7):1807–1814. https://doi.org/10.1016/j.neurobiolaging.2013.01.009
29. Muth K, Schonmeyer R, Matura S et al (2010) Mild cognitive impairment in the elderly is associated with volume loss of the cholinergic basal forebrain region. Biol Psychiatry 67(6):588–591. https://doi.org/10.1016/j.biopsych.2009.02.026
30. Moon WJ, Kim HJ, Roh HG et al (2008) Atrophy measurement of the anterior commissure and substantia innominata with 3T high-resolution MR imaging: does the measurement differ for patients with frontotemporal lobar degeneration and Alzheimer disease and for healthy subjects? AJNR Am J Neuroradiol 29(7):1308–1313. https://doi.org/10.3174/ajnr.A1103
31. Teipel SJ, Meindl T, Grinberg L et al (2011) The cholinergic system in mild cognitive impairment and Alzheimer's disease: an in vivo MRI and DTI study. Hum Brain Mapp 32(9):1349–1362. https://doi.org/10.1002/hbm.21111
32. Grothe M, Zaborszky L, Atienza M et al (2010) Reduction of basal forebrain cholinergic system parallels cognitive impairment in patients at high risk of developing Alzheimer's disease. Cereb Cortex 20(7):1685–1695. https://doi.org/10.1093/cercor/bhp232
33. Grothe M, Heinsen H, Teipel SJ (2012) Atrophy of the cholinergic basal forebrain over the adult age range and in early stages of Alzheimer's disease. Biol Psychiatry 71(9):805–813. https://doi.org/10.1016/j.biopsych.2011.06.019

34. Kilimann I, Grothe M, Heinsen H et al (2014) Subregional basal forebrain atrophy in Alzheimer's disease: a multicenter study. J Alzheimers Dis 40(3):687–700. https://doi.org/10.3233/jad-132345
35. Teipel S, Heinsen H, Amaro E Jr et al (2014) Cholinergic basal forebrain atrophy predicts amyloid burden in Alzheimer's disease. Neurobiol Aging 35(3):482–491. https://doi.org/10.1016/j.neurobiolaging.2013.09.029
36. Grothe MJ, Ewers M, Krause B et al (2014) Basal forebrain atrophy and cortical amyloid deposition in nondemented elderly subjects. Alzheimers Dement 10(5 Suppl):S344–S353. https://doi.org/10.1016/j.jalz.2013.09.011
37. Kerbler GM, Fripp J, Rowe CC et al (2015) Basal forebrain atrophy correlates with amyloid beta burden in Alzheimer's disease. Neuroimage Clin 7:105–113. https://doi.org/10.1016/j.nicl.2014.11.015
38. Schmitz TW, Nathan Spreng R (2016) Basal forebrain degeneration precedes and predicts the cortical spread of Alzheimer's pathology. Nat Commun 7:13249. https://doi.org/10.1038/ncomms13249
39. Grothe M, Heinsen H, Teipel S (2013) Longitudinal measures of cholinergic forebrain atrophy in the transition from healthy aging to Alzheimer's disease. Neurobiol Aging 34(4):1210–1220. https://doi.org/10.1016/j.neurobiolaging.2012.10.018
40. Grothe MJ, Heinsen H, Amaro E Jr et al (2016) Cognitive correlates of basal forebrain atrophy and associated cortical Hypometabolism in mild cognitive impairment. Cereb Cortex 26(6):2411–2426. https://doi.org/10.1093/cercor/bhv062
41. Ray NJ, Metzler-Baddeley C, Khondoker MR et al (2015) Cholinergic basal forebrain structure influences the reconfiguration of white matter connections to support residual memory in mild cognitive impairment. J Neurosci 35(2):739–747. https://doi.org/10.1523/jneurosci.3617-14.2015
42. Cantero JL, Zaborszky L, Atienza M (2016) Volume loss of the nucleus Basalis of Meynert is associated with atrophy of innervated regions in mild cognitive impairment. Cereb Cortex. https://doi.org/10.1093/cercor/bhw195
43. Grothe MJ, Scheef L, Bauml J et al (2016) Reduced cholinergic basal forebrain integrity links neonatal complications and adult cognitive deficits after premature birth. Biol Psychiatry. https://doi.org/10.1016/j.biopsych.2016.12.008
44. Kilimann I, Hausner L, Fellgiebel A et al (2017) Parallel atrophy of cortex and basal forebrain cholinergic system in mild cognitive impairment. Cereb Cortex 27(3):1841–1848. https://doi.org/10.1093/cercor/bhw019
45. Grinberg LT, Ferretti RE, Farfel JM et al (2007) Brain bank of the Brazilian aging brain study group - a milestone reached and more than 1,600 collected brains. Cell Tissue Bank 8(2):151–162. https://doi.org/10.1007/s10561-006-9022-z
46. Grinberg LT, Amaro E Jr, Teipel S et al (2008) Assessment of factors that confound MRI and neuropathological correlation of human postmortem brain tissue. Cell Tissue Bank 9(3):195–203. https://doi.org/10.1007/s10561-008-9080-5
47. Grinberg LT, Amaro Junior E, da Silva AV et al (2009) Improved detection of incipient vascular changes by a biotechnological platform combining post mortem MRI in situ with neuropathology. J Neurol Sci 283(1–2):2–8. https://doi.org/10.1016/j.jns.2009.02.327
48. Mesulam MM, Mufson EJ, Wainer BH et al (1983) Central cholinergic pathways in the rat: an overview based on an alternative nomenclature (Ch1-Ch6). Neuroscience 10(4):1185–1201
49. Reil JC (1809) Untersuchungen über den Bau des grossen Gehirns im Mensch. Arch Physiol (Halle) 9:136–146
50. Alheid GF, Heimer L (1988) New perspectives in basal forebrain organization of special relevance for neuropsychiatric disorders: the striatopallidal, amygdaloid, and corticopetal components of substantia innominata. Neuroscience 27:1–39
51. Heimer L, De Olmos J, Alheid GF et al (1991) "perestroika" in the basal forebrain: opening the border between neurology and psychiatry. Prog Brain Res 87:109–165
52. Lauer M, Heinsen H (1996) Cytoarchitectonics of the human nucleus accumbens. Journal of. Brain Res 37:243–254
53. Grinberg LT, Heinsen H (2007) Computer-assisted 3D reconstruction of the human basal forebrain. Dementia & Neuropsychologia 2:140–146
54. Heinsen H, Hampel H, Teipel SJ (2006) Nucleus subputaminalis: neglected part of the basal nucleus of Meynert - response to Boban et al.: computer-assisted 3D reconstruction of the nucleus basalis complex, including the nucleus subputaminalis (Ayala's nucleus). Brain 129(4):U1–U4
55. Ayala G (1915) A hitherto undifferentiated nucleus in the basal forebrain (nucleus subputaminalis). Brain 37:433–438
56. Simic G, Mrzljak L, Fucic A et al (1999) Nucleus subputaminalis (Ayala): the still disregarded magnocellular component of the basal forebrain may be human specific and connected with the cortical speech area. Neuroscience 89(1):73–89
57. Ashburner J (2007) A fast diffeomorphic image registration algorithm. NeuroImage 38(1):95–113. https://doi.org/10.1016/j.neuroimage.2007.07.007

58. Klein A, Andersson J, Ardekani BA et al (2009) Evaluation of 14 nonlinear deformation algorithms applied to human brain MRI registration. NeuroImage 46(3):786–802. https://doi.org/10.1016/j.neuroimage.2008.12.037
59. Eickhoff SB, Stephan KE, Mohlberg H et al (2005) A new SPM toolbox for combining probabilistic cytoarchitectonic maps and functional imaging data. NeuroImage 25(4):1325–1335. https://doi.org/10.1016/j.neuroimage.2004.12.034
60. Ashburner J (2009) Computational anatomy with the SPM software. Magn Reson Imaging 27(8):1163–1174. https://doi.org/10.1016/j.mri.2009.01.006
61. Boyes RG, Rueckert D, Aljabar P et al (2006) Cerebral atrophy measurements using Jacobian integration: comparison with the boundary shift integral. NeuroImage 32(1):159–169. https://doi.org/10.1016/j.neuroimage.2006.02.052
62. Malone IB, Leung KK, Clegg S et al (2015) Accurate automatic estimation of total intracranial volume: a nuisance variable with less nuisance. NeuroImage 104:366–372. https://doi.org/10.1016/j.neuroimage.2014.09.034
63. Avants BB, Epstein CL, Grossman M et al (2008) Symmetric diffeomorphic image registration with cross-correlation: evaluating automated labeling of elderly and neurodegenerative brain. Med Image Anal 12(1):26–41. https://doi.org/10.1016/j.media.2007.06.004
64. Callaert DV, Ribbens A, Maes F et al (2014) Assessing age-related gray matter decline with voxel-based morphometry depends significantly on segmentation and normalization procedures. Front Aging Neurosci 6:124. https://doi.org/10.3389/fnagi.2014.00124
65. Shen Q, Zhao W, Loewenstein DA et al (2012) Comparing new templates and atlas-based segmentations in the volumetric analysis of brain magnetic resonance images for diagnosing Alzheimer's disease. Alzheimers Dement 8(5):399–406. https://doi.org/10.1016/j.jalz.2011.07.002
66. Eggert LD, Sommer J, Jansen A et al (2012) Accuracy and reliability of automated gray matter segmentation pathways on real and simulated structural magnetic resonance images of the human brain. PLoS One 7(9):e45081. https://doi.org/10.1371/journal.pone.0045081
67. Gaser C (2009) Partial volume segmentation with adaptive maximum a posteriori (MAP) approach. NeuroImage 47:S121. https://doi.org/10.1016/s1053-8119(09)71151-6
68. Voevodskaya O, Simmons A, Nordenskjold R et al (2014) The effects of intracranial volume adjustment approaches on multiple regional MRI volumes in healthy aging and Alzheimer's disease. Front Aging Neurosci 6:264. https://doi.org/10.3389/fnagi.2014.00264
69. Boban M, Kostovic I, Simic G (2006) Nucleus subputaminalis: neglected part of the basal nucleus of Meynert. Brain 129(Pt 4):E42.; author reply E43. https://doi.org/10.1093/brain/awl025
70. Schliebs R, Arendt T (2006) The significance of the cholinergic system in the brain during aging and in Alzheimer's disease. J Neural Transm 113(11):1625–1644. https://doi.org/10.1007/s00702-006-0579-2
71. Rogers JD, Brogan D, Mirra SS (1985) The nucleus basalis of Meynert in neurological disease: a quantitative morphological study. Ann Neurol 17(2):163–170. https://doi.org/10.1002/ana.410170210
72. Samuel W, Terry RD, DeTeresa R et al (1994) Clinical correlates of cortical and nucleus basalis pathology in Alzheimer dementia. Arch Neurol 51(8):772–778
73. Iraizoz I, Guijarro JL, Gonzalo LM et al (1999) Neuropathological changes in the nucleus basalis correlate with clinical measures of dementia. Acta Neuropathol 98(2):186–196
74. Grothe MJ, Schuster C, Bauer F et al (2014) Atrophy of the cholinergic basal forebrain in dementia with Lewy bodies and Alzheimer's disease dementia. J Neurol 261(10):1939–1948. https://doi.org/10.1007/s00415-014-7439-z
75. Arendt T, Bigl V, Arendt A et al (1983) Loss of neurons in the nucleus basalis of Meynert in Alzheimer's disease, paralysis agitans and Korsakoff's disease. Acta Neuropathol 61(2):101–108
76. Teipel S, Raiser T, Riedl L et al (2016) Atrophy and structural covariance of the cholinergic basal forebrain in primary progressive aphasia. Cortex 83:124–135. https://doi.org/10.1016/j.cortex.2016.07.004
77. Wolf D, Grothe M, Fischer FU et al (2014) Association of basal forebrain volumes and cognition in normal aging. Neuropsychologia 53:54–63. https://doi.org/10.1016/j.neuropsychologia.2013.11.002
78. Olesen OF, Dago L, Mikkelsen JD (1998) Amyloid beta neurotoxicity in the cholinergic but not in the serotonergic phenotype of RN46A cells. Brain Res Mol Brain Res 57(2):266–274
79. Zheng WH, Bastianetto S, Mennicken F et al (2002) Amyloid beta peptide induces tau phosphorylation and loss of cholinergic neurons in rat primary septal cultures. Neuroscience 115(1):201–211
80. Boncristiano S, Calhoun ME, Kelly PH et al (2002) Cholinergic changes in the APP23 transgenic mouse model of cerebral amyloidosis. J Neurosci 22(8):3234–3243. doi:20026314

81. Bell KF, Ducatenzeiler A, Ribeiro-da-Silva A et al (2006) The amyloid pathology progresses in a neurotransmitter-specific manner. Neurobiol Aging 27(11):1644–1657. https://doi.org/10.1016/j.neurobiolaging.2005.09.034
82. Nitsch RM, Slack BE, Wurtman RJ et al (1992) Release of Alzheimer amyloid precursor derivatives stimulated by activation of muscarinic acetylcholine receptors. Science 258(5080):304–307
83. Ramos-Rodriguez JJ, Pacheco-Herrero M, Thyssen D et al (2013) Rapid beta-amyloid deposition and cognitive impairment after cholinergic denervation in APP/PS1 mice. J Neuropathol Exp Neurol 72(4):272–285. https://doi.org/10.1097/NEN.0b013e318288a8dd
84. Kolisnyk B, Al-Onaizi M, Soreq L et al (2016) Cholinergic surveillance over hippocampal RNA metabolism and Alzheimer's-like pathology. Cereb Cortex. https://doi.org/10.1093/cercor/bhw177
85. Ovsepian SV, Herms J (2013) Drain of the brain: low-affinity p75 neurotrophin receptor affords a molecular sink for clearance of cortical amyloid beta by the cholinergic modulator system. Neurobiol Aging 34(11):2517–2524. https://doi.org/10.1016/j.neurobiolaging.2013.05.005
86. Anaclet C, Pedersen NP, Ferrari LL et al (2015) Basal forebrain control of wakefulness and cortical rhythms. Nat Commun 6:8744. https://doi.org/10.1038/ncomms9744
87. Lim YY, Maruff P, Schindler R et al (2015) Disruption of cholinergic neurotransmission exacerbates Abeta-related cognitive impairment in preclinical Alzheimer's disease. Neurobiol Aging 36(10):2709–2715. https://doi.org/10.1016/j.neurobiolaging.2015.07.009
88. Teipel SJ, Bruno D, Grothe MJ et al (2015) Hippocampus and basal forebrain volumes modulate effects of anticholinergic treatment on delayed recall in healthy older adults. Alzheimers Dement (Amst) 1(2):216–219. https://doi.org/10.1016/j.dadm.2015.01.007
89. Kendziorra K, Wolf H, Meyer PM et al (2011) Decreased cerebral alpha4beta2* nicotinic acetylcholine receptor availability in patients with mild cognitive impairment and Alzheimer's disease assessed with positron emission tomography. Eur J Nucl Med Mol Imaging 38(3):515–525. https://doi.org/10.1007/s00259-010-1644-5
90. Bohnen NI, Muller ML, Kuwabara H et al (2009) Age-associated leukoaraiosis and cortical cholinergic deafferentation. Neurology 72(16):1411–1416. https://doi.org/10.1212/WNL.0b013e3181a187c6
91. Bohnen NI, Kaufer DI, Hendrickson R et al (2005) Cognitive correlates of alterations in acetylcholinesterase in Alzheimer's disease. Neurosci Lett 380(1–2):127–132. https://doi.org/10.1016/j.neulet.2005.01.031
92. Wolf D, Bocchetta M, Preboske GM et al (2017) Reference standard space hippocampus labels according to the EADC-ADNI harmonized protocol: utility in automated volumetry. Alzheimers Dement. https://doi.org/10.1016/j.jalz.2017.01.009
93. Frisoni GB, Jack CR Jr, Bocchetta M et al (2015) The EADC-ADNI harmonized protocol for manual hippocampal segmentation on magnetic resonance: evidence of validity. Alzheimers Dement 11(2):111–125. https://doi.org/10.1016/j.jalz.2014.05.1756

Part V

Patient Benefit and Ethical Considerations

Check for updates

Chapter 16

Clinical Meaningfulness of Biomarker Endpoints in Alzheimer's Disease Research

Kok Pin Ng, Tharick A. Pascoal, Xiaofeng Li, Pedro Rosa-Neto, and Serge Gauthier

Abstract

Advancement in biomarker research has enabled the in vivo detection of Alzheimer's disease (AD) pathophysiology, including amyloid plaques, neurofibrillary tangles, and neuronal degeneration. AD biomarkers play an important role in characterizing the trajectory of AD and have been incorporated in the research criteria for AD diagnosis. The presence of abnormal biomarkers for AD pathology in cognitively normal individuals has further led to the proposal of a preclinical stage in the AD spectrum. With the emerging conceptual framework that intervention in the early stages of the disease offers the greatest chance of success in preventing or delaying AD progression, recent clinical trials are now focusing on individuals with preclinical AD. While there are clear benefits from the use of biomarkers in research settings such as the enrichment of clinical trial population to confirm the presence of target brain pathology and target engagement by the intervention, the role of biomarkers in the clinical setting is less clear, especially in asymptomatic individuals. Potential ethical issues also arise with the use of biomarkers due to the conflict between the principles of benefits and not doing harm. In fact, a unique set of ethical issues arises in asymptomatic individuals, such as the disclosure of genetic mutation status, and abnormal biomarker results when their diagnostic validity is uncertain. In this chapter, we will discuss the issues and clinical meaningfulness of biomarkers in AD research. Specifically, we will focus on the potential benefits and ethical considerations when genetics and biomarkers for amyloid, tau, and neurodegeneration are used in the early stages of AD.

Key words Alzheimer's disease, Dementia, Mild cognitive impairment, Biomarker, Cerebrospinal fluid, Imaging, Tau, Amyloid-beta, Ethics, Genetics, Early diagnosis, Prognosis

1 Introduction

Amyloid plaques and neurofibrillary tangles are the core histopathological features of Alzheimer's disease (AD) [1]. Advancement in biomarker research has enabled the characterization of the trajectories of these pathologies in vivo across the AD spectrum [2] and has deepened the understanding of AD pathophysiology. As such, biomarkers have been incorporated in the research criteria for AD diagnosis, which aimed to detect both the earliest stages of AD and the full spectrum of the disease [3, 4].

Robert Perneczky (ed.), *Biomarkers for Preclinical Alzheimer's Disease*, Neuromethods, vol. 137,
https://doi.org/10.1007/978-1-4939-7674-4_16, © Springer Science+Business Media, LLC 2018

The identification of AD biomarkers crossing pathological threshold in cognitively normal individuals and presymptomatic autosomal-dominant mutation carriers has further led to the conceptual framework which proposes a preclinical stage in both sporadic and familial AD [5–7]. In fact, recent therapeutic trials have focused on preclinical AD individuals defined using abnormal AD biomarkers [8], given that early intervention may offer the greatest chance of treatment success. For example, solanezumab and gantenerumab are being studied in asymptomatic and very mildly symptomatic autosomal-dominant mutation carriers in the Dominantly Inherited Alzheimer Network Trials Unit (DIAN-TU) prevention trial [9], while solanezumab is being studied in asymptomatic or mildly symptomatic older individuals who have biomarker evidence of brain amyloid deposition in the Anti-Amyloid Treatment in Asymptomatic AD (A4) study [10].

There are benefits of AD biomarkers for both patients and families suffering from AD. With genetic counseling, genetic testing plays an important role in individuals with a positive family history of dementia, such as alleviating uncertainty and anxiety and allowing one to make plans for the future [11]. For the clinicians, AD biomarkers may also improve clinical diagnostic certainty, especially in patients with atypical or early-onset presentations [12]. This may allow an earlier diagnosis and, hence, earlier access to emerging research studies and clinical trials. AD biomarkers such as amyloid positron emission tomography (PET) may also be used in the selection of patients with the target pathology to enrich clinical trial population with individuals with a higher probability of AD-related clinical progression [13].

However, ethical issues arise with the use of AD biomarkers, due to a conflict between the principles of beneficence and nonmaleficence [14]. As current treatment frameworks focus on the earlier stages of the disease, which involves asymptomatic individuals, ethical issues such as the disclosure of biomarker result and its social implications must be considered. For example, *APOE* ε4 represents a higher risk for AD, and *APOE* ε4 carriers may benefit from preventive or disease-modifying clinical trials [15]. But given that *APOE* ε4 is neither necessary nor sufficient for developing AD, there may be ethical concerns regarding the disclosure of this genetic status, especially in asymptomatic individuals positive for *APOE* ε4 as they may face unnecessary social discrimination and psychological distress when they may never develop AD dementia.

In this chapter, we will discuss the issues and clinical meaningfulness of biomarkers in AD research. Specifically, we will focus on the potential benefits and ethical considerations when genetics, amyloid, tau, and neurodegeneration biomarkers are used to define the preclinical stage of AD.

2 Definition of Preclinical AD

The National Institute on Aging–Alzheimer's Association (NIA-AA) proposed an operational three-stage research criteria in 2011 [16]. Stage 1 begins when asymptomatic cerebral amyloidosis is diagnosed using amyloid biomarkers (PET and cerebrospinal fluid (CSF)). This is followed by stage 2, when neurodegeneration markers diagnosed using CSF tau, [^{18}F]flurodeoxyglucose (FDG) PET, or magnetic resonance imaging (MRI) are present in addition to cerebral amyloidosis in asymptomatic individuals. Stage 3 is defined when the above individuals develop subtle cognitive or behavioral decline.

The international working group (IWG-2) defined two preclinical states in 2014 [6]. The presymptomatic state is defined as asymptomatic individuals who are destined to develop AD, because they carry an autosomal-dominant mutation. The asymptomatic at-risk state is defined as asymptomatic individuals who have either decreased CSF $A\beta_{42}$ together with increased CSF total tau (t-tau) or phosphorylated tau (p-tau) or increased retention of amyloid measured with amyloid PET.

Most recently, the definition of preclinical AD has been updated based on the stratification of high or low risk for progression to clinical AD [7]. In preclinical AD when both amyloid and tau biomarkers are beyond pathologic thresholds, the risk of clinical progression to AD is particularly high. In asymptomatic at risk for AD when either amyloid or tau biomarker is beyond pathologic threshold, the risk of progression to clinical AD is less likely.

3 Ethical Principles

Ethics regulation has played an important role in AD research as it protects the interest of AD patients, especially when their autonomy and mental capacity may be impaired in the later stages of the disease. Ethical issues often arise due to the tension between the principle of beneficence, such as the development of an effective treatment, and non-maleficence, by not doing harm in this vulnerable population [14]. These ethical issues have focused on the decision to disclose the diagnosis and the potential social stigma following the disclosure of the diagnosis. In addition, the benefits of symptomatic treatment in the mild to moderate disease stages, decision to stop treatment in the severe stage of the disease, and the competence of the patients to give consent to research and therapy are often debated.

However, as the AD research landscape currently shifts toward preventive and disease-modifying treatments in the preclinical stage of the disease, a different set of ethical issues has arisen.

Although there is no doubt regarding the autonomy and competency of the cognitively normal individuals participating in AD research, researchers have to bear in mind the unique ethical issues, such as the disclosure of AD diagnosis, the uncertainty of a diagnosis based on biomarkers without full validation, and the social stigma of having a preclinical AD, mild cognitive impairment (MCI), and AD diagnosis [14]. These ethical issues arising from the use of specific AD biomarkers will be elaborated in the subsequent paragraphs.

4 Genetic Biomarkers

4.1 Autosomal-Dominant AD Genes (APP, PS1, PS2)

Familial AD represents <1% of AD, and patients often present with an early-onset (before 65 years old) of cognitive symptoms. In familial AD, there is overproduction of Aβ, either due to *APP* gene mutation or mutations in genes encoding presenilin 1 (PSEN1) or presenilin 2 (PSEN2), which are key components of the γ-secretase complexes that cleave and release Aβ. Given the autosomal-dominant inheritance of these genes, individuals with a positive family history of first-degree relatives with these genetic mutations have a 50% chance of having the mutation. Also, with a penetrance of almost 100%, the chance of progression to AD in these mutation carriers is almost 100%.

4.1.1 Benefits

In individuals with a positive family history of autosomal-dominant AD (ADAD), the knowledge of their carrier status of the genetic mutation may alleviate the uncertainty and anxiety of developing AD [17] and allow them to make important plans for their future, such as financial and family arrangements and when to stop working, as the likelihood of progression to AD in mutation carriers is almost 100%.

After testing for the mutation status, these individuals with or without ADAD genetic mutation may participate and contribute to AD research, such as the Dominantly Inherited Alzheimer Network (DIAN) study [18]. Given that AD pathophysiology is postulated to develop many years before the clinical onset of dementia [19], the DIAN aims to study early pathophysiological changes in the asymptomatic mutation carriers, with the family members without the mutation serving as a comparison group. In addition, presymptomatic mutation carriers will also have access to therapeutic clinical trials, such as the DIAN-TU [9]. DIAN-TU is a preventive study that investigates interventions such as solanezumab and gantenerumab in asymptomatic and mildly symptomatic mutation carriers, as it is postulated that these individuals may benefit from Aβ-targeting interventions. It is hypothesized that treatment in the preclinical stage of AD will be more effective and have a higher chance of slowing cognitive decline in these individuals [7].

4.1.2 Ethical Issues

The presymptomatic ADAD mutation carriers represent a vulnerable group of individuals who has full autonomy and competency. Hence, genetic counseling plays a key role in protecting their interest, and ample time must be allowed for making a decision to undergo genetic testing. In fact, a joint practice guideline for genetic counseling and testing has been proposed to guide clinicians in the effective counseling on hereditary risk and genetic testing [11]. In addition, a structured three-phase framework has also been proposed for the disclosure of AD diagnosis, which is especially important in this group of asymptomatic individuals with full insight [20]. In the "before phase," the key objective is to determine whether one wants to know the diagnosis, their coping strategies and their psychological profiles, the time and place where disclosure will take place, and the information that will be conveyed. In the "during phase," the individual's understanding of the disease should be established, and confidentiality has to be maintained unless consent is obtained to disclose the diagnosis to any family member. The subsequent phase focuses on ensuring that the individual understands the information that is delivered and follow-up meeting and support are provided.

The association of MCI and AD diagnosis with social stigmatization and discrimination is well established [21]. This social stigmatization and discrimination will likely be faced by ADAD mutation carriers, who are destined to develop AD. Given a recent shift of research from clinical stages to presymptomatic disease stages, the knowledge of the mutation status and its impact on social stigmatization in the cognitively normal population are still elusive. Hence, it is important that this sensitive information be kept confidential as they may have serious implications on employment and insurance. This is especially important when a mutation carrier has intact cognition and mental capacity and is perfectly well to work and contribute to the society.

While the potential benefits of ADAD genetic testing in gaining access to clinical trials have been discussed, the therapeutic risk-benefit ratio of the clinical trial in presymptomatic ADAD carriers has to be considered. For example, the use of a novel anti-amyloid treatment, although desirable in individuals with higher risk of AD, may be less acceptable in healthy asymptomatic individuals, due to potential side effects such as amyloid-related imaging abnormalities [22]. In this regard, it is important to ensure that the interventions being studied must pass through rigorous clinical trials that evaluate drug safety and effectiveness [9].

4.2 APOE Gene

There are three polymorphic alleles in the *APOE* gene, ε2, ε3, and ε4, with a frequency of 8.4%, 77.9%, and 13.7%, worldwide. However, the frequency of ε4 allele is increased in AD at an estimate of 40% [23]. The ε4 allele is the strongest genetic risk factor for sporadic AD, leading to an earlier age of onset in a gene dose-dependent manner [15]. The frequency of AD and the mean age of cognitive decline are 91% and 68 years old in ε4 homozygotes, 47% and 76 years old in ε4 heterozygotes, and 20% and 84 years old in ε4 non-carriers, respectively [15, 24]. *APOE* ε4 is associated with impaired memory performance and increased risk of memory decline in middle-aged (40–59 years) and elderly (60–85 years) people with MCI, as well as increased risk of progression from MCI to AD [15].

4.2.1 Benefits

In a population-based study, physical inactivity, dietary fat intake, alcohol drinking, and smoking at midlife are associated with the risk of dementia and AD, especially among the *APOE* ε4 carriers [25]. Type 2 diabetes is also a risk factor for AD, and the association between diabetes and AD is particularly important among the *APOE* ε4 carriers, which is supported by clinical and neuropathological data [26]. Hence, lifestyle modifications such as healthy diet, physical activity, and optimization of cardiovascular risk factors may modify dementia risk, particularly among *APOE* ε4 carriers. In this regard, a study which followed cognitively normal individuals for 9 years to detect incident dementia cases found that high education, active leisure activities, or maintaining vascular health may lower dementia risk related to *APOE* ε4 [27].

APOE genotyping has also been included in clinical trials to enrich the study population with individuals with a high risk of developing AD. For example, the Alzheimer's Prevention Initiative (API) *APOE* ε4 trial, which aims to prevent or delay AD using either the active amyloid immunotherapy CAD106 or the beta-secretase inhibitor 1 (BACE1) inhibitor CNP520 versus placebo, recruits cognitively normal individuals with homozygous ε4, who have the highest risk of progression to AD.

4.2.2 Ethical Issues

However, as *APOE* ε4 is neither necessary nor sufficient for developing AD, genetic testing for *APOE* ε4 is controversial, and genetic counseling for the individual is important to discuss the benefits and risks prior to testing. Although there is no effective prevention or treatment when one is tested positive for *APOE* ε4, these individuals may benefit from participating in preventive trials. It is also important to emphasize that the confidentiality of *APOE* ε4 status should be maintained, given that *APOE* ε4 is increasingly tested in the research setting. This will protect the interest of the individual as *APOE* ε4 carriers may face social stigmatization and discrimination.

There have been concerns that disclosure of the genetic screening results may lead to negative emotions among the *APOE* ε4 carriers. In the Risk Evaluation and Education for AD (REVEAL) study which examines the effects of *APOE* genotype status disclosure to asymptomatic individuals with a first-degree relative with AD, no difference was found in terms of anxiety and depression in individuals who are informed that they are *APOE* ε4 carriers, compared to non-carriers, and those not informed of their *APOE* status [28]. However, given the limitations of the study, the authors concluded that larger studies that follow subjects for more than 1 year will be required to detect uncommon and long-term effects, such as delayed emotional repercussions and injudicious life decisions.

5 Amyloid Biomarkers

Low CSF $A\beta_{42}$ and high uptake of amyloid PET tracers are biomarkers of amyloid fibrillary deposition [2]. There is a close correlation between in vivo amyloid biomarkers and brain amyloid plaques assessed in postmortem studies [29–31]. Longitudinal studies have suggested that AD biomarkers, beginning with amyloid deposition, become abnormal sequentially in a prolonged preclinical period [19]. In fact, CSF $A\beta_{42}$ levels decline up to 20 years before symptom onset [32]. There is high concordance between CSF $A\beta_{42}$ and amyloid PET in MCI and AD patients [33]. In addition, amyloid PET provides information regarding regional distribution and longitudinal changes of amyloid deposition. However, 10–30% of cognitively normal people have a positive amyloid PET scan [34], and pathological amyloid deposition in the brain is also shown in non-demented individuals [35]. Hence, amyloid pathology is proposed to be necessary, but not sufficient to cause progression to a symptomatic stage of the disease [2].

5.1 Benefits

Given a high degree of pathological heterogeneity underlying the clinical diagnosis of MCI [1], AD biomarkers, including amyloid, have been recently incorporated in the research criteria to improve the certainty of making a diagnosis of MCI due to AD [36]. An abnormal amyloid PET scan increases the likelihood of AD, while a negative amyloid PET scan provides useful information in ruling out AD etiology in atypical or early-onset presentations [12]. Hence, amyloid PET is considered appropriate in individuals with atypical presentation or dementia of uncertain etiology [37]. Amyloid PET in MCI may also provide useful prognostic information. A study showed that PIB-positive MCI patients are significantly more likely to convert to AD than PIB-negative patients, and faster converters have higher PIB retention levels at baseline [38].

Amyloid biomarkers have also been incorporated in the preclinical AD research criteria, which, together with tau, identify individuals with the highest risk of progression to AD dementia [7]. In this regard, amyloid biomarkers may guide study eligibility and target engagement in disease-modifying clinical trials, especially in early asymptomatic stage, as these biomarkers reflect in vivo biological process that occurs many years before symptom onset. In the phase 3 clinical trial with anti-Aβ monoclonal antibody bapineuzumab, up to 36.1% of study participants have negative amyloid PET scans [39]. Since then, subsequent clinical trials such as the EXPEDITION 3 with solanezumab and PRIME with aducanumab have added amyloid biomarkers in the inclusion criteria to enrich their study population. The A4 trial, which aims to delay progression from normal cognition to MCI or dementia using solanezumab, also recruits asymptomatic and mildly symptomatic individuals with biomarker evidence of brain amyloid deposition [10]. Amyloid biomarkers are used in outcome measures to demonstrate target engagement. For example, following treatment with aducanumab, brain Aβ plaques as measured by florbetapir PET imaging are reduced in a dose- and time-dependent fashion [13].

5.2 Ethical Issues

While it is expected that AD biomarkers, including amyloid, will be able to identify preclinical individuals who will progress to AD, current biomarker follow-up studies provide elusive information regarding biomarker signatures of those who will progress to dementia [40, 41]. The uncertainty of AD diagnosis based on biomarkers is further demonstrated in MCI individuals, where 50% have amyloid pathology, but only 40% of them develop dementia in 3 years [38]. Hence, the clinical use of amyloid biomarkers is currently not recommended in asymptomatic individuals [37].

The disclosure of amyloid status to study participants is also a key ethical issue. Among investigators, there are debates as to whether the results of amyloid biomarkers should be revealed to the participants. The Alzheimer's Disease Neuroimaging Initiative (ADNI) study has maintained a position where biomarker results, including amyloid, are not disclosed to study participants, due to uncertainty of the prognostic value of the biomarker [42]. However, this stance is likely to change once clinical utility associated with biomarkers is better established. In fact, recent clinical trials, such as the A4 which studies disease-modifying agents in amyloid-positive individuals, will, by study design, result in asymptomatic participants being informed of their amyloid imaging results [10]. Given that the implications of amyloid status on employment, health insurance, and social stigmatization are not fully understood, there is a need to accelerate the development of a best practice framework for disclosure of amyloid status and to identify individuals at risk of potential devastating reactions.

While amyloid PET has a clear benefit in research, its clinical cost-effectiveness is less clear [43]. There remains an uncertainty regarding the validity of amyloid PET, such as the presence of false positives and false negatives, the adjustment of threshold for amyloid positivity due to age and genetic factors, and the optimal threshold for determining scan positivity using a quantitative approach [44].

Similarly, interlaboratory variations of CSF analysis including $A\beta_{42}$ using current immunoassays are of a concern [45]. In the absence of disease-modifying therapy for AD at present, the important goal of a diagnostic evaluation is to identify treatable conditions or modifiable comorbidities. In this regard, amyloid PET adds little to the diagnostic workup of patients with a classic amnestic AD phenotype [37] and does not offer additional information on treatable or modifiable conditions such as cerebrovascular disease. As such, there remains a strong debate regarding potential cost-savings versus increasing cost when utilizing amyloid PET in dementia evaluation [46]. However, this is likely to change once there are effective disease-modifying therapies and therefore the clinical necessity of molecular imaging to better identify the therapeutic targets.

6 Tau Biomarkers

Neurofibrillary tangles (NFT) are intraneuronal aggregates of hyperphosphorylated and misfolded tau proteins and become extraneuronal when neurons with tangles die [1, 35]. There is a strong correlation between cortical NFT and cognitive decline [47, 48]. Elevated CSF p-tau and high uptakes of tau PET tracers are biomarkers of NFT. CSF p-tau is specific for the burden of AD-type tau pathology [49, 50], and its levels correlate with the severity of tau pathology in postmortem studies [51]. Tau PET imaging constitutes a new frontier in AD research [52], and current evidence suggests that tau imaging agents are able to demonstrate the topography of tau spread consistent with the description at autopsy by Braak in the typical AD spectrum [53, 54].

6.1 Benefits

Tau biomarkers have been incorporated in the research diagnostic criteria which enables the detection of AD pathophysiology at the early stages of the disease [4, 36]. Increased CSF p-tau is more sensitive and specific than CSF t-tau and $A\beta_{42}$ in discriminating AD from normal aging and non-AD dementia [55], and a high CSF p-tau/$A\beta_{42}$ ratio also provides high diagnostic accuracy in differentiating AD from normal controls and from subjects with non-AD dementia [56]. High uptake of tau tracers in the temporal lobes is able to distinguish AD from healthy individuals [57], and in amyloid-positive individuals, increasing levels of tau binding are associated with higher cognitive impairment [58].

Studies and clinical trials have tested interventions that target pathological tau aggregation [59, 60], and similar to amyloid biomarkers, it is likely that tau biomarkers will play a key role in future clinical trials and observational studies. Although tau PET imaging is still in the early stages of development, it is anticipated that tau imaging agents will enable the staging of the disease, selection of patients appropriate for a given tau therapy, confirmation of target engagement, and monitoring treatment efficacy for AD [54].

6.2 Ethical Issues

Given that tau PET imaging is still in the early stages of development, there has been limited clinical and research data for its current use compared to amyloid PET imaging. In fact, off target binding of tau tracers has been reported [61, 62], and its implications on the interpretation of scan findings will need to be carefully studied, given the potential use of tau PET imaging in future clinical trials as a marker of target engagement. In this regard, the disclosure of tau PET imaging results is also a potential ethical issue, given the uncertainty of the validity of this biomarker.

7 Neurodegeneration Biomarkers

Neurodegeneration represents the progressive loss of structure or function of neurons [63]. High CSF t-tau, decreased FDG PET uptake, and atrophy on structural MRI are considered biomarkers of neuronal degeneration. CSF t-tau levels correlate with neuronal tissue damage and reflect the intensity of neuronal degeneration in AD [49, 64]. FDG PET measures cerebral metabolic rates of glucose, which is postulated to be a proxy for synaptic activity. AD is characterized by a regional hypometabolism in the parietotemporal cortex, medial temporal lobes, and posterior cingulate cortex [63]. MRI measures brain atrophy, which is associated with neuronal degeneration and tissue loss. Vulnerable brain regions for atrophy in AD include the hippocampus and entorhinal cortex and are able to predict progression from MCI to AD dementia [65].

7.1 Benefits

Neurodegeneration biomarkers play an important role in the clinical evaluation of cognitive symptoms. From 36 studies with about 2500 AD patients and 1400 controls, the pooled sensitivity and specificity for CSF t-tau in differentiating AD from controls are 81% and 90%, respectively [49]. In a longitudinal study of MCI patients, a combination of CSF t-tau and $A\beta_{42}$ at baseline gave a sensitivity of 95% and a specificity of 83% for detection of incipient AD [66], further supporting the diagnostic use of CSF t-tau in the MCI stage. CSF t-tau levels also correlate with poor cognitive performance in AD [67]. FDG PET has high sensitivity in distinguishing AD from controls and individuals with higher versus lower AD risk and has good quantitative and topographical correlation with clinical progression [68]. FDG also improves accuracy in distinguishing frontotemporal dementia (FTD) from AD [69]. Atrophy of medial temporal structures in MRI is a valid diagnostic marker at the MCI stage, and hippocampal atrophy is able to distinguish AD from individuals with no cognitive impairment using visual rating scales [70]. The rates of change in structural measures, such as the entorhinal cortex, hippocampus, and temporal lobes, also correlate closely with changes in cognitive performance. However, it is important to note that medial temporal atrophy by itself lacks specificity to exclude other dementia, and its combination with other biomarker modalities such as CSF enhances the accuracy of making a diagnosis of AD [65].

MRI, which is widely available and devoid of radiation risk, has the added benefit of assessing comorbidities that increase AD risk, such as white matter hyperintensities, a marker of small vessel cerebrovascular disease, or other causes of cognitive symptoms, such as a space-occupying lesion. This adds valuable information during clinical evaluation, which aid in the treatment management of the patient, such as optimization of cardiovascular risk factors.

7.2 Ethical Issues

While neurodegeneration biomarkers support the diagnosis of AD, it is important to note that these biomarkers are least specific for AD compared to biomarkers for amyloid and tau pathologies. Brain atrophy in MRI and hypometabolism in FDG PET scan involving regions associated with AD can also occur in a variety of other disorders [71]. For example, atrophy in the anterior, medial, and basal temporal lobes can occur in non-AD tauopathies such as progressive supranuclear palsy, as well as in cerebrovascular disease, epilepsy, hippocampal sclerosis, primary age-related tauopathy, and argyrophilic grain disease. Hypometabolism in the temporoparietal region can also be found in corticobasal degeneration and cerebrovascular disease [50]. The non-specificity of neurodegeneration biomarkers has to be taken into account when explaining the biomarker results to patients as this can lead to inaccurate diagnosis, wrong treatment, and unnecessary anxiety.

8 Summary

Biomarkers play a key role in AD research which advanced our understanding of the AD pathophysiology. With the finding of abnormal AD biomarkers in cognitively normal individuals, a conceptual framework of a preclinical stage of AD has been proposed. While there are definite benefits linked to the use of biomarkers to define this early disease stage in research setting, a unique set of ethical issues have arisen as a result of a conflict between beneficence and non-maleficence. It is important to keep in mind these ethical concerns so as to protect the interests of research participants, especially those in the asymptomatic stages of the disease.

Acknowledgments

Kok Pin Ng is supported by the National Medical Research Council, Research Training Fellowship Grant (Singapore).

Pedro Rosa-Neto and Serge Gauthier are funded by the Canadian Institutes for Health Research.

References

1. Serrano-Pozo A, Frosch MP, Masliah E, Hyman BT (2011) Neuropathological alterations in Alzheimer disease. Cold Spring Harb Perspect Med 1:a006189
2. Jack CR, Knopman DS, Jagust WJ et al (2013) Tracking pathophysiological processes in Alzheimer's disease: an updated hypothetical model of dynamic biomarkers. Lancet Neurol 12:207–216
3. Dubois B, Feldman HH, Jacova C et al (2007) Research criteria for the diagnosis of Alzheimer's disease: revising the NINCDS-ADRDA criteria. Lancet Neurol 6:734–746
4. McKhann GM, Knopman DS, Chertkow H et al (2011) The diagnosis of dementia due to Alzheimer's disease: recommendations from the national institute on aging-Alzheimer's association workgroups on diagnostic guidelines for Alzheimer's disease. Alzheimers Dement 7:263–269
5. Sperling RA, Aisen PS, Beckett LA et al (2011) Toward defining the preclinical stages of Alzheimer's disease: recommendations from the national institute on aging and the Alzheimer's association workgroup. Alzheimers Dement 7:1–13
6. Dubois B, Feldman HH, Jacova C et al (2014) Advancing research diagnostic criteria for Alzheimer's disease: the IWG-2 criteria. Lancet Neurol 13:614–629
7. Dubois B, Hampel H, Feldman HH et al (2016) Preclinical Alzheimer's disease: definition, natural history, and diagnostic criteria. Alzheimers Dement 12:292–323
8. Folch J, Petrov D, Ettcheto M et al (2016) Current research therapeutic strategies for Alzheimer's disease treatment. Neural Plast 2016:1–15
9. Bateman RJ, Benzinger TL, Berry S et al (2017) The DIAN-TU next generation Alzheimer's prevention trial: adaptive design and disease progression model. Alzheimers Dement 13:8–19
10. Sperling RA, Aisen PS (2016) Anti-amyloid treatment of asymptomatic AD: A4 and beyond. Alzheimers Dement 12:P326–P327
11. Goldman JS, Hahn SE, Williamson Catania J et al (2011) Genetic counseling and testing for Alzheimer disease: joint practice guidelines of the American college of medical genetics and the national society of genetic counselors. Genet Med 13:597–605
12. Morris E, Chalkidou A, Hammers A et al (2016) Diagnostic accuracy of (18)F amyloid PET tracers for the diagnosis of Alzheimer's disease: a systematic review and meta-analysis. Eur J Nucl Med Mol Imaging 43:374–385
13. Sevigny J, Chiao P, Bussière T et al (2016) The antibody aducanumab reduces Aβ plaques in Alzheimer's disease. Nature 537:50–56
14. Gauthier S, Leuzy A, Racine E, Rosa-Neto P (2013) Diagnosis and management of Alzheimer's disease: past, present and future ethical issues. Prog Neurobiol 110:102–113
15. Liu C, Kanekiyo T, Xu H, Bu G (2013) Apolipoprotein E and Alzheimer disease: risk, mechanisms and therapy. Nat Rev Neurol 9:106–118
16. Sperling RA, Aisen PS, Beckett LA et al (2011) Toward defining the preclinical stages of Alzheimer's disease: recommendations from the national institute on aging- Alzheimer's association workgroups on diagnostic guidelines for Alzheimer's disease. Alzheimers Dement 7:280–292
17. Steinbart EJ, S D MZ et al (2001) Impact of DNA testing for early-onset familial Alzheimer disease and frontotemporal dementia. Arch Neurol 58:1828
18. Bateman RJ, Xiong C, Benzinger TLS, Fagan AM, Goate A, Fox NC, Marcus DS, Cairns NJ, Xie X, Nlazey TM, Holtman DM, Santacruz A, Buckles V, Oliver A, Moulder K, Aisen PS, Ghetti B, Klunk WE, McDade E, Martins RN, Masters CL, Mayeux R, Ringman JM, Rossor MN, Schofield PR, Sperling RA, Salloway S, Morris JC (2013) Clinical and biomarker changes in dominantly inherited Alzheimer's disease. N Engl J Med 367:795–804.
19. Bateman RJ, Xiong C, Benzinger TLS et al (2012) Clinical and biomarker changes in

dominantly inherited Alzheimer's disease. N Engl J Med 367:795–804
20. Pepersack T (2008) Disclosing a diagnosis of Alzheimer's disease. Rev Med Brux 29:89–93
21. Nicole L Batsch, Mary S Mittelman (2012) World Alzheimer report 2012 overcoming the stigma of dementia. https://www.alz.org/documents_custom/world_report_2012_final.pdf. Accessed 8 Apr 2017
22. Peters KR, Lynn Beattie B, Feldman HH (2013) A conceptual framework and ethics analysis for prevention trials of Alzheimer disease. Prog Neurobiol 110:114–123
23. Farrer LA, Cupples LA, Haines JL et al (1997) Effects of age, sex, and ethnicity on the association between apolipoprotein E genotype and Alzheimer disease. A meta-analysis. APOE and Alzheimer disease meta analysis consortium. JAMA 278:1349–1356
24. Corder EH, Saunders AM, Strittmatter WJ et al (1993) Gene dose of apolipoprotein E type 4 allele and the risk of Alzheimer's disease in late onset families. Science 261:921–923
25. Kivipelto M, Rovio S, Ngandu T et al (2008) Apolipoprotein E ε4 magnifies lifestyle risks for dementia: a population-based study. J Cell Mol Med 12:2762–2771
26. Peila R, Rodriguez BL, Launer LJ (2002) Type 2 diabetes, APOE gene, and the risk for dementia and related pathologies the Honolulu-Asia aging study. Diabetes 51:1256–1262
27. Ferrari C, W-L X, Wang H-X et al (2013) How can elderly apolipoprotein E ε4 carriers remain free from dementia? Neurobiol Aging 34:13–21
28. Green RC, Roberts JS, Cupples LA et al (2009) Disclosure of *APOE* genotype for risk of Alzheimer's disease. N Engl J Med 361:245–254
29. Strozyk D, Blennow K, White LR, Launer LJ (2003) CSF Abeta 42 levels correlate with amyloid-neuropathology in a population-based autopsy study. Neurology 60:652–656
30. Ikonomovic MD, Klunk WE, Abrahamson EE et al (2008) Post-mortem correlates of in vivo PiB-PET amyloid imaging in a typical case of Alzheimer's disease. Brain 131:1630–1645
31. Sojkova J, Driscoll I, Iacono D et al (2011) In vivo fibrillar β-amyloid detected using [11C] PiB positron emission tomography and neuropathologic assessment in older adults. Arch Neurol 68:232–240
32. Fagan AM, Xiong C, Jasielec MS et al (2014) Longitudinal change in CSF biomarkers in autosomal-dominant Alzheimer's disease. Sci Transl Med 6:226ra30
33. Leuzy A, Chiotis K, Hasselbalch SG et al (2016) Pittsburgh compound B imaging and cerebrospinal fluid amyloid-β in a multicentre European memory clinic study. Brain 139:2540–2553
34. Chételat G, La Joie R, Villain N et al (2013) Amyloid imaging in cognitively normal individuals, at-risk populations and preclinical Alzheimer's disease. Neuroimage Clin 2:356–365
35. Price JL, Morris JC (1999) Tangles and plaques in nondemented aging and "preclinical" Alzheimer's disease. Ann Neurol 45:358–368
36. Albert MS, DeKosky ST, Dickson D et al (2011) The diagnosis of mild cognitive impairment due to Alzheimer's disease: recommendations from the national institute on aging-Alzheimer's association workgroups on diagnostic guidelines for Alzheimer's disease. Alzheimers Dement 7:270–279
37. Johnson KA, Minoshima S, Bohnen NI et al (2013) Appropriate use criteria for amyloid PET: a report of the amyloid imaging task force, the society of nuclear medicine and molecular imaging, and the Alzheimer's association. Alzheimers Dement 9:E1–E16
38. Okello A, Koivunen J, Edison P et al (2009) Conversion of amyloid positive and negative MCI to AD over 3 years: an 11C-PIB PET study. Neurology 73:754–760
39. Salloway S, Sperling R, Fox NC et al (2014) Two phase 3 trials of Bapineuzumab in mild-to-moderate Alzheimer's disease. N Engl J Med 370:322–333
40. Hertze J, Minthon L, Zetterberg H et al (2010) Evaluation of CSF biomarkers as predictors of Alzheimer's disease: a clinical follow up study of 4.7 years. J Alzheimers Dis 21:1119–1128
41. Bertens D, Knol DL, Scheltens P, Visser PJ (2015) Temporal evolution of biomarkers and cognitive markers in the asymptomatic, MCI, and dementia stage of Alzheimer's disease. Alzheimers Dement 11:511–522
42. Shulman MB, Harkins K, Green RC, Karlawish J (2013) Using AD biomarker research results for clinical care: a survey of ADNI investigators. Neurology 81:1114–1121
43. Leuzy A, Zimmer ER, Heurling K et al (2014) Use of amyloid PET across the spectrum of Alzheimer's disease: clinical utility and associated ethical issues. Amyloid 21:143–148
44. Laforce R, Rabinovici GD (2011) Amyloid imaging in the differential diagnosis of dementia: review and potential clinical applications. Alzheimers Res Ther 3:11
45. Blennow K, Dubois B, Fagan AM et al (2015) Clinical utility of cerebrospinal fluid biomarkers

in the diagnosis of early Alzheimer's disease. Alzheimers Dement 11:58–69

46. Caselli R, Woodruff B (2016) Clinical impact of amyloid positron emission tomography—is it worth the cost? JAMA Neurol 73:1396–1398
47. Sabbagh MN, Cooper K, DeLange J et al (2010) Functional, global and cognitive decline correlates to accumulation of Alzheimer's pathology in MCI and AD. Curr Alzheimer Res 7:280–286
48. Nelson PT, Alafuzoff I, Bigio EH et al (2012) Correlation of Alzheimer disease neuropathologic changes with cognitive status: a review of the literature. J Neuropathol Exp Neurol 71:362–381
49. Blennow K, Hampel H (2003) CSF markers for incipient Alzheimer's disease. Lancet Neurol 2:605–613
50. Jack CR, Hampel HJ, Universities S et al (2016) A/T/N: an unbiased descriptive classification scheme for Alzheimer disease biomarkers. Neurology 87:539–547
51. Tapiola T, Alafuzoff I, Herukka S-K et al (2009) Cerebrospinal fluid β-amyloid 42 and tau proteins as biomarkers of Alzheimer-type pathologic changes in the brain. Arch Neurol 66:734–746
52. Villemagne VL, Fodero-Tavoletti MT, Masters CL, Rowe CC (2015) Tau imaging: early progress and future directions. Lancet Neurol 14:114–124
53. Braak H, Braak E (1995) Staging of Alzheimer's disease-related neurofibrillary changes. Neurobiol Aging 16:271–278. discussion 278–284
54. Johnson KA, Schultz A, Betensky RA et al (2016) Tau positron emission tomographic imaging in aging and early Alzheimer disease. Ann Neurol 79:110–119
55. Koopman K, Le Bastard N, Martin J, Nagels G (2009) Improved discrimination of autopsy-confirmed Alzheimer's disease (AD) from non-AD dementias using CSF P-tau 181P. Neurochem Int 55:214–218
56. Maddalena A, Papassotiropoulos A (2003) Biochemical diagnosis of Alzheimer disease by measuring the cerebrospinal fluid ratio of phosphorylated tau protein to β-amyloid peptide42. Arch Neurol 60:1202–1206
57. Harada R, Okamura N, Furumoto S et al (2016) 18F-THK5351: a novel PET radiotracer for imaging neurofibrillary pathology in Alzheimer's disease. J Nucl Med 57:208–214
58. Pontecorvo MJ, Devous MD Sr, Navitsky M et al (2017) Relationships between flortaucipir PET tau binding and amyloid burden, clinical diagnosis, age and cognition. Brain 140:748–763
59. Gauthier S, Feldman HH, Schneider LS et al (2016) Efficacy and safety of tau-aggregation inhibitor therapy in patients with mild or moderate Alzheimer's disease: a randomised, controlled, double-blind, parallel-arm, phase 3 trial. Lancet 388:2873–2884
60. Yanamandra K, Jiang H, Mahan TE et al (2015) Anti-tau antibody reduces insoluble tau and decreases brain atrophy. Ann Clin Transl Neurol 2:278–288
61. Ng KP, Pascoal TA, Mathotaarachchi S et al (2017) Monoamine oxidase B inhibitor, selegiline, reduces 18 F-THK5351 uptake in the human brain. Alzheimers Res Ther 9:25
62. Vermeiren C, Mercier J, Viot D et al (2015) T807, a reported selective tau tracer, binds with nanomolar affinity to monoamine oxidase a. Alzheimers Dement 11:P283
63. Mosconi L (2013) Glucose metabolism in normal aging and Alzheimer's disease: methodological and physiological considerations for PET studies. Clin Transl Imaging 1:217–233
64. Ost M, Nylén K, Csajbok L et al (2006) Initial CSF total tau correlates with 1-year outcome in patients with traumatic brain injury. Neurology 67:1600–1604
65. Frisoni GB, Fox NC, Jack CR et al (2010) The clinical use of structural MRI in Alzheimer disease. Nat Rev Neurol 6:67–77
66. Hansson O, Zetterberg H, Buchhave P et al (2006) Association between CSF biomarkers and incipient Alzheimer's disease in patients with mild cognitive impairment: a follow-up study. Lancet Neurol 5:228–234
67. Skillbäck T, Farahmand BY, Rosén C et al (2015) Cerebrospinal fluid tau and amyloid-$\beta_{1\text{-}42}$ in patients with dementia. Brain 138:2716–2731
68. Mosconi L, Berti V, Glodzik L et al (2010) Pre-clinical detection of Alzheimer's disease using FDG-PET, with or without amyloid imaging. J Alzheimers Dis 20:843–854
69. Foster NL, Heidebrink JL, Clark CM et al (2007) FDG-PET improves accuracy in distinguishing frontotemporal dementia and Alzheimer's disease. Brain 130:2616–2635
70. Scheltens P, Leys D, Barkhof F et al (1992) Atrophy of medial temporal lobes on MRI in "probable" Alzheimer's disease and normal ageing: diagnostic value and neuropsychological correlates. J Neurol Neurosurg Psychiatry 55:967–972
71. Fotuhi M, Do D, Jack C (2012) Modifiable factors that alter the size of the hippocampus with ageing. Nat Rev Neurol 8:189–202

Chapter 17

The Ethics of Biomarker-Based Preclinical Diagnosis of Alzheimer's Disease

Alexander F. Kurz and Nicola T. Lautenschlager

Abstract

In older adults who are developing Alzheimer-typical pathology but do not experience any symptoms—a condition termed "preclinical" Alzheimer's disease—the individual prognostic value of current biomarkers is low. Only a minority of those who have positive findings on biomarkers in the cerebrospinal fluid or upon brain imaging experience cognitive decline within 3–5 years. On the other hand, the majority of people who have negative biomarker results remain cognitively intact. This may be expected in research settings where alternative causes of cognitive impairment are infrequent. The main reasons for the poor predictive performance of biomarkers in asymptomatic individuals are the dynamic nature of the assessments with abnormality evolving over time and the large interindividual variability of the threshold for clinical manifestation. We do not recommend biomarker assessment and disclosure of biomarker results to asymptomatic individuals outside of research protocols or clinical trials at this point in time. More needs to be learned about the interplay between biomarkers, risk factors, and protective factors. Also, studies are required with extended follow-up intervals on less selected participant groups to determine the long-term predictive potential of biomarkers in presymptomatic Alzheimer's disease.

Key words Alzheimer's disease, Biomarkers, Preclinical, Ethics

1 Introduction

The clinical symptoms of Alzheimer's disease (AD) including mild cognitive impairment and dementia are late manifestations of the neurodegeneration. It is believed that the pathological changes begin decades before the diagnosis of AD can be established on clinical grounds [1, 2]. The stage of the neurodegeneration when pathological changes accumulate but are not clinically expressed is termed "preclinical AD." The subsequent stage of "prodromal AD" is characterized by subtle deficits in cognitive areas such as attention, information processing speed, working memory, episodic memory, and executive function, often labeled as "mild cognitive impairment" (MCI) [3]. The prodromal stage precedes the clinical diagnosis of dementia for several years [4]. In parallel,

Robert Perneczky (ed.), *Biomarkers for Preclinical Alzheimer's Disease*, Neuromethods, vol. 137,
https://doi.org/10.1007/978-1-4939-7674-4_17, © Springer Science+Business Media, LLC 2018

it has been shown that key features of the pathology including deposition of β-amyloid, aggregation of tau, metabolic dysfunction of neurons, and neuronal loss can be identified by measurements in the cerebrospinal fluid (CSF), positron emission tomography (PET), and, at a more advanced stage, magnetic resonance imaging (MRI) independently of clinical symptoms. These assessments are collectively referred to as "biomarkers." An additional class of diagnostic indicators is genetic tests for rare pathogenic mutations. Since the latter reveal the predisposition for developing the disease rather than the presence of brain changes, they are usually not included in the group of biomarkers. A large number of drug trials conducted during the past 20 years have uniformly shown that at the stage of overt clinical symptoms, treatment effects are disappointing. One of the explanations for this is that the damage to neuronal networks has advanced too far. Taken together, these observations have given rise to the hope that intervening with the neurodegenerative process at the preclinical or prodromal stage might have a greater impact on the evolution of pathology, on cognitive and functional decline, and ultimately on the development of dementia and disability. This novel scenario of AD management would require the combination of excellent diagnostic services, precise biomarkers, and effective disease-modifying treatments. In this chapter we focus on with the ethical issues that are associated with the assessment of biomarkers in asymptomatic individuals.

2 Methods

We conducted a selective literature search using the databases CINAHL, EMBASE, MEDLINE, and PubMed and the search terms Alzheimer's disease, biomarker, early diagnosis, and ethics.

3 Results

3.1 What Kind of Information Do Biomarkers Provide?

Two types of biomarkers can be distinguished: one category indicating β-amyloid pathology (low β-amyloid concentration in the CSF and evidence of β-amyloid deposits on imaging) and another category demonstrating neuronal injury and neurodegeneration (high total tau or phosphorylated tau in the CSF, glucose hypometabolism in typical locations, accumulation of tau, and hippocampal atrophy on imaging). When using and interpreting biomarkers, particularly in asymptomatic people, their specific nature and associated problems need to be observed.

- Biomarkers of both categories capture different single elements of AD neurodegeneration [5, 6]. These features are not specific of AD but also occur in other brain diseases such as

Lewy body disease, frontotemporal degenerations, Creutzfeldt-Jakob disease, and traumatic brain injury and even in old age itself [7–9]. Therefore, biomarkers should not be looked at in isolation.

- Biomarker findings have been found to evolve in a temporal sequence. In most cases, indicators of β-amyloid pathology are the first to become abnormal [10–12]. However, neurodegeneration in terms of atrophy of the mediotemporal lobe may occur in the absence of detectable β-amyloid accumulation [13, 14]. This implies that AD does not always present the complete "signature," and findings may be conflicting [15]. No algorithm has been developed yet to deal with this situation [16].
- Conflicting results may not only be encountered between but also within biomarker categories. The concordance of assessments is not perfect, and the biomarkers of one category are not interchangeable. This is particularly true for the indicators of neuronal injury and neurodegeneration [17–21].
- There is no close association between the amount of pathological changes in the brain and the severity of clinical symptoms. This especially applies to β-amyloid burden. The clinical manifestation of pathology is counterbalanced by resilience or "reserve" factors such as education, occupational attainment, lifetime mental activity, and other factors [22, 23]. The relationship between these factors and the biomarkers is largely unexplored [24].

For these reasons, information provided by biomarkers is particularly difficult to interpret in asymptomatic people. In individuals who already experience symptoms, e.g., are diagnosed with MCI, normal biomarker findings make the presence of AD pathology less likely and point to an alternative cause of cognitive deterioration [25]. Conversely, abnormal biomarker results support the diagnosis of prodromal AD [26, 27] and predict progression to dementia with high accuracy [19, 28]. The scenario in asymptomatic individuals is different. Here, normal biomarker findings may indicate that there is no brain disease and no increased risk of cognitive decline and dementia. However, results will also be normal if the pathological changes are still below detection threshold. The biomarkers may become abnormal several years later, and clinical deterioration will eventually occur. Abnormal biomarker results in asymptomatic individuals are associated with an increased risk of future cognitive decline [29], particularly if they refer to both categories [30]. However, they do not predict whether or when the clinical manifestations of AD will develop during life. Thus, in asymptomatic individuals, even if they harbor the disease, a much wider interindividual variability can be expected regarding the distance to clinical manifestation than in people who already have symptoms, and therefore predictions of cognitive deterioration will necessarily be uncertain.

Predictive testing (i.e., in asymptomatic individuals) for pathogenic mutations in the amyloid precursor and presenilin genes differs from CSF and imaging biomarkers in several respects. It affects a small minority of individuals, since only 1–5% of AD follows an autosomal dominant pattern of inheritance [31]. The results of these tests are invariable over time and have a very high predictive value. Moreover, they have implications not only for the tested individuals but also for their blood-related family members, as their disease risk will increase dramatically by positive findings. The ε4 variant of the apolipoprotein E gene is a susceptibility gene. Testing has neither diagnostic nor predictive relevance and therefore is not recommended for clinical practice [32].

3.2 How Well Do Biomarkers Predict Cognitive Decline?

Among cognitively healthy older adults, 25–45% have abnormal CSF or imaging biomarker values [33]. Amyloid positivity on imaging increases with age from 5% in those 50–60 years to 46% in people aged 81 years and older [34]. Abnormal CSF biomarker results are associated with an increased likelihood of cognitive decline within a follow-up interval of 3 years. However, even in a research setting, using the best biomarker combination and optimal cutoff values, individual prediction only achieved 65% accuracy, with sensitivity and specificity values being 74% and 57%, respectively. These figures indicate that one third of cognitive decliners were not identified, while almost half of the predictions were false positive [35]. In a longitudinal study at Washington University, cognitively healthy adults over the age of 65 years who were positive for CSF ß-amyloid showed clinical deterioration over 5 years in only 13%. If they also had a positive biomarker for neuronal injury, the progression rate increased to 25% [1]. Thus, a considerable number of healthy older individuals with abnormal CSF biomarker findings will not experience cognitive deterioration within the next couple of years. Similarly, abnormal findings on ß-amyloid imaging are generally associated with greater decline in memory and other domains and indicate an increased risk of dementia on a group level [36]. However, such results do not predict with certainty if and when someone will develop clinical symptoms [37–40]. In the Australian Imaging, Biomarkers and Lifestyle (AIBL) study, only 26% of cognitively healthy individuals who had an AD-like result on ß-amyloid imaging progressed to MCI or dementia within 3 years. If a positive ß-amyloid scan was combined with reduced hippocampal volume on MRI, the percentage of cognitive decliners increased to 47%. Among participants with normal imaging findings in both modalities, 92% remained cognitively intact [41]. These findings show that within a follow-up period of several years, less than half of cognitively healthy older adults who have abnormal biomarker findings will experience cognitive decline. Most but not all of those who have normal biomarker results are likely to remain cognitively stable. This may be expected in research settings where the a priori probability of alternative causes of cognitive decline is low. Studies on less selected participants with longer-follow-up intervals are needed to determine the longer-term predictive value of biomarkers in asymptomatic individuals.

3.3 What Are the Benefits of Biomarker Information?

The greatest benefit of biomarker assessment in an asymptomatic individual would be if, in the case of abnormal findings, a treatment could be initiated which delays the progression of disease and postpones the onset of cognitive impairment. However, such interventions are currently not available. A potential benefit may be seen in the participation in a clinical trial which evaluates the efficacy of a treatment for secondary prevention. Due to high costs and methodological difficulties, we expect that only few studies of this kind, e.g., the A4 trial, will be conducted and also will be an option only for a small minority due to strict inclusion and exclusion criteria. Moreover, potential participants need to be aware that treatments in drug trials are experimental and may not be effective [42]. An important benefit for the worried well, particularly for family members of people who have or have had dementia, is the assurance they might get from negative biomarker findings that they are not at risk. However, due to the dynamic nature of the biomarkers, normal findings may not remain normal over time. Thus, in asymptomatic individuals, the benefit of normal CSF and imaging biomarker results would only materialize upon repeated assessment, which appears to be an unlikely scenario. In the absence of any preventive intervention, the benefits of biomarker determination may refer to future planning and lifestyle modification. Positive biomarker findings might promote providing wills and advance directives or affect financial planning and decisions whether to continue working or when and where to retire [43]. They may also stimulate people to pursue a healthier lifestyle or engage in risk-lowering behaviors such as enhanced cognitive or physical activity [16]. However, one could take the position that these steps should be part of everyone's plan for their later years whether or not there is an increased likelihood of cognitive decline [44].

3.4 What Are the Risks Associated with Biomarker Information?

There are two kinds of risks associated with biomarker assessment. The first kind are minor adverse effects that may be associated with the assessment procedure, i.e., the invasiveness of lumbar puncture for CSF measurements, and the radiation exposure associated with PET scanning. The second kind of risk is related to the information provided by the biomarker results. In the absence of preventive or treatment options, this information is a potentially harmful knowledge. It may cause psychological distress including anxiety, depression or even suicidal ideation, disruption of family dynamics, and worries about loss of or failure to obtain insurance or other social or economical benefits such as employment [45, 46] or lead to social stigmatization. Out of fear, presymptomatic persons may also be more vulnerable to therapeutic misconception and might falsely expect a therapeutic benefit of biomarker research [47]. Higher education has been found to be associated with consideration of suicide in response to (nongenetic) biomarker information [48]. Negative psychological outcomes are particularly likely if people have the misconception that biomarker findings provide a diagnosis and not only risk information, without predicting if or when cognitive deterioration will occur. Other risks of biomarker information include stigmatization and misuse of the information by third parties.

3.5 Whether and How Should Biomarker Results Be Disclosed?

Asymptomatic individuals who are interested in an assessment of AD biomarkers need to know that CSF indicators and β-amyloid imaging provide information about the presence of amyloid deposition in the brain which is associated with an increased risk of cognitive deterioration and ultimate development of dementia, but is frequent in old age and occurs in several brain disorders such as Lewy body disease, and thus does not establish a diagnosis [29]. In like manner, elevated CSF levels of tau and increased tau PET tracer take-up indicate an ongoing neurodegeneration but do not distinguish between the different causes [49]. Applicants for a biomarker evaluation should also consider the dynamic nature of these indicators. A negative or indetermined result at one time might turn into a positive result a few years later. Moreover, because of the individually variable threshold of clinical manifestation which may be related to brain reserve, these biomarkers cannot tell whether and when cognitive impairment will eventually occur even in people where evidence for AD pathology is found. In contrast, the rare genetic mutations are associated with an almost 100% risk of developing the pathology and clinical manifestations of AD, but there is also considerably interindividual variation of onset and symptomatology [31, 50]. In light of these limitations, asymptomatic subjects need to weigh potential benefits and risks of obtaining biomarker information. In research settings, disclosure of the apolipoprotein E genotype [51] or of amyloid PET results provoked little change in levels of depressive, anxiety, and stress symptoms. Reported risk of self-harm [52] and catastrophic reactions following such disclosure are rare [53]. However, these findings may not apply to biomarker findings which have a much greater impact on individual prognosis. An international working group recommends not to disclose biomarker information to asymptomatic individuals, except when well-informed subjects request the information, in cases of high social responsibility (e.g., pilots, bus drivers) or in case of inclusion in research protocols or clinical trials [2]. Regarding predictive genetic testing, most persons at risk for a dominantly inherited mutation do not want to know their risk marker status [54], and very few actually choose to be tested [55].

Predictive genetic testing should include pre- and posttest counseling as well as baseline neurologic and neuropsychological or psychiatric evaluation [56, 57].

3.6 What Will Change with the Advent of Disease-Modifying Treatment?

Novel treatments are currently being evaluated which address the amyloid and tau components of AD. Examples are passive or active immunization which inhibits the aggregation of the two proteins and/or removes the proteins from the brain or enzyme blockers that reduce the production of the proteins. Expected efficacy is a slowing of neurodegeneration, and the hope is that early use of these treatments will delay the onset of cognitive impairment and disability in daily living. Biomarkers will then be the gatekeepers to treatment and determine the access to treatment. However, these novel treatments are not without side effects and most probably will not be tolerated by everyone. In addition, the indication (when to start treatment, when to stop) and reimbursement by health insurances is unknown to date. Furthermore, a proportion of people who might be eligible for treatment may not wish to be treated, balancing the benefit of postponing the onset of symptoms in an uncertain future with the procedural burden and cost of treatment [58]. It is therefore likely that only a minority of people who are at risk, i.e., who have no symptoms but ongoing neurodegeneration, will access these novel treatments. The majority of those tested positively for the presence of AD pathology will not gain access to disease-slowing treatment but will have the information of a fatal brain disease. For them and their families, practical coping strategies and concrete problem solutions need to be delineated [47].

4 Discussion

In the field of AD research, biomarkers can be viewed as reflections of brain pathology. In asymptomatic individuals their major role is to predict or exclude future cognitive decline and development of dementia. However, the prognostic potential of current assessments is low, and they must not be interpreted mechanistically. People with positive biomarker findings may remain free of symptoms for an extended period of time, because there is a wide interindividual variability regarding the threshold of clinical manifestation. Negative biomarker results may convey false certainty, because they gradually evolve over time, and findings may change upon repeated measurement. To achieve a better estimation of cognitive decline in asymptomatic individuals who are interested in learning about their prognosis, the role of co-pathologies including vascular abnormalities and depression, the contribution of risk factors, but also the impact of protective factors such as brain reserve [59] need to be better understood [24, 60, 61]. Since there are no pharmacological interventions to date that would slow the neurodegeneration and postpone the onset of cognitive and functional impairment, the potential benefits of biomarker assessment are limited to lifestyle modification. However, most changes regarding healthier living, financial planning, or advance directives should be part of everybody's life plan [44, 48]. On the other hand, biomarker assessments are associated with significant risks, primarily regarding the potentially negative psychological, social, and economical repercussions of the information provided. The potentially negative effects beyond the emotional burden are related to the impact on such issues as insurance and employment. We assume that the advent of novel disease-modifying treatments will change the risk-benefit ratio of biomarker assessment only for a minority of asymptomatic individuals. Most probably, access to these treatments will be limited, costs will be high, and many will decide against participation in view of a significant procedural burden, long treatment duration, or significant side effects. Furthermore, there will be a large group of individuals who cannot be treated because of contraindications or poor tolerance. They will not have the benefit of the intervention but only the burden of knowing their risk.

In summary, at this point in time, we do not recommend biomarker assessment and disclosure of biomarker results to asymptomatic individuals outside of research protocols or clinical trials. If in research contexts biomarker assessments are employed, participants need to be comprehensively informed about the nature of the information they might receive and about the consequences this information possibly has for their future planning, and they should be given the choice of knowing or not knowing the biomarker results.

References

1. Vos SJB, Xiong C, Visser PJ et al (2013) Preclinical Alzheimer's disease and its outcome: a longitudinal cohort study. Lancet Neurol 12:957–965
2. Dubois B, Hampel H, Feldman HH et al (2016) Preclinical Alzheimer's disease: definition, natural history, and diagnostic criteria. Alzheimers Dement 12:292–323
3. Petersen RC (2011) Mild cognitive impairment. N Engl J Med 364:2227–2234
4. Ritchie K, Carrière I, Berr C et al (2016) The clinical picture of Alzheimer's disease in the decade before diagnosis: clinical and biomarker trajectories. J Clin Psychiatry 77:e305–e311
5. Roberts JS, Dunn LB, Rabinovici GD (2013) Amyloid imaging, risk disclosure and Alzheimer's disease: ethical and practical issues. Neurodegener Dis Manag 3:219–229
6. Witte MM, Foster NL, Fleisher AS et al (2015) Clinical use of amyloid-positron emission tomography neuroimaging: practical and bioethical considerations. Alzheimers Dement 1:358–367
7. Jack CR, Albert M, Knopman DS et al (2011) Introduction to revised criteria for the diagnosis of Alzheimer's disease: national institute on aging and the Alzheimer association workgroups. Alzheimers Dement 7:257–262
8. Arlt S (2013) Non-Alzheimer's disease–related memory impairment and dementia. Dialogues Clin Neurosci 15:465–473
9. Villemagne VL, Burnham S, Bourgeat P et al (2013) Amyloid beta deposition, and cognitive decline in sporadic Alzheimer's disease: a prospective cohort study. Lancet Neurol 12:357–367

10. Han SD, Gruhl J, Beckett L et al (2012) Beta amyloid, tau, neuroimaging, and cognition: sequence modeling of biomarkers for Alzheimer's disease. Brain Imaging Behav 6:610–620
11. Jack CR, Knopman DS, Jagust WJ et al (2013) Tracking pathophysiological processes in Alzheimer's disease: an updated hypothetical model of dynamic biomarkers. Lancet Neurol 12:207–216
12. Sutphen CL, Jasielec MS, Shah AR et al (2015) Longitudinal cerebrospinal fluid biomarker changes in preclinical Alzheimer disease during middle age. JAMA Neurol 72:1029–1042
13. Wirth M, Villeneuve S, Haase CM et al (2013) Association between Alzheimer disease biomarkers, neurodegeneration, and cognition in cognitively normal older people. JAMA Neurol 70:1512–1519
14. Knopman DS, Jack CR, Lundt ES et al (2015) Role of beta-amyloidosis and neurodegeneration in subsequent changes in mild cognitive impairment. JAMA Neurol 72:1475–1483
15. Alexopoulos P, Werle L, Roesler J et al (2016) Conflicting cerebrospinal fluid biomarkers and progression to dementia due to Alzheimer's disease. Alzheimers Res Ther 8:51
16. Porteri C, Frisoni GB (2014) Biomarker-based diagnosis of mild cognitive impairment due to Alzheimer's disease: how and what to tell. A kickstart to an ethical discussion. Front Aging Neurosci 6:41
17. Landau SM, Lu M, Joshi AD et al (2013) Comparing PET imaging and CSF measurements of A beta. Ann Neurol 74:826–836
18. Toledo JB, Weiner MW, Wolk DA et al (2014) Neuronal injury biomarkers and prognosis in ADNI subjects with normal cognition. Acta Neuropathol Commun 2:26
19. Leuzy A, Carter SF, Chiotis K et al (2015) Concordance and diagnostic accuracy of 11C PIB PET and cerebrospinal fluid biomarkers in a sample of patients with mild cognitive impairment and Alzheimer's disease. J Alzheimers Dis 45:1077–1088
20. Leuzy A, Chiotis K, Hasselbalch SG et al (2016) Pittsburgh compound B imaging and cerebrospinal fluid amyloid-beta in a multicentre European memory clinic study. Brain 139:2540–2553
21. Vos SJB, Gordon BA, Su Y et al (2016) NIAA-AA staging of preclinical Alzheimer disease: discordance and concordance of CSF and imaging biomarkers. Neurobiol Aging 44:1–8
22. Valenzuela MJ (2008) Brain reserve and the prevention of dementia. Curr Opin Psychiatry 21:296–302
23. Murray AD, Staff RT, McNeil CJ et al (2011) The balance between cognitive reserve and brain imaging biomarkers of cerebrovascular and Alzheimer's diseases. Brain 134:3687–3696
24. Almeida RP, Schultz SA, Austin BP et al (2015) Cognitive reserve modifies age-related alterations in CSF biomarkers of Alzheimer's disease. JAMA Neurol 72:699–706
25. Doraiswamy PM, Sperling RA, Johnson K et al (2014) Florbetapir F 18 amyloid PET and 36-month cognitive decline: a prospective multicenter study. Mol Psychiatry 19:1044–1051
26. McKhann GM, Knopman DS, Chertkow H et al (2011) The diagnosis of dementia due to Alzheimer's disease: recommendations from the national institute on aging and the Alzheimer's association workgroup. Alzheimers Dement 7(3):263–269
27. Janelidze S, Zetterberg H, Mattsson N et al (2016) CSF Abeta 42/Abeta 40 and Abeta42/−abeta38 ratios: better diagnostic markers of Alzheimer's disease. Ann Clin Transl Neurol 3:154–165
28. Counts SE, Ikonomovic MD, Mercao N et al (2017) Biomarkers for the early detection and progression of Alzheimer's disease. Neurotherapeutics 14:35–53
29. Chételat G, La Joie R, Villain N et al (2013) Amyloid imaging in cognitively normal individuals, at-risk populations and preclinical Alzheimer's disease. NeuroImage Clin 2:356–365
30. Soldan A, Pettigrew C, Cai Q et al (2016) Hypothetical preclinical Alzheimer disease groups and longitudinal cognitive change. JAMA Neurol 73:698–705
31. Goldman JS (2012) New approaches to genetic counseling and testing for Alzheimer's disease and frontotemporal degeneration. Curr Neurol Neurosci Rep 12:502–510
32. Hauser PS, Ryan RO (2013) Impact of apolipoprotein E on Alzheimer's disease. Curr Alzheimer Res 10:809–817
33. Randall C, Mosconi L, de Leon M et al (2014) Cerebrospinal fluid biomarkers of Alzheimer's disease in cognitively healthy elderly. Front Biosci 18:1150–1173
34. Fleisher AS, Chen K, Liu X et al (2013) Apolipoprotein E e4 effects on florbetapir positron emission tomography in healthy aging and Alzheimer disease. Neurobiol Aging 34:1–12
35. Steenland K, Zhao L, Goldstein F et al (2014) Biomarkers for predicting cognitive decline in those with normal cognition. J Alzheimers Dis 40:587–594

36. Johnson KA, Minoshima S, Bohnen NI et al (2013) Appropriate use criteria for amyloid PET: a report of the amyloid imaging task force, the society of nuclear medicine and molecular imaging, and the Alzheimer's association. Alzheimers Dement 9: e-1–e-15
37. Morris JC, Roe CM, Grant EA et al (2009) PIB imaging predicts progression from cognitively normal to symptomatic Alzheimer's disease. Arch Neurol 66:1469–1475
38. Villemagne VL, Pike KE, Chételat G et al (2011) Longitudinal assessment of Abeta and cognition in aging and Alzheimer disease. Ann Neurol 69:181–192
39. Harkins K, Sankar P, Sperling R et al (2015) Development of a process to disclose amyloid imaging results to cognitively normal older adult research participants. Alzheimers Res Ther 7:26
40. Rabinovici GD, Karlawish J, Knopman D et al (2016) Testing and disclosure related to amyloid imaging and Alzheimer's disease: common questions and fact sheet summary. Alzheimers Dement 12:510–515
41. Rowe CC, Bourgeat P, Ellis KA et al (2013) Predicting Alzheimer disease with beta-amyloid imaging: results from the Australian imaging, biomarkers, and lifestyle study of ageing. Ann Neurol 74:905–913
42. Howe E (2013) Clinical implications of the new diagnostic guidelines for dementia. Innov Clin Neurosci 10:32–38
43. Baum ML (2016) Patient requests for off-label bioprediction of dementia. Camb Q Healthc Ethics 25:686–690
44. Gordon M (2013) Identification of potential or preclinical cognitive impairment and the implications of sophisticated screening with biomarkers and cognitive testing: does it really matter? Biomed Res Int 2013:976130
45. Karlawish J (2011) Addressing the ethical, policy, and social challenges of preclinical Alzheimer disease. Neurology 77: 1487–1493
46. Kim SYH, Karlawish J, Berkman BE (2015) Ethics of genetic and biomarker test disclosures in neurodegenerative disease prevention trials. Neurology 84:1488–1494
47. Schicktanz S, Schweda M, Ballenger JF et al (2014) Before it is too late: professional responsibilities in late-onset Alzheimer's research and pre-symptomatic prediction. Front Hum Neurosci 8:921
48. Caselli RJ, Langbaum J, Marchant GE et al (2014) Public perceptions of presymptomatic testing for Alzheimer's disease. Mayo Clin Proc 89:1389–1396
49. Villemagne VL, Fodero-Tavoletti MT, Masters CL et al (2015) Tau imaging: early progress and future directions. Lancet Neurol 14:114–124
50. Tanzi RE (2012) The genetics of Alzheimer disease. Cold Spring Harb Perspect Med 2:a006296
51. Bemelmans SASA, Tromp K, Bunnik EM et al (2016) Psychological, behavioral and social effects of disclosing Alzheimer's disease biomarkers to research participants: a systematic review. Alzheimers Res Ther 8:46
52. Lim YY, Maruff P, Getter C et al (2016) Disclosure of positron emission tomography amyloid imaging results: a preliminary study of safety and tolerability. Alzheimers Dement 12:454–458
53. Paulsen JS, Nance M, Kim JI et al (2013) A review of quality of life after predictive testing for and earlier identification of neurodegenerative diseases. Prog Neurobiol 110:2–28
54. Hooper M, Grill JD, Rodriguez-Agudelo Y et al (2013) The impact of the availability of prevention studies on the desire to undergo predictive testing in persons at-risk for autosomal dominant Alzheimer's disease. Contemp Clin Trials 36:256–262
55. Steinbart EJ, Smith CO, Poorkay P et al (2001) Impact of DNA testing for early-onset familial Alzheimer disease and frontotemporal dementia. Arch Neurol 58:1828–1831
56. Cassidy MR, Roberts JS, Bird TD et al (2008) Comparing test-specific distress of susceptibility versus deterministic genetic testing for Alzheimer's disease. Alzheimers Dement 4:406–413
57. Goldman JS, Rademakers R, Huey ED et al (2011) An algorithm for genetic testing of frontotemporal lobar degeneration. Neurology 76:476–483
58. Molinuevo JL, Cami J, Carné S et al (2016) Ethical challenges in preclinical Alzheimer's disease observational studies and trials: results of the Barcelona summit. Alzheimers Dement 12:614–622
59. Soldan A, Pettigrew C, Li S et al (2013) Relationship of cognitive reserve and CSF biomarkers to emergence of clinical symptoms in preclinical Alzheimer's disease. Neurobiol Aging 34:2827–2834
60. Donovan NJ, Hsu DC, Dagley AS et al (2015) Depressive symptoms and biomarkers of Alzheimer's disease in cognitively normal older adults. J Alzheimers Dis 46:63–73
61. Garrett MD, Valle R (2016) A methodological critique of the national institute of aging and Alzheimer's association guidelines for Alzheimer's disease, dementia, and mild cognitive impairments. Dementia 15:239–254

Chapter 18

Shared Decision-Making and Important Medical and Social Decisions in the Context of Early Diagnosis of Alzheimer's Disease

Katharina Bronner and Johannes Hamann

Abstract

Shared decision-making is a model of medical decision-making, which aims at reducing the informational and decisional power asymmetry between physician and patient to obtain jointly reached agreements on medical treatment. SDM can also be applied in the case of early stage of Alzheimer's disease, as long as individuals' with dementia decisional capacity still exists. Because of the loss of cognitive functions in the course of the disease, upcoming important issues like medical treatment and legal and social issues should be made as soon as possible after diagnosis, so that persons with dementia can be involved in the process of decision-making.

Key words Alzheimer's disease, Social aspects, Early diagnosis, Dementia, Shared decision-making

1 Introduction

In recent years many efforts have been undertaken to improve early diagnosis of Alzheimer's disease (AD). This issue of AD has raised awareness due to its high prevalence and incidence. In Germany, for example, there are 1.6 million individuals with dementia, thereof approximately 70% affected by Alzheimer's disease, and 300,000 new cases a year. In addition, demographic development with an increase of the elderly population will lead to an increase of persons affected, i.e., approximately 3 million in 2050 in Germany [1] and 131 [2] to 135 [3] million worldwide.

AD causes great expenditures that range from approximately 15,000 Euro per individual and year in the case of mild dementia to approximately 42,000 Euro in the case of severe dementia. About 75% of total cost origins from costs of care and nursing [4].

Thus, on the one side, dementia burdens persons with AD and their caregivers and families and on the other side society and healthcare systems, because enormous financial resources must be made available [2, 3], which are limited in a healthcare system based on

Robert Perneczky (ed.), *Biomarkers for Preclinical Alzheimer's Disease*, Neuromethods, vol. 137,
https://doi.org/10.1007/978-1-4939-7674-4_18, © Springer Science+Business Media, LLC 2018

social principles [5]. As a consequence persons with AD are threatened by a shortage of resources because of their higher age, nonexisting curative therapy, and lacking lobby. Additionally, dementia is rather stigmatized in the population because of the lacking chances of recovery. In general population AD is the best known, most frequent, and most feared neurodegenerative disorder [3]. In a survey of a German health insurance company, 51% of respondents were afraid of dementia [6].

Despite the discrepancy between already made progresses regarding molecular genetics, laboratory, neuropsychological and instrument-based diagnostics, and the still stagnant therapeutic possibilities, it is recommended to make the diagnosis of AD as soon as possible. One rationale behind this recommendation is that diagnosis is the basis for therapy and thorough examination of persons with cognitive decline may help detecting treatable reasons of dementia symptoms, for example, depression [7].

The other rationale is founded in the progressive cognitive decline throughout the course of AD. Thus, in later stages of AD persons with AD inevitably loose decisional capacity, making it reasonable to make important medical and social decisions early, as long as persons with AD themselves still have the ability to make them. These decisions include medical treatment, legal issues, and coping with illness [8]. "Shared decision-making" may be a feasible approach to facilitate joint decision-making between persons with dementia, their carers, and physicians [9].

2 Models of Medical Decision-Making

2.1 Four Models of Physician-Patient Relationship

In their seminal paper, Emanuel and Emanuel [10] describe four models of the doctor-patient relationship. In the *paternalistic model* patients get selected information and instructions from the physician, which they are expected to follow. The physician makes decisions about medical procedures like clinical diagnostics and therapy all alone. And there is no explicit patient involvement in decision-making. In the *informative model* the physician provides the patient with all relevant specialized information without giving an advice. The patient then comes to a decision all alone, which the physician is expected to implement. In the *interpretative model*, the physician provides the relevant information, inquires moral values and objectives, and proposes appropriate measures of therapy. The patient decides alone; however, the physician is involved in decision-making. The *deliberative model* assumes that the physician discusses the best treatment option with the patient. He knows about the living conditions and also the moral values and objectives of the patient. He tries to convince the patient to accept the best option, but if the patient disagrees, the physician doesn't try to overrule him. Decisions are made by the patient and physician together.

2.2 Shared Decision-Making

Shared decision-making (SDM) has been located right in the middle of the paternalistic and the informative model [11]. Already in 1997 four key features were postulated: (1) At least two persons are participating, physician and patient; (2) both sides share information with each other; (3) both sides take steps to actively engage in a process of decision-making; (4) both sides accept the decision and actively contribute to its implementation [9]. Since the 1990s enormous political efforts have been undertaken to foster research on and implementation of SDM in various healthcare systems [12–14].

Research has shown that SDM has the potential to improve patients' treatment satisfaction as well as their adherence with therapies [15, 16].

3 Decision-Making and Decision-Making Capacity

3.1 Decision-Making in Mental Illness

In mental health, patient participation and SDM have not yet been fully implemented, although the majority of individuals with mental illnesses express a clear wish for information and participation in medical decisions and although there is growing evidence that SDM may positively influence treatment outcomes [17]. Meanwhile several clinical guidelines explicitly recommend patient participation in all medical decisions for the majority of mental illnesses, such as depression or schizophrenia (e.g., National Institute for Health and Care Excellence (NICE) [18, 19]).

However, in mental health it may be a special challenge to engage individuals into medical decision-making, especially due to the cognitive limitations, lack of insight, and reduced decisional capacity that all may be a result of the mental illness [20].

3.2 Decision-Making in AD

The challenge of reduced decisional capacity is of special concern in AD, since AD is per definition a continually progressive disease with loss of cognitive functions, and as individuals' decision-making capacity decreases, the more the disease is progressing [21]. However, individual's decisional capacity for medical as well as social decisions can still be present in early stages of AD [22]. This pattern is also reflected in the guidelines of the National Institute for Health and Care Excellence (NICE) that on the one side emphasize the individual's autonomy as a major aim but on the other side judge the reduced cognitive capacity as a limiting factor for participating [23]. Therefore, clinical guidelines do not demand patient participation in decision-making for individuals with AD as explicitly as for other mental illnesses. Nevertheless self-determination and autonomy in general are important issues for persons with AD. In 2006 the US Alzheimer's Association founded an initiative with the objective of promoting public relation and lobby work for persons with early AD [24]. In Europe, Alzheimer Europe and its member organizations support autonomy and self-determination of persons with AD [25].

3.3 Decision-Making in Early-Stage Dementia

Individuals in early stages of dementia are in most cases still capable of making decisions [26] and actually want to have a say in decisions concerning their own treatment [27]. Thereby, participation preferences of persons with dementia were especially marked in decisions about social issues, for example, future housing or restitution of the driver's license. On the contrary, individuals with dementia prefer to hand over decisions about medical issues to their physicians. From the persons' with dementia point of view, relatives should play a minor part in medical as well as in social decisions [28]. It can be assumed, however, that individuals' participation in decision-making does not take place, because psychiatrists tend to exclude cognitively impaired individuals in decision-making [29], i.e., when they are worried about the decrease of decisional capacity in the progression of dementia [26, 30]. Additionally many persons with AD are prevented from the process of decision-making, because their relatives tend to take over control [27].

In most studies on decision-making in early-stage dementia, more emphasis was placed on the relationship between individuals with dementia and relatives and relatives and physicians than on the issue of shared decision between individuals with dementia and physicians [21, 31, 32].

3.4 Decision-Making at the Stage of MCI and at the Preclinical Stage of AD

For individuals at the stage of MCI and even more for individuals in a preclinical stage of AD, decisional capacity is in most cases not a problematic issue in decision-making. Here it is rather a complicated matter regarding the often debatable decisions (e.g., for or against excessive diagnostic testing) that can or should be made at these stages of the disease.

Thus, before dementia can be diagnosed, many individuals pass through a period of mild cognitive impairment (MCI) [33]. This diagnosis has actually no therapeutic consequences, because there is no medical indication for the use of the licensed antidementives for prevention at the stage of MCI and at the preclinical stage of AD, respectively [34, 35]. Moreover, there is an ongoing debate about whether persons without any symptom of dementia should participate in a predictive examination of AD by use of diagnostic methods of early detection [36]. Although studies have finally identified relevant cardiovascular risk factors for development of a dementia (smoking, hypertension, obesity, diabetes mellitus, and physical inactivity) [37, 38] and there are hints for protecting dementia by special nutrition [39] and sport activity [40], there are no specific recommendations for risk reduction of AD but only generally effective measures, which have a preventive effect on other diseases, too ("what is good for the heart, is good for the brain") [41]. Predictive diagnostics can also be adverse for individuals. Experiences with cancer screening have shown that early detection is in no way without side effects [42]. Using current methods there is, except the rare heritable form of AD [43], no individual prediction of 100%, but only statements regarding one's individual risk for developing AD [44]. So there may be false-positive or false-negative results [45] and therefore, because of the lack of therapeutic options, a risk of a worsening of the quality of life by developing anxieties and depression.

The German Alzheimer Society therefore does not recommend early detection in persons without any symptoms of dementia. By explicit request of individuals, there is a need for a competent medical consultation and a comprehensive education about the consequences by well-informed specialists. The wish for participation in clinical trials to support dementia research is an additional reason for an early detection [46].

4 Important Medical and Social Decisions After Early Diagnosis of Alzheimer's Disease

The diagnosis of early AD is a major turning point in the individual's and relative's life and often challenges the future life of whole families. Because of the progressive course with an inevitable loss of cognitive abilities including decisional capacity, important decisions should be made as soon as possible after diagnosis to ensure that persons with dementia can fully participate in decision-making, if they want to do so. For individuals with AD, spouses, and physicians, a wide range of medical and social decisions has been reported to be of importance [8]. Some of these decisions refer to the current situation of the individual with AD, while others rather deal with issues of the future. As relatives have a disposition to take over more and more responsibility and decisions, even if persons with AD still have decisional capacity [21], the decisions mentioned below should ideally be made within a joint decision process including persons with AD, their carers, and physicians.

4.1 Medical Treatment

Although there isn't an effective therapy yet, which can cure or at least stop the disease process, there are drugs with disease-delaying properties and drugs for accompanying symptoms such as depression, anxiety, and aggression [47]. It is also possible for persons with AD to participate in clinical trials. In addition, psychosocial interventions are available, which can increase individuals' with dementia and relatives' well-being and quality of life.

4.1.1 Drug Treatment

Currently there is no causal medical therapy for AD. There are hints that early drug treatment may have positive effects on the disease process [48]. Licensed antidementives are able to delay dementia symptoms for all stages of AD [49] and also have an effect on mental and behavioral symptoms [50] in some cases. However, over the course of time, all individuals with AD will develop severe dementia with the need for care, and, like all drugs, antidementives may have side effects, especially on the gastrointestinal system. On the part of the physicians, comprehensive information should take place about the potential benefits and side effects using a shared decision-making approach. In addition, the benefit-risk profile should be observed carefully, especially when prescribed for early stages of AD [7].

In addition to treatment with licensed drugs, persons with dementia may participate in clinical trials, which are performed at specialized and certificated centers. The European parent organization Alzheimer Europe has published information and a register of current clinical trials [51].

4.1.2 Psychosocial Interventions

Besides drug treatment psychosocial interventions are part of persons' with dementia and relatives' support in dementia treatment. This includes, for example, cognitive training, occupational therapy, physical activity, and art and sensory therapy. When making the decision whether or not to accept support with psychosocial therapies, providers and individuals with dementia are faced with conflicting evidence on these therapies because of the heterogeneous studies and partially poor study designs [52]. The effects measured are often small and not constant over several studies. Nevertheless, various psychosocial therapies were recommended, because they have shown beneficial effects [7, 23], for example, cognitive training for depression, cognition, and quality of life [53]; occupational therapy for apathy, quality of life, cognition [54], and functional decline [55]; and physical activity for basic life skills [56]. In addition, support for relatives with the aim of improving the individuals' with dementia situation yields effects regarding behavioral symptoms and depression [51]. Since mental and physical health of the caregivers is often affected [57], structured offers can help to prevent illnesses and reduce burden [58]. Persons with dementia and relatives, which are integrated in dementia networks, are more frequently in contact with physicians and get more frequently drug-free therapy and psychosocial interventions [59].

4.2 Legal Issues

Over the course of the disease, individuals with dementia lose their capacity to conclude legal transactions or to provide a declaration of will. Therefore, all legal issues that could be of importance in the early or later stages of AD should be discussed with persons with dementia as long as they are definitely capable to do so. Written powers and advance directives help persons of authority to make arrangements in lieu of the individual with dementia in later stages of the disease and avoid lengthy legal disputes. However, it is advisable to prove and certify the individual's with dementia ability to make a will even in early stages of the disease by a medical specialist, mostly a psychiatrist.

4.2.1 Healthcare Proxy, Last Will, Advance Healthcare Directive, and Advance Care Planning

After diagnosis of AD, it is the right time, according to experts, to settle legal issues [8], for example, to arrange the legacy; the healthcare proxy, which designates a person for arranging personal and economic affairs, if AD is progressing; and the advance healthcare directive, which among others defines the extent of medical procedures acceptable to the person with dementia. Because of the increased accident risk of dementia, it may be advisable to contract homeowner's insurance or to request a disabled person's pass, which in some countries includes financial relief [60] and advanced worker protection [61].

Again, last will, healthcare proxy, and advance healthcare directives can only be done by persons who are still capable to consent and who can sign personally.

In recent years the concept of advance care planning (ACP) [62] has gained more and more attention. It addresses persons in critical stages of illness and may remedy shortcomings of advance healthcare directives, which persons often feel overburdened to complete. Moreover, an advance healthcare directive is often vaguely formulated; it may not be available at the critical moment and therefore will be disregarded. ACP provides personal assistance and support in filling in individual forms. In addition an emergency guide is linked, which specifies how to act in a case of emergency [63].

4.2.2 Car Driving

Another important "legal" issue is car driving. Here the persons' with dementia wish for autonomy may conflict with his or her ability to keep driving motor vehicles. Although AD at early stages is not necessarily associated with the loss of capacity to drive [64], reaction capacity and speed estimation can be affected. In some European countries [65] there are obligatory regular medical examinations at a certain age, which confirm or reject the ability to drive. In other countries the responsibility is with persons concerned, their relatives, or their physicians. Here individuals with dementia should be informed by their physicians at the time of diagnosis that AD leads to loss of the driving ability during the course of the disease. In case of doubt, a medical-psychological assessment can give some guidance in the decision-making process. If AD progresses and individuals' ability to drive is impaired, it may be required that another person informs authorities to revoke the persons' with dementia driving license.

4.3 Social Issues and Coping with Illness

4.3.1 Coping with Illness

The progression of AD symptoms raises a lot of social issues. After diagnosis the question arises, how individuals with AD and their relatives will deal with the illness. Many persons have the desire to "carry on as always" [66] and not to further look into AD. This also touches the issue of informing the social environment about the disease, which individuals with dementia want to impart or not. Many individuals with dementia and relatives feel to have no reason to be concerned with future planning, as long as dementia symptoms aren't well pronounced [8], because they aren't fully aware of the importance of timely decisions [67]. Even GPs have problems to communicate to individuals with dementia the importance of future planning because of the absence of decisional capacity in the later course of the disease [68].

4.3.2 Assistance by Family Members and External Help

During the course of the disease, individuals with dementia lose more and more abilities and become dependent on support. In most cases, relatives take over the additional assistance needed, which is both expected and claimed from persons with dementia and is taken for granted from relatives, who adapt their living conditions for that. Both sides rather disapprove the possibility of external help. In the course of AD, many caring relatives become overburdened with this situation [8], have a worse quality of life, risk their own health, and cannot ensure adequate care with the best will in the world. Therefore professionals in dementia recommend to plan external help and to extend the social network as early as possible [69].

4.3.3 Living Conditions

The issue of keeping up living conditions (especially housing conditions) or changing them is also a difficult one, which from the point of view of professionals in dementia is often repressed and ignored. Most individuals with AD express the wish to stay in their familiar environment as long as possible. Their relatives confirm this and often promise that individuals shall stay at home even in the later stages of AD [8]. During the course of AD, living at home may, however, become more and more difficult, especially for persons with dementia living alone without families or lack of assistance by family. Therefore, persons with dementia and their relatives should timely be informed about alternative living conditions such as assisted community living or care homes for dementia.

5 Implementation of SDM in Early AD

To implement SDM for these decisions in the early course of AD, several challenges have to be overcome: First, the model of SDM has been shown to increase patient participation in decision-making, but it does not clearly address decision-making of triads (of patients, carers, and physicians) [70]. Second, the decisions to be made are complex, and—in contrast to "classical" decisions often described in decision aids—scientific evidence might neither be available nor helpful in case of most decisions (e.g., housing conditions). Third, some individuals with early AD might already exhibit cognitive deficits, making it necessary to account for this limitation. Finally, many individuals with dementia and their carers tend to postpone decisions. Any implementation strategy will therefore have to motivate them not to procrastinate.

We therefore suggest that a complex intervention consisting of elements of SDM (e.g., physicians' communicative competencies for SDM [71], decision aids [72] (mainly as an aid to memory), and advance care planning (i.e., a guide who helps patient-carer dyads through the decision-making process)) might be suited to the persons' with AD interests. Due to the complexity and quantity of decisions, a multi-session approach will be necessary. Thus, with the help of a decision guide and a decision aid containing the most important decision to be made, the different options and the potential pros and cons patient-carer dyads may be prepared for decisions or decisional talks with their physicians in charge.

6 Conclusions

Alzheimer's disease is a progressive disease which is related to high burden for the individuals with dementia, their relatives, and the society. In the course of the disease, persons with AD lose their decisional capacity and consequently the possibility to make decisions. In the early stages of AD, decisional capacity is, however, still present. Thus, shortly after diagnosis important decisions should be made to allow participation of the persons affected as long as they still have decisional capacity. However, many individuals with dementia postpone these decisions not to burden themselves or others. This often leads to decision-making in later stages of the disease without the individual with dementia, and decisions might therefore be not in line with the individuals' initial preferences. Shared decision-making (SDM) and advance care planning might be suitable approaches to foster early and joint decision-making of individuals with dementia, their carers, and physicians.

References

1. Deutsche Alzheimer Gesellschaft (2017) Die Häufigkeit der Demenzerkrankungen. https://www.deutsche-alzheimer.de/fileadmin/alz/pdf/factsheets/infoblatt1_haeufigkeit_demenzerkrankungen_dalzg.pdf. Accessed 25 April 2017
2. Prince MJ et al (2015) The world Alzheimer report 2015, The global impact of dementia: an analysis of prevalence, incidence, cost and trends. https://www.alz.co.uk/research/world-report-2015. Accessed 26 April 2017
3. WHO (2016) Dementia fact sheet April 2016. http://www.who.int/mediacentre/factsheets/fs362/en/. Accessed 26 April 2017
4. Leicht H et al (2011) Net costs of dementia by disease stage. Acta Psychiatr Scand 124:384–395
5. BMBF (2016) Rahmenprogramm der Gesundheitsforschung der Bundesregierung
6. Forsa (2013) .Angst vor Krankheiten. https://www.dak.de/dak/download/forsa-umfrage-demenz-1331362.pdf. Accessed 26 April 2017
7. Deutsche Gesellschaft für Psychiatrie, Psychotherapie und Nervenheilkunde (DGPPN); Deutsche Gesellschaft für Neurologie (Hrsg.) (2016) S3-Leitlinie "Demenzen". http://www.dgppn.de/fileadmin/user_upload/_medien/download/pdf/kurzversion-leitlinien/S3-LL-Demenzen-240116-1.pdf. Accessed 27 April 2017
8. Bronner K et al (2016) Which medical and social decision topics are important after early diagnosis of Alzheimer's disease from perspec-

tives of people with Alzheimer's disease, spouses and professionals? BMC Res Notes 9:149
9. Charles C et al (1997) Shared decision-making in the medical encounter: what does it mean? (or it takes at least two to tango). Soc Sci 44:681–692
10. Emanuel EJ, Emanuel LL (1992) Four models of the physician-patient relationship. JAMA 267:2221–2226
11. Charles C et al (1999) What do we mean by partnership in making decisions about treatment? BMJ 319:780–782
12. Coulter A et al (2017) Shared decision making in the UK: moving towards wider uptake. Z Evid Fortbild Qual Gesundhwes 123-124:99–103. https://doi.org/10.1016/j.zefq.2017.05.010
13. Spatz ES et al (2017) Shared decision making as part of value based care: new U.S. policies challenge our readiness. Z Evid Fortbild Qual Gesundhwes 123-124:104–108. https://doi.org/10.1016/j.zefq.2017.05.012
14. Martin H et al (2017) The long way of implementing patient-centered care and shared decision-making in Germany. Z Evid Fortbild Qual Gesundhwes 123-124:46–51. https://doi.org/10.1016/j.zefq.2017.05.006
15. Stacey D et al (2017) Decision aids for people facing health treatment or screening decisions. Cochrane Database Syst Rev 4:CD001431
16. Hamann J et al (2006) Partizipative Entscheidungsfindung. Implikationen des Modells "Shared Decision Making" für Psychiatrie und Neurologie. Nervenarzt 9: 1071–1078
17. Reichhart T et al (2008) Patientenbeteiligung in der Psychiatrie–eine kritische Bestandsaufnahme. Psychiatr Prax 35:111–121
18. NICE (2009) Depression in adults: recognition and management: NICE Clinical Guideline [CG90]. https://www.nice.org.uk/guidance/cg90/resources/depression-in-adults-recognition-and-management-pdf-975742636741. Accessed 27 April 2017
19. NICE (2014) Psychosis and schizophrenia in adults: prevention and management: NICE Clinical Gudeline [CG178]. https://www.nice.org.uk/guidance/cg178/resources/psychosis-and-schizophrenia-in-adults-prevention-and-management-pdf-35109758952133. Accessed 27 April 2017
20. Hamann J, Heres S (2014) Adapting shared decision making for individuals with severe mental illness. Psychiatr Serv 65:1483–1486
21. Hirschman KB et al (2004) How does an Alzheimer's disease patients' role in medical decision making change over time? J Geriatr Psychiatry Neurol 17:55–60
22. Okonkwo OC et al (2008) Medical decision-making capacity in mild cognitive impairment: a 3-year longitudinal study. Neurology 71:1474–1480
23. NICE (2006) Dementia: supporting people with dementia and their carers in health and social care: NICE Clinical Guideline [CG42]. https://www.nice.org.uk/guidance/cg42/resources/dementia-supporting-people-with-dementia-and-their-carers-in-health-and-social-care-pdf-975443665093. Accessed 27 April 2017
24. Alzheimer's Association Early-Stage Initiative (2006) History of the Alzheimer's association early-stage initiative. http://www.alz.org/documents/national/EDInitiative.pdf. Accessed 27 April 2017
25. Alzheimer Europe (2006) Paris declaration 2006. http://www.alzheimer-europe.org/Policy-in-Practice2/Paris-Declaration-2006. Accessed 27 April 2017
26. Moye J et al (2006) Neuropsychological predictors of decision-making capacity over 9 months in mild-to-moderate dementia. J Gen Intern Med 21:78–83
27. Hirschman KB et al (2005) Do Alzheimer's disease patients want to participate in a treatment decision, and would their caregivers let them? Gerontologist 45:381–388
28. Hamann J et al (2011) Patient participation in medical and social decisions in Alzheimer's disease. J Am Geriatr Soc 59:2045–2052
29. Hamann J et al (2009) Psychiatrists' use of shared decision making in the treatment of schizophrenia: patient characteristics and decision topics. Psychiatr Serv 60:1107–1112
30. Vollmann J et al (2004) Mental competence and neuropsychologic impairments in demented patients. Nervenarzt 75:29–35
31. Karlawish JHT et al (2000) Caregivers' preferences for the treatment of patients with Alzheimer's disease. Neurology 55:1008–1014
32. Karlawish JHT et al (2002) Relationship between Alzheimer's disease severity and patient participation in decision about their medical care. J Geriatr Psychiatry Neurol 15:68–72
33. Peterson RC et al (2001) Current concepts in mild cognitive impairment. Arch Neurol 58:1985–1992
34. Raschetti R et al (2007) Cholinesterase inhibitors in mild cognitive impairment: a systematic review of randomised trials. PLoS Med 4:e338
35. DeKosky ST (2008) Ginkgo Biloba for prevention of dementia: a randomized controlled trial. JAMA 300:2253–2262
36. Rosen C et al (2013) Fluid biomarkers in Alzheimer's disease–current concepts. Mol Neurodegener 8:20

37. Kivipelto M, Soloman A (2008) Alzheimer's disease–the ways of prevention. J Nutr Health Aging 12:89S–94S
38. Alonso A et al (2009) Cardiovascular risk factors and dementia mortality: 40 years of follow-up in the seven countries study. J Neurol Sci 280:79–83
39. Feart C et al (2009) Adherence to a Mediterranean diet, cognitive decline, and risk of dementia. JAMA 302:638–648
40. Scarmeas N et al (2009) Physical activity, diet, and risk of Alzheimer disease. JAMA 302:627–637
41. Sindi S, Mangialasche F, Kivipelto M (2015) Advances in the prevention of Alzheimer's disease. F1000Prime Rep 7:50. 10.12703/P7-50
42. Marckmann G, in der Schmitten J (2014) Krebsfrüherkennung aus Sicht der Public-Health-Ethik. Bundesgesundheitsbl 57:327. https://doi.org/10.1007/s00103-013-1913-0
43. Williamson J, Goldman J, Marder KS (2009) Genetic aspects of Alzheimer disease. Neurologist 15:80–86
44. Jessen F, Dodel R (2014) Prädiktion der Alzheimer Demenz. Nervenarzt 85: 1233–1237
45. Quian W et al (2016) Misdiagnosis of Alzheimer's disease: inconsistencies between clinical diagnosis. Alzheimers Dement 12:293
46. Deutsche Alzheimer Gesellschaft (2013) Empfehlungen zum Umgang mit Frühdiagnostik bei Demenz. https://www.deutsche-alzheimer.de/fileadmin/alz/pdf/empfehlungen/empfehlungen_fruehdiagnostik_demenz_dalzg.pdf. Accessed 2 May 2017
47. Robert PH et al (2005) Grouping for behavioral and psychological symptoms in dementia: clinical and biological aspects. Consensus paper of the European Alzheimer disease consortium. Eur Psychiatry 20:490–496
48. Winblad B et al (2006) 3-year study of donepezil therapy in Alzheimer's disease: effects of early and continuous therapy. Dement Geriatr Cogn Disord 21:353–363
49. NICE (2011) Donepezil, galantamine, rivastigmine and memantine for the treatment of Alzheimer's disease. Technology appraisal guidance [TA217]. https://www.nice.org.uk/guidance/ta217/resources/donepezil-galantamine-rivastigmine-and-memantine-for-the-treatment-of-alzheimers-disease-pdf-82600254699973. Accessed 2 May 2017
50. McShane R, Areosa Sastre A, Minakaran N (2006) Memantine for dementia. Cochrane Database Syst Rev 19:CD003154
51. Alzheimer Europe (2016) Clinical trials watch. http://www.alzheimer-europe.org/Research/Clinical-Trials-Watch. Accessed 3 May 2017
52. Institut für Qualität und Wirtschaftlichkeit im Gesundheitswesen (IQWiG) (2009) Nichtmedikamentöse Behandlung der Alzheimer Demenz. Abschlussbericht A05-19D(Version 1.0 Stand: 13.01.2009), Köln
53. Pinquart M, Forstmeier S (2012) Effects of reminiscence interventions on psychosocial outcomes: a meta-analysis. Aging Ment Health 16:541–558
54. Korczak D, Habermann C, Braz S (2013) The effectiveness of occupational therapy for persons with moderate and severe dementia. GMS Health Technol Assess 9:1–7
55. McLaren AN, Lamantia MA, Callahan CM (2013) Systematic review of non-pharmacologic interventions to delay functional decline in community-dwelling patients with dementia. Aging Ment Health 17:655–666
56. Forbes D et al (2013) Exercise programs for people with dementia. Cochrane Database Syst Rev 12:CD006489
57. Schölzel-Dorenbos CJ et al (2009) Quality of life and burden of spouses of Alzheimer disease patients. Alzheimer Dis Assoc Disord 23:171–177
58. Chien LY et al (2011) Caregiver support groups in patients with dementia: a meta-analysis. Int J Geriatr Psychiatry 26:1089–1098
59. Wolf-Ostermann K et al (2017) Users of regional dementia care networks in Germany: first results of the evaluation study DemNet-D. Z Gerontol Geriatr 50:21–27
60. Sozialgesetzbuch (SGB) Neuntes Buch (IX) - Rehabilitation und Teilhabe behinderter Menschen - (Artikel 1 des Gesetzes v. 19.6.2001, BGBl. I S. 1046) § 145 Unentgeltliche Beförderung, Anspruch auf Erstattung der Fahrgeldausfälle
61. Sozialgesetzbuch (SGB) Neuntes Buch (IX) - Rehabilitation und Teilhabe behinderter Menschen - (Artikel 1 des Gesetzes v. 19.6.2001, BGBl. I S. 1046) § 85 Erfordernis der Zustimmung
62. Emanuel LL, von Gunten CF, Ferris FD (2000) Advance care planning. Arch Fam Med 9:1181–1187
63. Coors M, Jox R, in der Schmitten J (Hrsg) (2015) Advance care planning: Von der Patientenverfügung zur gesundheitlichen Vorausplanung. Kohlhammer, Stuttgart
64. Wolter DK (2014) Beginnende Demenz und Fahreignung. Teil 2: Das Assessment und seine praktischen Konsequenzen. Z Gerontol Geriat 47:345–355

65. Fastenmeier W, Gstalter H (2014) Fahreignung älterer Kraftfahrer im internationalen Vergleich. Literaturrecherche, Analyse und Bewertung. Forschungsbericht Nr. 25. Unfallforschung der Versicherer, Berlin
66. Von Kutzleben M et al (2012) Community-dwelling persons with dementia: what do they need? What do they demand? What do they do? A systematic review on the subjective experiences of persons with dementia. Aging Ment Health 16:378–390
67. Hirschman KB, Kapo JM, Karlawish JH (2008) Identifying the factors that facilitate or hinder advance planning by persons with dementia. Alzheimer Dis Assoc Disord 22:293–298
68. DeVleminck A et al (2014) Barriers to advance care planning in cancer, heart failure and dementia patients: a focus group study on general practitioners' views and experiences. PLoS One 9:e84905
69. Callahan CM et al (2009) Integrating care for older adults with cognitive impairment. Curr Alzheimer Res 6:368–374
70. Morant N, Kaminskiy E, Ramon S (2016) Shared decision making for psychiatric medication management: beyond the micro-social. Health Expect 19:1002–1014
71. Towle A, Godolphin W (1999) Framework for teaching and learning informed shared decision making. BMJ 319:766–771
72. O'Connor AM et al (2009) Decision aids for people facing health treatment or screening decisions. Cochrane Database Syst Rev 8:CD001431

Index

Robert Perneczky (ed.), *Biomarkers for Preclinical Alzheimer's Disease*, Neuromethods, vol. 137,
https://doi.org/10.1007/978-1-4939-7674-4, © Springer Science+Business Media, LLC 2018

Printed in the United States
By Bookmasters